DNA: Forensic and Legal Applications

DNA: Forensic and Legal Applications

Lawrence Kobilinsky
John Jay College of Criminal Justice
The City University of New York

Thomas F. Liotti
Law Offices of Thomas F. Liotti

Jamel Oeser-Sweat
Law Offices of Jamel Oeser-Sweat

A John Wiley & Sons, Inc Publication

Published by John Wiley & Sons, Inc., Hoboken, New Jersey.
Published simultaneously in Canada.

Library of Congress Cataloging-in-Publication Data:

Kobilinsky, Lawrence
DNA : forensic and legal applications / Lawrence Kobilinsky, Thomas F. Liotti, Jamel Oeser-Sweat.
p. cm.
Includes bibliographical references and index.
ISBN 0-471-41478-6 (cloth)
1. DNA fingerprinting - - United States. 2. Evidence, Expert - - United States.
3. Forensic genetics - - United States. I. Liotti, Thomas F. II. Oeser-Sweat, Jamel. III. Title.

KF9666.5.K63 2005
343.73′067 - - dc22

2004003677

Printed in the United States of America

10 9 8 7 6 5 4 3 2

Contents

Foreword

The Banbury Center of Cold Spring Harbor Laboratory was the place to be in November 1988, if you were interested in forensic applications of DNA fingerprinting. Molecular geneticists, population geneticists, forensic scientists, lawyers, and bioethicists came to Banbury for what turned out to be an historic meeting. It was only three years earlier that Alec Jeffreys had published an esoteric study describing a curious DNA sequence that he had found in the myoglobin gene. It ceased to be an arcane topic when Jeffreys and his colleagues showed how this DNA sequence could be used for individual identification and demonstrated its use in an immigration case. And when, in 1987, Jeffreys was called in to assist in a notorious murder case, DNA fingerprinting attracted worldwide attention.

In the two years prior to the Banbury Center meeting, DNA fingerprinting was taken up eagerly by the forensic community, eager to have an identification method that was of far greater applicability than fingerprinting (traces of DNA are left everywhere); appeared to be of far greater precision (identifications could be made with figures like one in ten billion); and it came with the cachet of molecular genetics. But by 1988 controversies had arisen and DNA fingerprinting was under attack; the numerical estimates of identity were challenged; the competence of forensic laboratories doing DNA testing was called into question; and concerns were raised about the collection, retention, and use of DNA samples.

Alec Jeffreys was one of the participants who came to Banbury that November, together with others who were to become well known in the coming years for their involvement with DNA profiling: Bruce Budowle, Tom Caskey, Rockne Harmon, John Hicks, Eric Lander, Henry Lee, Peter Neufeld, and Barry Scheck, to name a few. The meeting proved historic because it was the first to subject the forensic practice of DNA fingerprinting to critical scrutiny. It was controversial, and discussions were frequently heated, but it initiated a comprehensive review of DNA fingerprinting including studies by independent bodies such as the National Academy of Sciences. These in turn led to improvements in technique; refinements in statistical analysis; the establishment of quality controls; and a change in way DNA evidence is regarded in the courtroom. Indeed, it may yet be that the rigorous scrutiny of the theory and practice of DNA fingerprinting will come to be applied to other forms of forensic evidence.

Scientific findings move slowly from the laboratory bench to application in the wider world, so the rapidity with which DNA fingerprinting moved from Jeffreys' laboratory to become the gold standard in evidence is unprecedented. And, just as the DNA double helix became emblematic of scientific research, so, through exposure on television programs such as *CSI,* DNA fingerprinting has become an icon for forensic science.

Last year was the 50th anniversary of the DNA double helix, and there was a grand, black tie gala at the Waldorf. One speaker was Marvin Anderson, who had been sentenced to 210 years, and had served 15 years, for a crime he did not commit. DNA evidence had exonerated him, and now he was addressing a gathering of Nobel laureates, eminent scientists, and celebrities. He came down from the stage to the table where Alec Jeffreys was sitting and the two embraced. A fitting and moving testimony to the power of DNA fingerprinting.

DNA: Forensic and Legal Applications is a comprehensive and invaluable guide to the field, covering topics ranging from collecting samples in the field to presenting the complex results to a jury. We are sure that it will play its part in promoting this most powerful tool in the forensic scientist's armamentarium.

James D. Watson
Jan A. Witkowski
Cold Spring Harbor Laboratory

Left to Right: Dr. Jan Witkowski, Jamel Oeser-Sweat, Dr. Lawrence Kobilinsky, Dr. James Watson, Thomas F. Liotti

Preface

The recently developed techniques that permit human identification by analysis of specific regions of DNA within the human genome have emerged as powerful evidentiary tools for the criminal justice system. The realization that a person can be "individualized" by analyzing his or her DNA has been heralded as one of the greatest revelations of the twentieth century. The number of clinical, scientific, and forensic uses for DNA grows with each passing day. As scientists strive to elucidate the many mysteries locked in the code that comprises DNA, we begin to understand why nature has made it the medium for storage of its blueprint for life, the genetic code. In DNA lies not merely the story of our evolution but also who we are, what lies in store for us in the future, and perhaps even the reason for our very existence.

Forensic DNA analysis has had a major impact on our criminal justice system and on the law during the last decade of the twentieth century. It has been employed in criminal law to help prove guilt or innocence, in family law to prove paternity, and in immigration law to prove blood relationships to establish citizenship. Its usefulness as a human identification tool is clear. Accordingly, in recent years, our legal system has given forensic DNA analysis the credibility that nature has given it as the blueprint of life. However, the examination of DNA can become compromised when environmental factors intervene, leading to deterioration, destruction, or contamination of the evidence, or when human error results in incorrect conclusions. These factors are crucial in determining what *weight* to give DNA evidence. Determining whether these factors exist and, if so, the extent to which they have corrupted the evidence or compromised the analysis are important tasks for a lawyer. A lawyer must not only understand what is advantageous about the science of DNA but what can go wrong and how to detect and prevent procedural errors. Attorneys facing trials in which DNA evidence will be offered must understand the underlying science and technology on which DNA testing is based. This book guides attorneys and judges through the complexities of the biochemical sciences to help them understand the methodology of DNA analysis. It will provide them with this knowledge so that, at trial, they can ask appropriate questions and understand the responses that are given. This book has been written for students of science and law, for criminal justice practitioners, and for those forensic scientists who do not currently work in the field of DNA identification but who seek to learn more about the scientific and legal procedures involved. It is assumed that the reader knows very little about DNA and written in a style that is easy to read and comprehend.

The first chapter will provide the reader with the background necessary to understand the science underlying the common tests employing DNA. While much can be written regarding the chemistry, uses, and functions of DNA, this discussion is limited to an overview of its chemistry, structure, and its ability to replicate, which provides the information necessary to explore more advanced topics including the molecular biology and forensic applications of DNA in later portions of the book.

The second chapter provides the reader with information on the techniques employed by criminalists on the path from crime scene to final result; evidence is recognized at a crime scene and samples are collected, documented, packaged and brought to the laboratory, and analysis begins. It explores several issues that are relevant to each of the above procedures, as well as the impact of environmental factors, contamination, aging, and so forth on DNA evidence and test results. It also reviews methods of chemical and/or physical identification of common items of biological evidence.

The third chapter familiarizes the reader with procedures used to analyze biological evidence to determine its origin and if an association can be made between a suspect, victim, and/or crime scene. The different kinds of human DNA identification tests are reviewed, beginning with DNA fingerprinting, which was developed in the mid-1980s and used effectively into the mid-1990s. The chapter continues with a discussion of tests based on reverse dot blot technology, amplified fragment-length polymorphism (AmpFLP) analysis, and finally, the two current state-of-the-art techniques, known as PCR-STR analysis and mitochondrial DNA sequencing. Each of these tests differs markedly from the other, and each is a product of scientific and technological advancements. The advantages, drawbacks, and significance of each procedure are described. The chapter introduces the reader to several important issues related to the interpretation of test data. Some background knowledge of biochemistry and the molecular biology of DNA is necessary to understand the specific details of these technical procedures. It is helpful to have some knowledge of basic statistics and the laws of probability to appreciate the significance of the test results.

The first part of Chapter 4 discusses human genetics, population genetics, and statistics. Mendel's laws of heredity and the Hardy–Weinberg equilibrium are explained. Population genetics provides the scientific foundation for using DNA testing to develop genetic profiles whose frequencies of occurrence are so rare that each can be considered unique to an individual.

In certain cases, where the quality or quantity of nuclear DNA is limited, mitochondrial DNA testing is conducted. In the case of sexual assault, Y-chromosomal

STR DNA testing can be advantageous. These tests as well as single-nucleotide polymorphism analysis are also discussed in Chapter 4.

A discussion of the importance of having a national database of digitized DNA profiles, CODIS, to help solve crimes where biological evidence has been found but where no suspects have been identified by eyewitness or police investigation is also included in Chapter 4. The CODIS database has been especially helpful when serial criminals perpetrate their crimes across state lines. A discussion of several statistical considerations in genetic testing, including the power of discrimination, probability of exclusion, and the likelihood ratio, are also included. There is also a discussion of how to treat evidence that is composed of a mixture of DNA from a number of sources. Regarding the use of DNA in civil law, we provide a discussion of how paternity is currently established by accredited testing laboratories.

The fourth chapter also discusses the need for quality control and quality assurance in the forensic laboratory. Quality assurance is demonstrated by the laboratory's accreditation, certification of its personnel, and proficiency testing. The chapter closes with a review of four DNA reports, the first dealing with RFLP, the second with PCR-based testing including HLA DQA1, Polymarker, and D1S80, the third, describing results of a PCR-STR analysis, and the fourth a paternity report based on DNA testing.

The fifth chapter will provide those who work in the criminal justice system, but who have little or no science background, the ability to understand and interpret DNA evidence with respect to past and present law. This chapter bridges the disciplines of science and the law by focusing on the admissibility of scientific evidence. It shows how the law applies to the evidence that has been collected and analyzed, and the findings and subsequent summary report issued by the laboratory. The chapter includes a discussion of the Federal Rules of Evidence, the *Daubert, Joiner*, and *Kumho Tire* decisions of the U.S. Supreme Court, as well as the judicial gatekeeping function of judges. It outlines what types of evidence are or are not admissible and relevant cases are discussed. The various aspects of a legal proceeding are detailed, including arraignment, grand jury, discovery, preparation for trial, jury selection and voir dire, opening statements, direct examination, cross-examination, and closing arguments.

The sixth chapter of this book introduces the reader to the concepts and procedures of challenging or defending DNA evidence. It opens with the importance of developing a strategy of how to have DNA evidence admitted at trial. It also explains how to make the best use of an expert witness by thoroughly and properly preparing him or her to testify. There is a discussion of how to introduce the expert and his or her credentials to the court to be deemed qualified as an expert witness. There is a list of issues that should be brought out during direct examination of the expert, allowing the jury to learn about the DNA testing, experimental observations, and, finally, the expert's conclusions.

The chapter then addresses the defense effort to mitigate the effects of DNA evidence being introduced by the prosecution. There are a number of potential routes to attack the admissibility of all or part of the DNA evidence including (a) expert not qualified, (b) expert not certified, (c) laboratory not accredited, (d) lack of discovery, and (e) improperly obtained evidence. There may be a challenge to the statistics, to the database, or to "insufficient" quality control or quality assurance used in the laboratory, or to a perceived break in the chain of evidence. These arguments may be helpful to the attorney who has never before litigated a DNA trial and can be beneficial for the experienced attorney by its comprehensive review of important issues to be brought out in testimony. It concludes with a discussion of the type of summation that might be effective in trials of this sort.

The seventh and final chapter of this book discusses postconviction appeals based on analysis of existing DNA evidence. It discusses the role of the prosecutor and defense counsel in achieving a just solution for the innocent convict. It also discusses the legal standards governing postconviction testing. The Innocence Project has been highlighted as the first and most successful organized effort to exonerate innocent convicts. The chapter ends with a brief discussion about the future of DNA technology. It attempts to explain how new and improved technology will make analysis of evidence at the crime scene possible. Testing will be far more rapid, more economical, easier to perform, less labor intensive, and even more reliable. Today's technology sometimes fails to identify the source of evidence that had become seriously degraded or corrupted as a result of environmental insult. The same evidence could produce results using the technology of the future. In addition to the currently performed STR analysis, sequencing analysis and SNP detection technology are both likely to become more and more utilized by the forensic analyst. The impact of these changes on the criminal investigatory process and on existing national, state, and local DNA database collections is also explored.

This book is unprecedented in its merger of law with the science of DNA. Now for the first time in a single volume lawyers, judges, scientists, professors, students, and experts can find everything that they need in order to understand the forensic and legal applications of DNA. The science of DNA and its potential in and outside of the courtroom is unlimited. It has already changed the face of jurisprudence as perhaps the single most important development in science and in the law in the past half century. The evolution of the merger has just begun. The book takes the reader on a guided tour of what the future holds in store for all of us, but instead of being mere passive observers, each of us can now be active participants in the ever changing world of science and the law.

LAWRENCE KOBILINSKY
THOMAS F. LIOTTI
JAMEL OESER-SWEAT

Acknowledgments

The authors would like to thank our colleagues and friends who supported our efforts and engaged us in interesting discussions about how DNA is being used in forensic science and in the courtroom. We would like to thank our families for their encouragement and patience. We would especially like to thank Drs. John Butler and Peter Valone for their assistance and cooperation in allowing us to use a number of their figures in this book. We have found STRbase to be a wonderful source of invaluable information related to STR technology. STRbase (available online at http://www.cstl.nist.gov/div831/strbase/) was developed and continues to be maintained by Dr. Butler, Dr. Dennis J. Reeder and others at the National Institute of Standards and Technology as well as the National Institute of Justice. We would also like to thank Dr. Louis Levine for reviewing the scientific portion of our book and for making valuable suggestions on the topic of population genetics. In addition we would like to thank the following individuals who have extended to us their support and encouragement: Prof. Robert Rothchild, Prof. Susan Stabile, Prof. Ettie Ward, Dr. Philip Furmanski, Dr. John Sexton, Dr. Edward Bottone, Emile Nava, Robin Maynard, Polly Powell, Henry Lung, Jason Spector, George Phillips, Bhavini Shah, Judge Robert Kohm, Judge Jaime Rios, and Maria Albano. Finally, we would like to thank The Innocence Project at the Benjamin Cardozo School of Law, the Federal Bureau of Investigation, and the American Prosecutors Research Institute.

1

Biochemistry, Genetics, and Replication of DNA

1.1 EVOLUTION OF IDENTIFICATION: FROM FACES TO FINGERPRINTS TO DNA

When a crime has been committed, it is the job of the investigator to reconstruct the events leading up to and during the incident. The investigator will seek information from a number of sources including witnesses, physical evidence, and records.

People are a very good source of information, but their observations and reporting must be carefully evaluated. One can often learn a great deal by questioning the victim's family members, and associates, as well as strangers. Those who come into contact with a criminal suspect may include, among others, witnesses to or the victims of a crime. Investigators can use information obtained from such individuals to recreate the past and to solve a crime mystery. Before modern scientific and technological methods were developed to study physical evidence, this was one of the most important methods for solving crimes. Eyewitness testimony was the best way to identify the perpetrator. Indeed, two eyewitnesses were required to convict a person of a crime under Hebraic law (Deut. 17:6).

When eyewitnesses are lacking, physical evidence may be the only way to solve a crime. Materials found at a crime scene can be used to link or associate a suspect to it, and the information derived from evidence analysis can be used to exonerate or convict a suspect. Various types of physical evidence can be found at crime scenes. Shoe prints are used to show what type of shoes a suspect was wearing (Bodziak, 1990). Such evidence allows an investigator to identify the shoe as being part of a certain class. As a result, shoes belonging to other classes can be ruled out. For

DNA: Forensic and Legal Applications, by Lawrence Kobilinsky, Thomas F. Liotti, and Jamel Oeser-Sweat
ISBN 0-471-41478-6

example, if a shoe print found at a crime scene is from a particular brand of sneaker, this might be used to rule out suspects who are known to have been wearing some other brand of sneaker during the time that the crime was committed. Such information can also be used to individualize an item. One refers to a specimen from a known source as an *exemplar*. The pattern on the sole of a shoe obtained from a suspect, classified as an exemplar, can be matched to the pattern of a plaster cast of a print found at a crime scene (which we refer to as the *questioned print*). The comparison between exemplar and questioned specimen can result in what is known as a *physical match*. Experts can analyze the patterns found on the sole and print, and, as with fingerprints, certain distinguishing markings could prove a match, for example, if there is a deep scratch in the sole of the shoe, caused by wear, and the print found at the crime scene exhibits the identical scratch. Such a match is incontrovertible evidence of the origin of the questioned print. The same kind of analysis can be conducted on tire impressions or even on tool marks. Tool marks refer to the markings made on an object when a tool or other instrument is used to gain entry, i.e. to break open a locked closet or window. Another example of a physical match is a sheet of paper or fabric ripped in half. A comparison of the torn ends by microscopic analysis can reveal if the two halves were created from the same sheet. Both torn ends constitute a physical match or perfect fit. In some situations a relatively unique kind of evidence is found that associates a suspect and victim. Several examples follow of evidence that is uniquely important at a crime scene. A relatively rare type of carpet fiber is found on the body of a murder victim, and it is subsequently found to be identical to the fibers of a rug in the bedroom of a suspect. In such a scenario, the fibers become significant associative evidence.

During a 3-year period in Atlanta, Georgia (1979–1981), more than 25 young black male children were found murdered as a result of strangulation, blunt trauma, or asphyxiation. The evidence found on the clothing and bodies of the victims consisted primarily of fibers and hair. The hair was found to be canine and some of the fibers were found to be of an uncommon type. Two types of fiber were identified. The first was a violet acetate fiber, and the second was an unusual yellow-green nylon fiber that appeared to be trilobed when viewed in cross section. The latter appeared to be a carpet fiber, but the Georgia State Crime Laboratory was unable to identify the manufacturer. The police decided to set up a stakeout under a bridge on the Chattahoochee River where many of the bodies were found. On May 22, 1981, the police hearing a splash investigated a car that had just passed over the bridge, its driver 23-year-old Wayne B. Williams. They interrogated him and let him go. Two days later, the body of an ex-felon Nathanial Cater appeared downstream of the bridge, and the police refocused their attention on Williams. He was arrested on June 21 and charged with the murder of Cater. He was indicted a month later for the murders of Cater and another person, Jimmy Payne. Payne was also an ex-convict who was murdered by suffocation in mid-April of 1981. The list of child victims associated with the case was extensive. The key evidence at trial was the hair and fibers found on the victims. Experts testified that the trilobed fiber matched the carpet in Williams's home. Other fibers were matched to the trunk liner of the Williams family 1979 Ford as well as to a second family car, a 1970 Chevrolet. On February 27, 1982, Wayne Williams was convicted of the murders of Cater and Payne and sentenced to two terms of life imprisonment.*

Synthetic or natural fibers are sometimes transferred from person to person upon physical contact, making these forms of trace evidence important. This is an example of the Locard exchange principle described in more detail in Section 3.6.1. When two bodies come into contact, there is an exchange of material between them. Sometimes these materials are so miniscule as to go unnoticed; for example, when extremely small fibers are transferred from victim to suspect or vice versa. If these fibers are somewhat unique, for example, dyed using an unusual chemical compound, they may be useful in linking victim and suspect. The same is true of hair that is easily transferred on clothing from person to person. There are numerous characteristics of evidentiary hair both visual and microscopic that can be used to compare it to known hair specimens taken from a suspect or victim. Hair can also be examined using several DNA (deoxyribonucleic acid) techniques to determine its origin.

The O.J. Simpson prosecution team attempted to link him to the crime scene after one bloody glove was found at the scene and its mate was found behind his house (Schmalleger, 1996). These gloves were easily identified because of the palm vent, stitching, hem, and other characteristics. Nicole Brown Simpson, his former wife and murder victim, had bought him two pairs of these Aris Isotoner Lights, size extra-large gloves in December of 1990, about three and a half years before she was murdered. What makes these gloves even more significant is the fact that there were only about 200 pairs sold that year by Bloomingdales's department store. The defense argued that the pair of gloves in question was unlikely to be Simpson's when O.J. unsuccessfully tried, at the prosecutor's request, to put on the pair of gloves found at the crime scene. Many observers of the trial felt that the prosecution should never have allowed the defendant to take possession of the evidence, thereby allowing him to demonstrate to the jury that the gloves were too small to fit his hands. In fact, some felt that either the gloves had shrunk or that Simpson was trying to show that he could not get them on. Simpson had put on a pair of rubber gloves before trying to put his hands into the evidentiary gloves. This would have made it difficult to put the gloves on even if they actually had fit his hands.

Fingerprints are another type of physical evidence used for human identification and also to link an individual to a location or to a victim (Cole, 2001; Lambourne, 1984; Cowger, 1983; Rhodes, 1956). Fingerprints are impressions made from the *papillary ridges* on ones fingertips. These epidermal ridges are arranged in very different or unique patterns on each of our fingertips. These patterns do not change

*For further information about this case see Katherine Ramsland http://www.crimelibrary.com/criminal_mind/forensics/trace/II.html.

from the time of birth until the time of death. Fingerprints provide absolute proof of identity. The science of fingerprint identification is called *dactyloscopy*. Similar ridges can be found on the palms and on the toes and soles of the foot. Although a print pattern can be described as a loop, whorl, or arch, within each of these configurations, ridges can be arranged in a variety of ways including straight lines with forks that split like forks in a road. In 1880 Henry Faulds and William Herschel discovered that every individual has unique, permanent fingerprint patterns on the tips of his or her fingers. This discovery was subsequently verified by Sir Francis Galton, who proposed a system of fingerprint classification based on the patterns of loops (hairpin ridges), whorls (circular or spiral ridges), and arches (tent or mountain-like ridges). Sir Edward Henry developed a classification system based on the work of Galton. The Henry system of classification was published in 1900 and was used to collect and categorize fingerprints of criminals and is still in use today. Beside the Henry system there is also a widely used system called the NCIC (National Crime Information Center) classification system. The FBI (Federal Bureau of Investigation) classifies a total of eight different fingerprint patterns: (1) plain arch, (2) tented arch, (3) radial loop, (4) ulnar loop, (5) double whorl, (6) central pocket whorl, (7) plain whorl, and (8) accidental whorl.

The NCIC system classifies fingerprints with a string of 20 characters, 2 per finger. The numbering starts with the right thumb (1) and then the other fingers across the right hand to the little finger (5). It then moves to the thumb of the left hand (6) and across the left hand to the little finger (10). For example, the fingerprint classification (FPC) {PO PI 11 PO 12 15 CI CO CI 11} is translated using the following code:

01–49, ulnar loop
51–99, radial loop
AA, plain arch
TT, tented arch
PI, plain whorl—inner tracing
PO, plain whorl—outer tracing
PM, plain whorl—meet tracing
CO, central pocket whorl—outer tracing
CI, central pocket whorl—inner tracing
CM, central pocket whorl—meet tracing
DO, double whorl—outer tracing
DI, double whorl—inner tracing
DM, double whorl—meet tracing
XI, accidental whorl—inner tracing
XO, accidental whorl—outer tracing
XM, accidental whorl—meet tracing

Two consecutive numbers represents a loop pattern with the number indicating the number of ridges. The pattern may be classified as an ulnar loop or a radial loop. The loop may be slanted to the left or to the right, and the classification depends on whether the pattern is on the left or right hand. There are four different kinds of whorls and three different *tracings*. Tracings are classified as either inside or outside the main whorl pattern and depend on the position of the two deltas (triangle-like patterns) found in whorls.

The most common type of fingerprint pattern is the ulnar loop. Another fingerprint classification system based on Galton's work was introduced by Juan Vucetich, an Argentinian, in 1888, and this system is still used in many Latin American countries.

Although the impression left by the ridges on the ends of our fingertips when we handle objects are not always visible, they can be made visible through chemical enhancement. A fingerprint invisible to the naked eye is called a *latent print.* Latent prints are usually created upon touching an object or surface and at the same time depositing some naturally secreted or environmentally acquired material onto it. However, after one's finger contacts an object, a perfect fingerprint is not always left (Barnett and Berger, 1977). Sometimes only a portion of the pattern on the edge of one's finger remains; a partial fingerprint can be found in the spot that was touched. Other times no useful print information remains at the point of contact. Fingerprint examiners will compare latent prints to known or exemplar prints by examining specific identification points in the pattern that consist of either dots, islands, ridge endings, or bifurcations (branching of a single ridge into two ridges). Although an inked print can reveal up to 100 such points (minutiae), latent prints may have only a small fraction of these points. Some examiners require at least 7 or 8 identical points before they will state that the latent and exemplar have the same origin. However, there is no specific minimum number that will satisfy all examiners.

In the early 1970s, law enforcement began using computers to classify, store, and retrieve fingerprint data. Today, crime labs have Automated Fingerprint Identification Systems (AFIS) that scan a fingerprint image and convert the minutiae to digital information (Wilson, 1986). The computer also records relative position and orientation of the minutiae and therefore stores geometric data. Computer databases have been created to record the unique imprints of those in the population who have been arrested or who have provided their prints for employment (armed forces or security guards) or gun ownership. Using AFIS, law enforcement agencies have been able to store fingerprint digital data in large databases. Using these digitized files and a powerful search algorithm, prints obtained from new crime scenes can be compared to those of known offenders on file. The computer also determines the degree of correlation of the pattern, location, and orientation of the minutiae. It can compare hundreds of thousands of prints on file in a second or two. AFIS then prepares a list of those prints that come closest to matching the questioned print so that a fingerprint examiner can make the ultimate call of identification or not. In recent years, using fingerprints to run background checks on individuals attempting to gain employment in certain areas such as early childhood teaching has become common.

Forensic DNA testing has emerged as a highly effective way to identify the source of biological evidence with reliability equal to that of fingerprint identification (Neufield and Coleman, 1990). An individual's total genetic composition, in the form of DNA, is referred to as the human genome. Most of the genome is located in the nucleus of a cell, while the remainder is found in the subcellular organelle known as the mitochondrion. Differences in DNA make every individual unique, and that uniqueness can be attributed to differences in certain areas of the human genome. Portions of DNA are invariant from person to person while other portions differ. Most people thinking about an individual's identification will focus on differences in physical appearance such as height, weight, hair color, eye color, skin color, and so forth. However forensic DNA examiners study the differences in the sequence of subunits that make up the DNA molecule. It is known that the difference between two individuals is only 1 in 1000 building blocks. With the human genome consisting of approximately 3.1 billion building blocks, there are about 3.1 million differences in genome subunit sequence between any two persons. The one exception is the DNA of identical twins (or cloned animals), which is identical.

In recent years DNA analysis has been used the same way that fingerprints have been used to link individuals to crime scenes (Kelly et al., 1987; Jeffreys et al., 1985). One advantage of fingerprinting over DNA analysis lies in the fact that even though identical twins have identical genomes, they still have different fingerprints and can easily be distinguished in this way. DNA does not control this trait because the establishment of ridge patterns on the fingertips, palms of the hands, and soles of the feet is a developmental process that takes place as the embryo develops into a fetus and is not a DNA-coded trait, that is, fingerprint patterns are not related to a person's genetic blueprint.

The advantages of DNA analysis over fingerprint analysis are clear. Even if a surface is touched, a useful record of that contact is not always left behind. A latent print requires a suitable surface and certain conditions for a print to remain. As mentioned above, natural and environmentally derived materials present on the fingers result in fingerprints. If a surface is not smooth, or if it is porous, irregular, or rough, it is unlikely that a useful fingerprint can be obtained. If nothing is touched or gloves are worn, discovering any fingerprint whether whole or partial will be virtually impossible. However, DNA can be obtained from a site even if nothing has been touched. A hair fiber with or without its root intact might have fallen from one's scalp, a cigarette butt with saliva (containing epithelial cells) may have been discarded (Hochmeister et al., 1998), or an item of clothing such as a hat or glove worn by a suspect could be discovered. Today, technology is so advanced that exceedingly small amounts of biological substances (blood, semen, saliva, urine, etc.) generated during the commission of a crime can be DNA tested resulting in the identification and conviction of a suspect (Stouder et al., 2001; Erlich, 1989). The one requirement is that there must be a sufficient amount of DNA and that it be in relatively good enough condition to allow testing to be successful.

However, shifts in environmental conditions including high temperature and/or humidity can have an adverse impact on the extraction of high-quality DNA. When DNA becomes fragmented as a result of bacterial or fungal enzymes, it may become so degraded as to render it useless for forensic purposes. This concept is explained in greater detail in Section 3.2.1.

Official records, documents, and databases are a type of physical evidence that can be used in criminal investigation. DNA sequences of various individuals can be recorded in the same way as fingerprints, to create databases by which characteristics of unknown perpetrators of a crime could be matched. The federal government, through the FBI, and all 50 states, through their state crime laboratories, have established such databases. Some ethical and legal issues still remain unresolved. For example, there has been much discussion about whose DNA profile should be placed into these databases, how they can be protected from abuse, who should have access to this information, and the like. Another issue with respect to databases concerns what happens to the specimen (e.g., blood or buccal swab) after the forensic analyst has obtained from it the desired identifying information. Does the government have the right to retain these specimens or should they be destroyed or returned to the subject?

DNA IN ACTION: DNA USED TO IDENTIFY 9/11 VICTIMS

On September 11, 2001, the World Trade Center was hit and destroyed by two hijacked planes in what was later discovered to be a terrorist attack. Over the next month, thousands of people were reported missing. On October 24, 2001, New York City officials said that the number of missing was 4339. [We now know there were 2749 people killed and, as of this writing, 1538 victims identified.] New York City Mayor Rudolph Giuliani urged relatives of the missing to submit DNA samples, which ultimately came to dominate the identification process. The identification of victims using DNA was part of a 24-hr per day operation coordinated by the New York City Medical Examiner's Office.

James DeBlase was a 45-year-old bond broker for Cantor Fitzgerald and worked on the 105th floor of the World Trade Center. His wife, Marion DeBlase, did not hear from him subsequent to the World Trade Center attacks. She last spoke to her husband shortly after the hijacked airliner collided with Tower One of the World Trade Center. Visits to the site, later dubbed "Ground Zero," left her little hope that her husband or his remains would be found. In a newspaper article, Mrs. DeBlase was quoted as stating, "I assume that this is a very tedious and lengthy job, because especially if they're not finding whole persons ... I would love to think that I would have something, but it's horrible to think that's the point we have come to." Mrs. DeBlase submitted her husband's hairbrush and toothbrush. Her family also submitted cheek swabs from the couple's three sons, as well as cheek swabs from Mr. DeBlase's parents.

DNA was used to help identify the remains of Mr. DeBlase as well as almost 700 other employees of Cantor Fitzgerald who were lost in the attack. DNA was used when other identification technologies, such as the use of fingerprints, surgical scars, and dental records, could not be employed. "You have to come to some kind of closure somehow as each day goes by, but it's very difficult to come to terms with it when you have nothing to hold on to," according to Ms. DeBlase. DNA helped bring closure to countless families who suffered the loss of loved ones and might not have otherwise been able to find the closure they so desperately needed.

Source: Sara Kugler, DNA Identifies Trade Center Victims, October 24, 2001 (Associated Press).

1.2 DNA AND HEREDITY

1.2.1 A Look at DNA from the Outside In

Imagine the construction of a building. First, plans are drawn up by an architect. The plans are given to a builder who uses different materials to construct the building. The builder uses the plans to guide him in figuring out what goes where. Different symbols and shorthand are used in the plans to show what must go where. Using the plans in the proper environment at the proper site with the right materials, the builder can construct the desired structure.

Most of us only see the finished product of this complex process, namely the completed structure. This process is similar to the way our bodies are constructed. Our bodies are similar to the building, our cells analogous to the bricks. The plans are analogous to the DNA found in our cells.

1.2.2 DNA—The Chemistry

The acronym DNA is shorthand for deoxyribonucleic (dee-ahk-see-ry-boh-noo-klee-ik) acid. DNA contains the plans that are responsible for the construction of our cells, tissues, organs, and body. DNA houses the information required to produce proteins. Proteins play a number of roles within cells as structural elements (keratin, actin, myosin, tubulin), hormones, antibodies, or enzymes. Less than 5% of our DNA contains the genes that code for the production of proteins. Enzymes are catalysts, chemicals that speed up the rate of a reaction but are not used up in the process. These proteins are necessary for the development and maintenance of cells that are then used to construct tissues. Tissues are groups of cells that function as a unit to form sheets or tubes (building elements). Tissues of different types can become physically and functionally related such that they give rise to organs, such as kidneys and livers. Thus, one can trace the pathway of life from the macromolecule known as DNA to the formation of cells, tissues, organs, and ultimately, to complete organisms.

To provide some perspective about the size of the DNA genome and where it is stored in the cell, imagine that you are able to travel within the cell (see Fig. 1.1). Cells are composed of nucleic acids [DNA and ribonucleic acid (RNA)], proteins, lipids, sugars, and a variety of other important molecules. Cells contain a large number of discrete membrane-bound structures, each with a unique and vital function. On this brief journey as you enter the outer cell membrane, you will encounter within the cytoplasm a large spherical nucleus, hundreds (and in some cases, thousands) of mitochondria, ribosomes, lysosomes, a Golgi body, interior membranes (endoplasmic reticulum) that provide channels within the cytoplasm and vacuoles, vessicles, and other structures related to cellular "feeding and drinking." The intracellular structures of human cells (animal and plant cells are eukaryotic) differ from those of bacterial cells (prokaryotic) that are much more simply organized. For a description of eukaryotic cell ultrastructure, see Figure 1.2. In virtually every human cell there is a double membrane-enclosed and somewhat

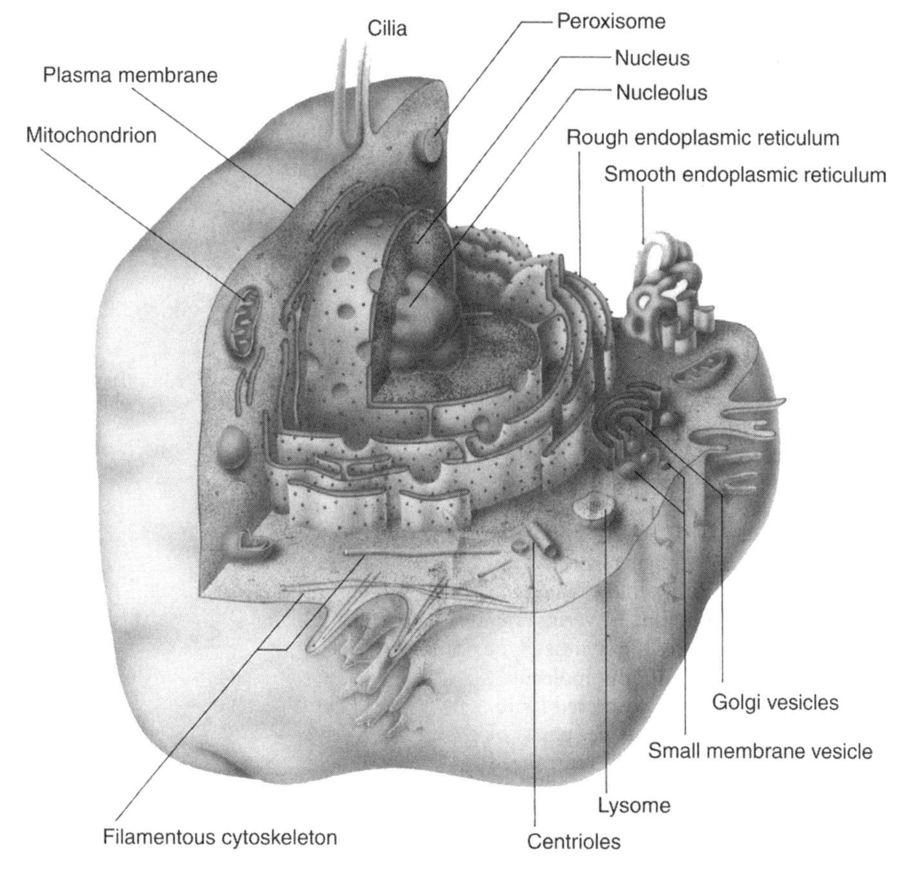

Fig. 1.1 *See color insert. Structure of the cell.*

spherical structure called the nucleus. It floats in a liquid medium called the cytoplasm that has characteristics of both a solution and a gel. Most of the cell's DNA is located in the nucleus and is known as genomic DNA. DNA within the nucleus is invisible to the naked eye as well as to the light microscope. It consists of 46 units. If the 46 molecules were linked end to end the resulting molecule would be 2 meters long, approximately 6 ft. However, it is so thin (20 Ångstrom units) that one requires a transmission electron microscope to magnify its image to visualize it.

$$1\,\text{Å} = 10^{-9}\,\text{m} = 1\,\text{billionth of a meter}$$

During the process of mitosis (or cell division), the DNA within the nucleus becomes coiled and supercoiled with the cooperation of certain proteins and takes on a unique appearance, while at the same time the nuclear membrane starts to break down in preparation for cell division. These supercoiled structures are called *chromosomes*. Human cells have 22 pairs of nonsex chromosomes (*autosomes*) plus

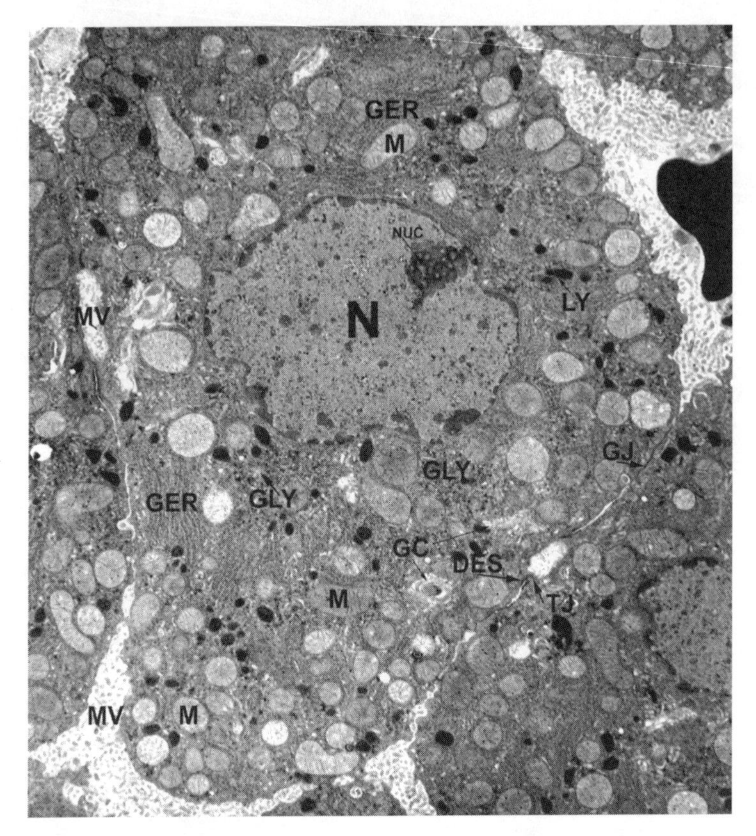

Fig. 1.2 Mammalian liver cell exhibiting the microvillus border (MV) of the cell into the sinusoidal space and the bile canaliculi. Also visible are tight junctions (TJ) and desmosomes (DES) adjacent to the bile canaliculi, and long gap junctions (GJ) between adjacent cells closer to the sinusoidal spaces. These cells have large round nuclei (N) with a single prominent nucleolus (NUC). The cytoplasm contains glycogen (GLY) and many organelles including mitochondria (M), Golgi Complexes (GC), granular endoplasmic reticulum (GER), and lysosomes (LY). Courtesy of Dr. Ronald Gordon, Mt. Sinai School of Medicine, New York, NY.

2 chromosomes that are important for sex determination ($44 + 2 = 46$ in all). The *diploid* number for the human species is 46, the total complement of chromosomes. Only our *gametes* (spermatozoa and ova) have half of that number, or 23. These cells are said to be *haploid. Sex chromosomes* are called X and Y, with females having 2 X chromosomes (XX) and males having 1 X and 1 Y (XY) chromosome. One member of each pair is maternal (originates from the mother) and the other is paternal (originates from the father). A chromosome is thus composed of a very long and thin thread of DNA combined with various kinds of protein and RNA. These molecules are intimately involved with chromosome structure and with its function. It is rather remarkable that the DNA within the 46 chromosomes of a diploid cell, when extended and placed end to end will reach a length of 6 ft, yet all of it can be packaged into a cell nucleus that is only about $1/1000$ of an inch in diameter.

The 46 chromosomes consist of approximately 6.1 picograms (pg) of DNA. (Note: One picogram is 10^{-12} g or 1/1,000,000,000,000 of a gram). Sperm and egg cells contain just half this amount. DNA is packaged in this way so that during mitosis (cell division) and during spermatogenesis (sperm formation) and oogenesis (egg formation) in the male and female, respectively, DNA can be moved from one location to another without loss of content and in a highly efficient manner. Imagine the dilemma posed by taking a structure so long and so thin and trying to move it from intracellular site to site every time a cell is to divide. When a cell divides to form two "new" daughter cells, it is vital that the 46 chromosomes first duplicate and then separate properly so that the two new cells each have the same chromosomes and DNA that the parent cell had. The cell has solved this problem by constructing DNA packages that are moved about the cell and parceled out to daughter cells by microtubules that self-assemble from the cellular protein tubulin at the appropriate time.

If we were to look at any particular gene site on a chromosome, that location would be called a locus. The nuclear genes which are located at specific sites (*loci*) along these chromosomes and that are inherited from one's mother and father are not always the same. The two chromosomes in a pair may contain genes at a particular locus that are different. Different forms of a gene are called alleles. Each chromosome contains hundreds or thousands of loci. Each locus consists of 2 alleles. If we were to compare the same locus on a pair of chromosomes, we would be able to determine whether or not the pairs are different forms of the same gene or whether they are identical. If the genes in the pair are structurally or functionally different, then the individual is said to be heterozygous at this locus. If they are identical, then the individual is said to be homozygous at this locus. An individual may be homozygous at one locus and heterozygous at another. If a person is heterozygous at a certain locus, we may or may not see these differences in the physical features of that person. However, when we look at the genetic composition of an individual, we are examining that individual's *genotype*. When we examine the physical features of a person, we are examining that person's *phenotype*.

If a chromosome was uncoiled and freed of its attached proteins and RNA, you would then see the naked DNA molecule. The "thread" actually consists of two strands that are twisted together to form a helical structure (Watson and Crick, 1952). The molecule is actually a double helix since the two strands are wound around each other, each turning in a right-handed helix but oriented in opposite directions. The two strands of DNA making up this double helix are *antiparallel* relative to each other, which means that they are aligned in opposite directions (one positioned 5′ to 3′ and the other positioned 3′ to 5′). The numbers 3′ and 5′ refer to positions of atoms on the ribose sugar portion of the molecule, and the directionality of a strand is signified by using these numbers (described further below). This double helical molecule looks like a twisting, turning ladder. When the two threads that make up the double helix come apart, the molecule is said to have become *denatured*, and the new configuration is known as single stranded (as opposed to double stranded). If you looked at the double helix, you would probably notice that it is made up of a string of nitrogenous bases held together by a backbone

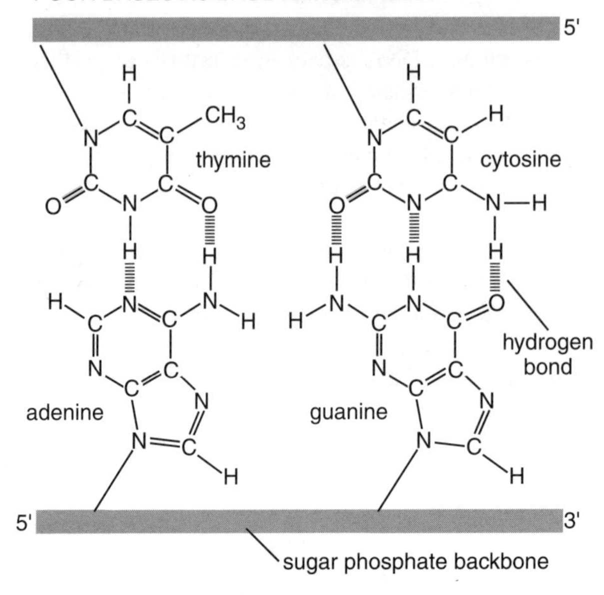

Fig. 1.3 Four nitrogenous bases found in DNA. Bases are chemically linked to backbone of sugar–phosphate groups. Two strands of DNA thus form the double helix. Reprinted with permission of CRC Press, LLC, Forensic DNA Technology by M.A. FArley and J.J. Harrington. Copyright CRC Press.

composed of repeating sugar–phosphate units. Four different chemical groups, or bases, are chemically linked like puzzle pieces to form the DNA helix. The DNA double helix is held together by hydrogen bonds that form between the nitrogenous bases on each strand. The bonded bases form the rungs of the ladder (see Fig. 1.3). The fundamental unit that is used to construct the macromolecule, DNA, is called the *nucleotide*, which consists of a 5-carbon sugar, a phosphate group, and a nitrogenous base (see Fig. 1.4). It takes approximately 3.1 billion of these units to construct the human genome (Primrose, 1998).

To understand how the double helix is held together, think of the four different building blocks of DNA, known as nucleotides. Each nucleotide contains a nitrogen-containing base along with a sugar and phosphate group of atoms. The first nucleotide contains the nitrogenous base *adenine [A]* (ad-uh-neen). The second contains the nitrogenous base *thymine [T]* (thy-meen). The third contains *cytosine [C]* (syt-uh-seen), and the fourth nucleotide contains the nitrogenous base *guanine [G]* (gwah-neen). A and G are called *purines* and C and T are called *pyrimidines* because of their structural differences. Even A and G (and C and T) have slightly different structures. As a result, only certain pieces can fit together in a stable form and associate chemically with one another. The nitrogenous bases between two strands connect as doublets, which are called base pairs (bp). The A piece only fits with the T piece and vice versa. The C piece only fits with the G piece

The components of nucleotides

Nucleotide = + sugar + phosphate

4 different dNTP's (deoxynucleoside triphosphate) :

deoxyadenosine triphosphate = dATP

deoxyguanosine triphosphate = dGTP

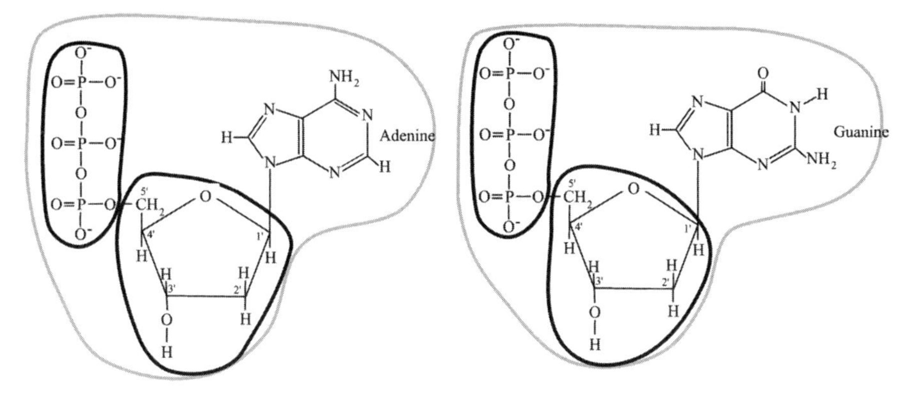

deoxycytidine triphosphate = dCTP

deoxythymidine triphosphate = dTTP

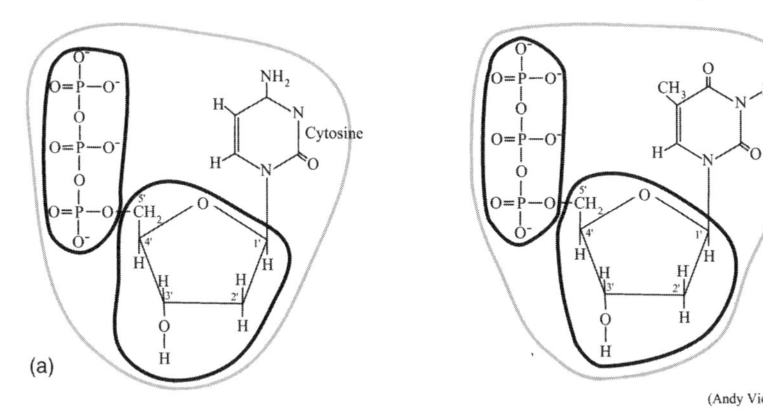

(Andy Vierstraete 1999)

Fig. 1.4 *(a) Fundamental unit of DNA is called a nucleotide. Each of these building blocks is composed of a 5-carbon sugar, a phosphate group, and a nitrogenous base. (b) Four nucleotides are linked together on a single strand of DNA.*

and vice versa. A purine can only fit easily with a pyrimidine. The bases are held together by *hydrogen bonds*, two of which hold the A-T pair and three hold the G-C pair (see Fig. 1.5). Hydrogen bonds are relatively weak and can easily be broken by raising the temperature from 68°C (room temperature) to 95°C or higher.

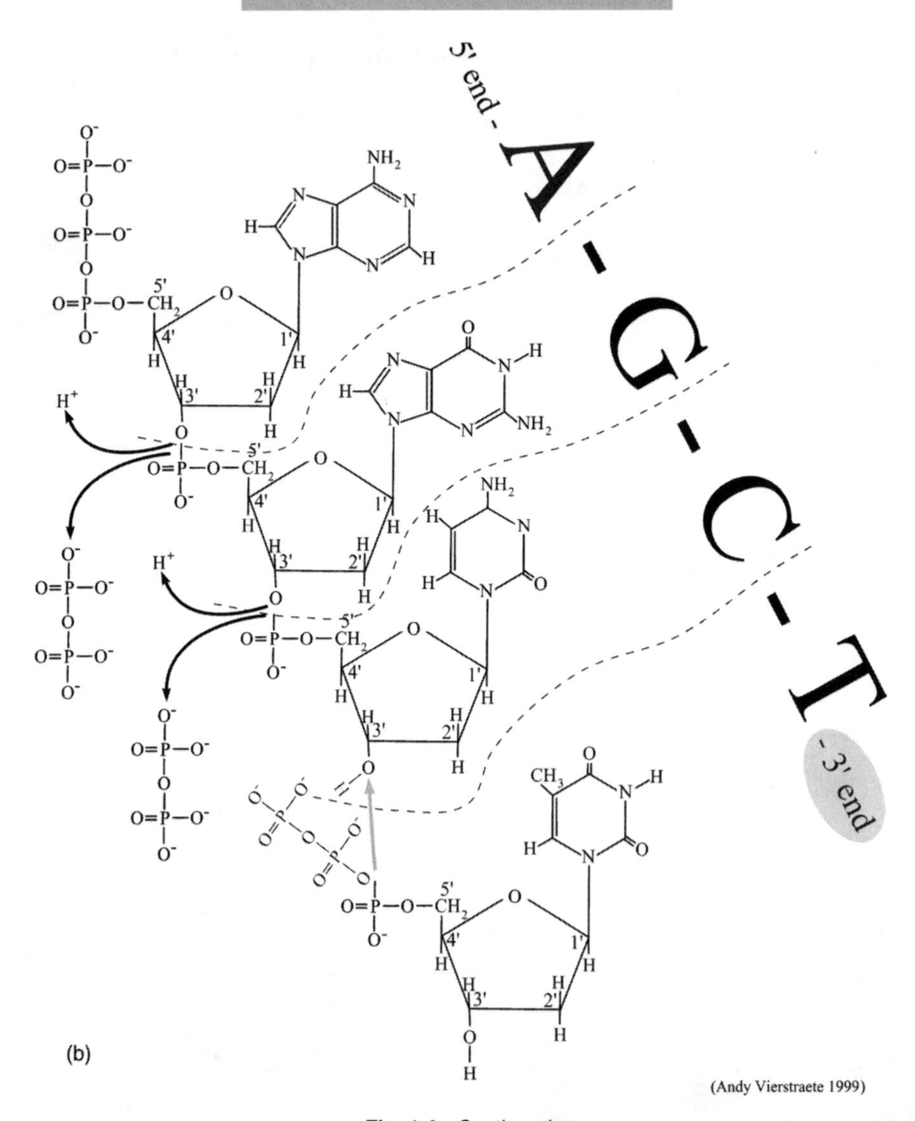

Fig. 1.4 Continued.

If the sequence of nucleotide bases on one DNA strand is specified, then the sequence of bases on the opposite strand can easily be determined. Hence, if one strand has the sequence 3′A-G-G-T-C-A-C5′, then the sequence on the second strand must be 5′T-C-C-A-G-T-G3′. This sequence-dependant association is known as *complementary binding*. The 3′ and 5′ designation at each end of the two strands indicates that each strand has directionality, and a sequence can be specified with

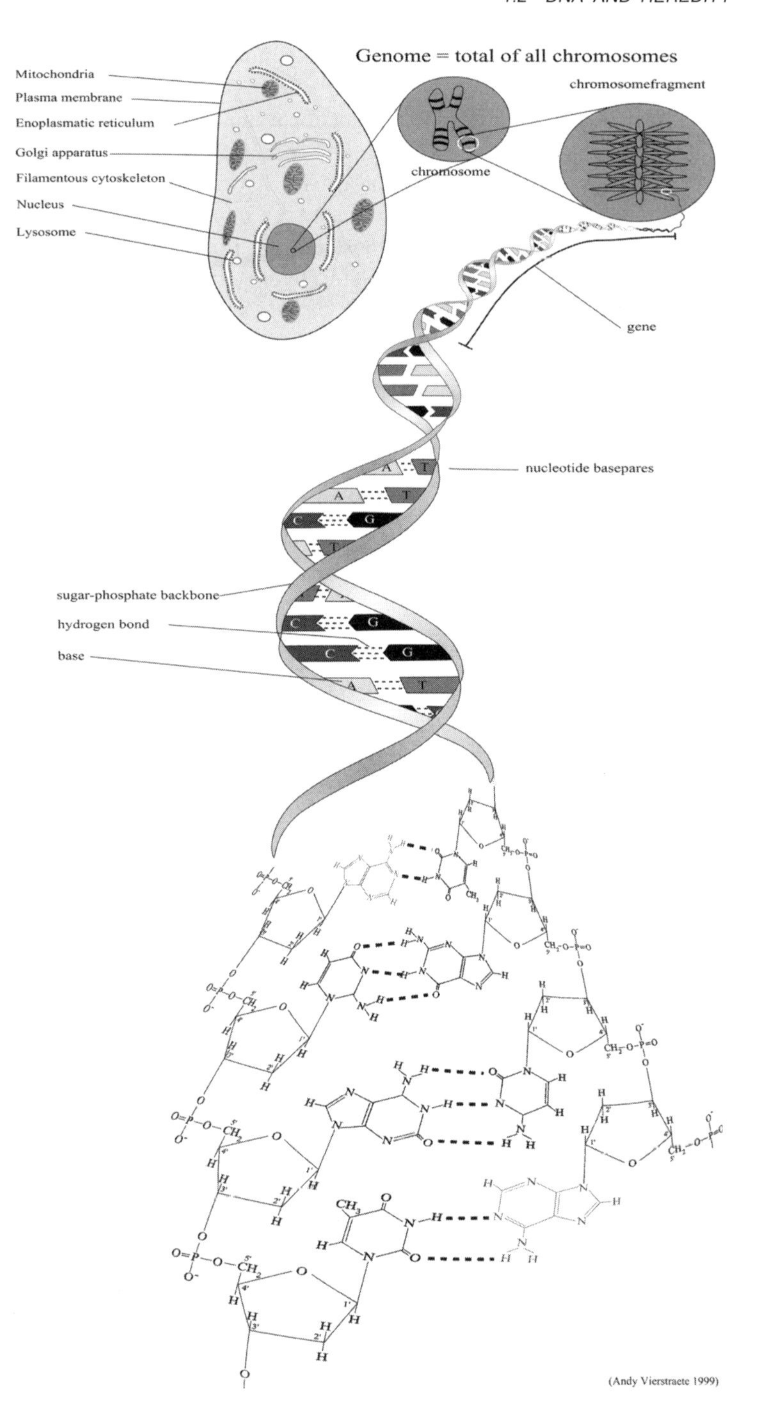

Fig. 1.5 *See color insert. Bases are held together by hydrogen bonds, two of which hold the A-T pair together while three bonds hold the G-C pair together.*

direction, for example, the 5′ to 3′ direction. The numbers correspond to the 3′ and 5′ carbon atoms within the deoxyribose sugar entity of the nucleotide mentioned above. Complementary binding is important for a number of reasons. The first is the way DNA is copied or replicated in cells. The method of copying DNA in a test tube using polymerase chain reaction (PCR) is very similar to the way DNA is replicated in living cells. Another reason it is very important to understand complementary binding and replication is because current forensic DNA tests are based upon replicating small amounts of template DNA within evidence, while others are based on complementary binding of "probe DNA" to target segments of DNA fragments within samples, and still other tests are based on both DNA replication and sequence detection.

1.2.3 Unique Sequence and Repetitious DNA

Although it has long been believed that the human genome contains the information for the synthesis of some 100,000–200,000 genes, it has recently been determined that there may be fewer than 30,000 genes responsible for the synthesis of the hundreds of thousands of proteins. In fact, some believe there are as few as 20,000 genes in the human genome. These genes are contained in approximately 5% of the 3.1 billion base pairs that constitute our nuclear genome. The proteins for which these genes code are important to the structure and function of cells and the maintenance of life. As described above, one of the major roles of proteins is to function as *enzymes.* In cells, reactions take place rapidly and efficiently at body temperature, 37°C, thanks to these enzymes. Similar reactions in test tubes in the absence of enzymes would require adding heat or another form of energy to activate the reaction. Cells can perform these reactions while at the same time maintaining their temperature at 98°F.

Sequence analysis of the human genome reveals a variety of DNA forms including (1) unique sequences, (2) moderately repetitious sequences, and (3) highly repetitious sequences. It is primarily the unique sequence DNA that is responsible for the production of most of the enzymes and structural proteins described above. Some of the highly repetitious sequences are responsible for the production (transcription) of the RNA molecules required for the synthesis (translation) of proteins on ribosomes. There is also a form of highly repetitious DNA known as *SINEs*, which stands for short interspersed nuclear elements (Fowler et al., 1988). The best known SINE is the polymorphic family known as *Alu insertions.* (Polymorphism refers to the existence of multiple alleles at a single locus and is described more fully below.) These units are approximately 300 bp long and are repeated up to a million times in the human genome. They are inserted at specific sites in different chromosomes in different individuals and inherited in a stable manner. Thus, some individuals have an Alu insert at a specific chromosomal site while others do not. The presence of the Alu insert increases the length of the insertion locus and can thus be detected by PCR followed by electrophoresis, a technique that separates DNA fragments of different sizes. (Electrophoresis is described more fully below.) These interesting DNA insertion fragments have not generated much interest in the forensic community largely because they are biallelic rather than multiallelic. This means that they are either present or they are not present at a specific locus. Because there are much better and more powerful ways to individualize evidence, Alu inserts have not found usage in forensic laboratories.

It is the moderately repetitious DNA that is of greatest interest to forensic analysts. It has been shown that 99.9% of human DNA is the same in every individual. In fact, every individual's DNA has a relatively small number of variations from that of others. It is that variation of 1 in every 1000 bases that allows us to dis-

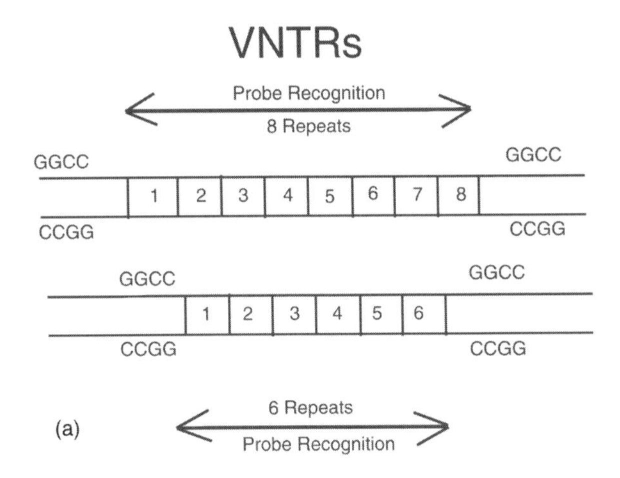

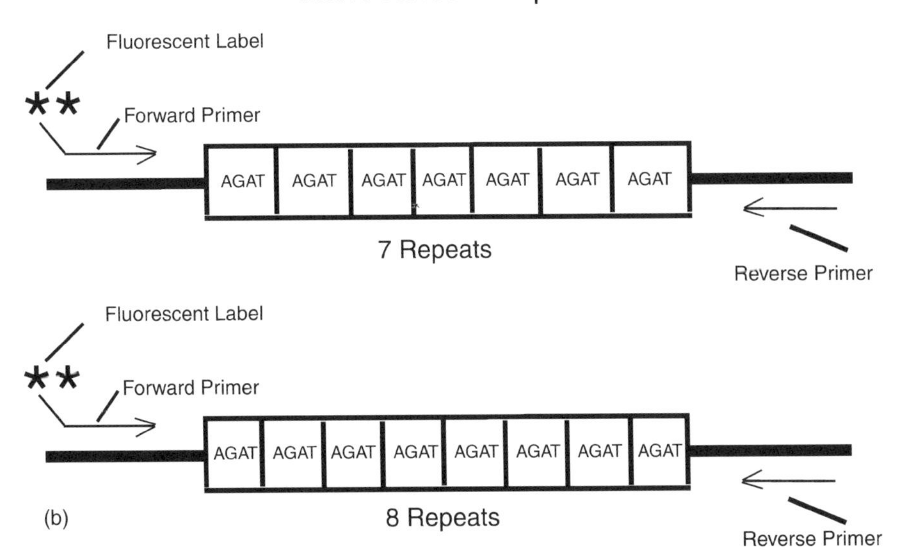

Fig. 1.6 *(a) Variable nucleotide tandem repeats (VNTRs). (b) Short tandem repeats (STRs).*

tinguish one individual from another through forensic genetic testing. These differences are responsible for the different phenotype and genotype profiles that people have.

There are several types of moderately repetitious DNA: minisatellite DNA includes *variable nucleotide tandem repeats* (VNTRs), microsatellite DNA includes *short tandem repeats* (STRs) (Fig. 1.6*a* and 1.6*b*). VNTRs are segments of DNA that consist of repeating *core* units made up of 7–25 bp that are connected adjacent to one another. These segments can consist of up to approximately 50 repeating core units and can reach lengths of several kilobase (1000 base) pairs. These regions were the first to be exploited through the technology of *restriction fragment length polymorphism* (RFLP—also known as "DNA fingerprinting") analysis for identification purposes (Jeffreys et al., 1985). The great benefit of using these VNTR regions is their very high degree of polymorphism due to the existence of so many alleles that can possibly be found at any one locus.

Short tandem repeats, or STRs, are similar to VNTRs in that they are tandemly connected core units each of which consists of 2–6 bp (Li et al., 1993). These core units can be repeated up to about 40 times at a locus but more commonly are repeated about 7–15 times. Because STRs are smaller in size (length) than VNTRs, they can be replicated and amplified more efficiently and accurately. STR testing has the advantage over VNTR testing in that less DNA is required and the analysis is completed much more quickly. Furthermore, if the DNA under study has become degraded, then testing the shorter STR fragments is much more likely to succeed than if one were to test the longer VNTR-containing fragments.

DNA fingerprinting techniques are based on the examination of VNTR loci, whereas the early PCR-based techniques are based on either sequence polymorphism (reverse dot blot technique) or on length polymorphism, such as *amplified fragment-length polymorphism* (AmpFLP). The current methods for examining nuclear DNA are based on the amplification and detection of STR-length polymorphisms. PCR has also been used to amplify polymorphic regions of mitochondrial DNA (mtDNA), which are found in the *hypervariable displacement (D) loop* of the mitochondrial genome. All of these techniques are described more fully in Chapter 3.

1.3 DNA REPLICATION

1.3.1 Replication in the Cell

There are some 200 or more different classifications of cells that make up the trillions of cells within our body. Due to aging and normal wear and tear in the body, cells die constantly and must be replaced. As cells replicate, so too must their internal components replicate, including the DNA found within the nucleus. Failure to copy DNA accurately results in a mutation or change in sequence. Although DNA repair enzymes that function to fix such unexpected changes exist, they are not infallible. In fact, unrepaired mutations occurring in the germ line (cells that produce gametes, spermatozoa, and ova) can be transmitted to offspring as heritable mutations. Not every mutation is harmful, but some can result in serious disease and some can have lethal consequences for the bearer of such altered genes. Clearly, DNA replication in cells must be efficient, faithful, and accurate.

In the laboratory, DNA can be copied using a mechanism similar to that used naturally, *in vivo*, by cells of the body. The early DNA analysis technology had several limitations. When using the procedure known as DNA fingerprinting (RFLP), a significant amount [100 nanograms (ng)] of nonfragmented DNA was required if the analysis was to be successful. Thus, insufficient quantity and poor quality of the sample often limited the ability to obtain useful identifying information. In some instances, biological evidence found at a crime scene can be exceedingly small in amount. In such a case, the analyst cannot obtain the genetic information that might be helpful in solving the crime. If we had the ability to repeatedly replicate (amplify) a small quantity of sample DNA and produce large numbers of identical copies, we then might be able to determine the genetic information present and compare it to that of the suspect. In essence we would like to copy specific regions of nuclear DNA much like a copy machine is used to duplicate a desired document. We do not need to copy the entire DNA of the sample (as the cell does *in vivo*), only the specific regions of the genome that are informative about human identity. In the mid-1980s Kary Mullis developed a procedure known as the polymerase chain reaction that mimics the replication of DNA *in vivo*. The procedure is composed of three distinct temperature-dependent stages in which the template molecule is first denatured and then primers are annealed to each of the resulting single strands. The primers identify the specific regions of the genome that are to be amplified. The polymerase enzyme extends the new strand by incorporating building blocks onto the primers. The building blocks are laid down so that they are complementary to the sequence on the template strand. In the final stage of the cycle the enzyme ligates (ties together) the nucleotide building blocks that have been incorporated into the new strand. Repeated thermal cycling results in hundreds of millions or even billions of amplified products that can subsequently be detected and analyzed to produce useful identifying information. Multiplex amplification results in the analysis of multiple loci simultaneously, thereby conserving evidence, generating a great deal of information from minute amounts of evidence, and saving time, labor, and money (Kimpton et al., 1993). The polymerase chain reaction is described in detail in Section 3.3.

1.3.2 Cloning (Gene Amplification)

For a number of applications of importance to molecular biologists, it became important to obtain large amounts of a particular protein. Rather than attempting to isolate relatively large quantities of substances normally present in only very small amounts in certain tissues/organs, a very important technique was developed in the early 1980s that enabled scientists to amplify genes of interest and to use this amplified DNA to produce large amounts of the needed protein. *Gene amplification (cloning)* is a method used to amplify segments of DNA that have been incorporated into plasmids, extrachromosomal circles of DNA that normally inhabit certain bacteria (Drlica, 1984). Some plasmids can even be found in a small number of eukaryotes or higher cells. Generally, bacteria do not require intracellular plasmids for their own existence unless the plasmid contains a gene that is vital for the host bacterias' survival under certain environmental conditions. Recombinant DNA technology methods have allowed biologists to insert DNA fragments of interest into plasmids (Watson et al., 1992). The plasmid serves as a vector that will enter the bacterial cell and form a stable relationship with the bacterium. This is called transformation. Plasmids such as pSC101 or ColE1 can replicate themselves autonomously, and, as they do so, the desired segment of DNA becomes amplified or cloned as well. Analysts use restriction enzymes to cut both the plasmid and the gene of interest in such a manner as to form sticky ends on both, mixing the two results in the desired combination. A DNA ligating enzyme covalently links the two components forming a larger plasmid containing the desired DNA fragment. Of course, many other methods are available to form the union of plasmid and target DNA. After extensive replication of the plasmid within the bacteria, there are many copies of the desired fragment available. The analyst, using the same restriction enzymes can now harvest large amounts of the desired DNA. This is one way of producing probes that are used in DNA fingerprinting (RFLP). This method of amplification is quite different from PCR, which is now routinely used for forensic DNA analysis and is described in Chapter 3.

UNIQUENESS OF EVERY INDIVIDUAL

The analysis of DNA evidence can be a very useful and powerful tool for the purpose of human identification. Human identity testing (Sajantila, 1998) has been so successful that additional fields have emerged for the identification of animals and plants where the issues of endangered species and breeding are important. Using various techniques, evidence found at the scene of a crime can be linked to an individual. The results of DNA testing are admissible in courtrooms throughout the United States, and experts can provide testimony about these procedure and their findings. Expert witnesses can express their opinions about the evidence unlike other witnesses.

With the knowledge of the chemistry of DNA and how it is replicated within the cell, it becomes easier to understand how it is used in human identification. Each person is different. The unique and distinct traits of individuals occur because, though much of our DNA is the same (99.9%), a portion of it is not. The genetic difference between one individual and another is due to differences in 1 million base pairs of DNA. This represents only 0.1% of the human genome or one out of every thousand bases. Some of those variable DNA sequences, which differ from person to person, are the targets that we wish to focus our attention on for individual identification. Scientists have been able to define and isolate a number of these regions (loci). The type of DNA that is examined is classified as moderately repetitious. Sometimes these regions consist of thousands of base pairs. DNA fingerprinting, or RFLP, can be used to analyze these loci. After isolating DNA from crime scene evidence, restriction enzymes are used to cut the DNA at sites that allow for an analysis of those specific loci that have been determined to be useful for identification. Alternately, after isolation of DNA from evidentiary samples, polymerase chain reaction can be used to make copies of specific loci and to examine them for genetic type at each locus. If a suspect then becomes available, that suspect's DNA can be matched against the DNA found at the crime scene. If the DNA matches, it is likely that the suspect was at the crime scene. The results of the comparative analyses can then be used at trial to demonstrate the probability that the suspect produced the biological evidence at the crime scene. DNA analysis has also been used to show that a defendant could not possibly be the source of evidence found at a crime scene.

The importance of using forensic DNA analysis in sexual assault cases cannot be overstated. DNA found on evidentiary swabs obtained from a rape victim can be used to show that a man accused of a rape had penetrated and ejaculated within the victim's vagina. In the same way, DNA analysis can exonerate a man who has been falsely accused of rape.

Anastasia was born in June 1901 into the Romanov family, which ruled Russia for many years. The czar was overthrown in 1916 and Anastasia was killed along with her family by a firing squad in 1918. The only member of the family to survive was the Dowager Empress Marie who settled in Denmark. Anna Anderson claimed to be the missing Anastasia, and many people who had known Anastasia swore that Anna was telling the truth. However, in the 1920s, a German court found her to be merely a Polish peasant girl despite her strong physical resemblance to the real Anastasia. Anna Anderson died in 1984 in Richmond, Virginia. In 1994, her DNA was compared with that of Anastasia's grand nephew, Prince Philip, Queen Elizabeth's husband, proving conclusively that Anna Anderson was not Anastasia. The comparison could be performed because of the familial relationship between the royal families of England, Austria, and Russia. Queen Victoria of England was the great grandmother of Anastasia.

The mass grave of Anastasia's family was discovered in Yekaterinberg in the Ural Mountains in 1991, after the collapse of the Soviet Union. Nine bodies were recovered including four males and five females along with the bones of a small dog thought to be owned by Anastasia. All the males were positively identified including Czar Nicholas II. However, supposedly six females were executed at Yekaterinberg in 1918. These were: Czarina Alexandra, her four daughters, Olga, Titiana, Marie, and Anastasia, and the family cook. This means that the body of one female is unaccounted for. What happened to Anastasia therefore remains a mystery.

Source: Anastasia Claimant a Fraud, Reuter, Oct. 5, 1994; DNA Test Shows Virginia Woman Wasn't Anastasia, Reuter, Oct. 4, 1994.

REFERENCES

Barnett, P. D. and R. A. Berger. The Effects of Temperature and Humidity on the Permanency of Latent Fingerprints. *J. Forensic Sci. Soc.*, **16**, 249 (1977).

Bodziak, W. J. *Footwear Impression Evidence.* Elsevier, New York, 1990. DeForest, P. R., R. E. Gaensslen, and H. C. Lee. *Forensic Science: An Introduction to Criminalistics.* McGraw-Hill, New York, 1983.

Cole, S. A. *Suspect Identities: A History of Fingerprinting and Criminal Identification.* Harvard University Press, Cambridge, MA, 2001.

Cowger, J. F. *Friction Ridge Skin: Comparison and Identification of Fingerprints.* Elsevier, New York, 1983.

DeForest, P. R., R. E. Gaensslen, and H. C. Lee. *Forensic Science: An Introduction to Criminalistics.* McGraw-Hill, New York, 1983.

Drlica, K. *Understanding DNA and Gene Cloning: A Guide for the Curious.* Wiley, New York, 1984.

Erlich, H. A. (ed.). *PCR Technology: Principles and Applications for DNA Amplification.* Stockton Press, New York, 1989.

Fowler, J. C. S., L. A. Burgoyne, A. C. Scott, and H. W. J. Harding. Repetitive Deoxyribonucleic Acid (DNA) and Human Genome Variation—A Concise Review Relevant to Forensic Biology. *J. Forensic Sci.*, **33**, 1111–1126 (1988).

Hochmeister, M. N., O. Rudin, and E. Ambach. PCR Analysis from Cigaret [sic] Butts, Postage Stamps, Envelope Sealing Flaps and Other Saliva-Stained Material. In *Forensic DNA Profiling Protocols*, Vol. 98. P.J. Lincoln and J. Thomson (eds.). Humana Press, Totowa, NJ, (1998), pp. 27–32.

Jeffreys, A. S., V. Wilson, and S. L. Thein. Individual-Specific Fingerprints of Human DNA. Nature **316**, 76–79 (1985).

Kelly, K. F., J. J. Rankin, and R. C. Wink. Method and Applications of DNA Fingerprinting: A Guide for the Nonscientist. *Criminal Law Rev.*, **105**, 108 (Feb., 1987).

Kimpton, C., P. Gill, A. Walton, A. Urquhart, E. Millican, and M. Adams. Automated DNA Profiling Employing Multiplex Amplification of Short Tandem Repeat Loci. *PCR Methods and Applications*, **3**, 13–22 (1993).

Lambourne, G. *The Fingerprint Story.* Harrap, London, 1984. [It was William Herschel who first used fingerprints for the purpose of identification of individuals.]

Li, H., L. Schmidt, M. H. Wei, T. Hustad, M. I. Lerman, B. Zbar, and K. Tory. Three Tetranucleotide Polymorphisms for Loci: D3S1352, D3S1358, D3S1359, *Hum. Mol. Genet*, **2**, 1327 (1993).

Neufield, P. J. and N. Coleman. When Science Takes the Witness Stand. *Sci. Am.*, **262**(5), 46–53 (1990).

Primrose, S. B. *Principles of Genome Analysis: A Guide to Mapping and Sequencing DNA from Different Organisms.* Blackwell Science, Malden, MA, 1998.

Rhodes, H. T. F. *Alphonse Bertillon, Father of Scientific Detection.* Harrap, London, 1956. [Alphonse Bertillon came up with the first system that allowed data on various people to be categorized in such a way that it was possible to quickly find a desired individual using a description.]

Sajantila, A. *Second European Symposium on Human Identification.* Promega, Madison, WI, 1998, pp. 1–5.

Schmalleger, F. *Trial of the Century: People of the State of California vs. Orenthal James Simpson.* Prentice Hall, Upper Saddle River, NJ, 1996.

Stouder, S. L., K. J. Reubush, D. L. Hobson, and J. L. Smith. Trace Evidence Scrapings: A Valuable Source of DNA? *Forensic Sci. Commun.*, **3**(4) (October, 2001).

Watson, J. D. and F. H. C. Crick. A Structure for Deoxyribose Nucleic Acid, *Nature* **171**, 737–738 (1952).

Watson, J. D., M. Gilman, J. Witkowski, and M. Zoller. *Recombinant DNA*, 2nd ed. W. H. Freeman, New York, 1992.

Wilson, T. Automated Fingerprint Identification Systems. *Law Enforcement Tech.*, Aug–Sept (17–20) 45–48, 1986.

BIBLIOGRAPHY

Berg, P. and M. Singer, *Dealing with Genes: The Language of Heredity.* University Science Books, 1992.

Cooper, G. M. *The Cell: A Molecular Approach.* ASM Press, Washington, D.C., 1997.

Hua, T., H. G., Reza, and L. Romney. New Suspect Charged as Man Held 17 years Is Freed; Crime: DNA Points Investigators to an Ex-Marine Already In Custody. Police Link Him To Six Slayings, *Los Angeles Times*, Vol. 115, p. 1, 1996.

Inman, K. and N. Rudin. *Principles and Practice of Criminalistics: The Profession of Forensic Science.* CRC Press Boca Raton, FL, 2001.

Kennedy, J. M. DNA Test Clears Man Convicted of Rape, *Los Angeles Times*, Vol 113, 1990.

King, P. H. Killing of Jogger Barbara Schoener by Mountain Lion Shocks Residents of Cool California, *Los Angeles Times*, Vol. 113, 1994.

Lodish, H., A. Berk, S. L. Zipursky, P. Matsudaira, D. Baltimore, and J. Darnell. *Molecular Cell Biology*, 4th ed. W.H. Freeman, New York, 2000.

Mays, G. L., N. Purcell, and L. T. Winfree. DNA (Deoxyribonucleic Acid) Evidence, Criminal Law, and Felony Prosecutions: Issues and Prospects, *Justice System J.*, **16**(1) 111 (1992).

Osterburg, J. W. and R. H. Ward. *Criminal Investigation: A Method for Reconstructing the Past.* Anderson, Cincinnati, Ohio 1992.

Rudin, N. DNA Untwisted: Correcting Some Misconceptions about Genetic Evidence, *L.A. Daily J.* Apr. 20, 1995, p. 6.

Schmidt, J. E. *Attorneys' Dictionary of Medicine and Word Finder*, Vol. 4. Matthew Bender, Albany N.Y. 1962, 1984.

State v. Pennington, 327 N.C. 89, 93, 393 S.E.2d 847, 850 (1990) [in which the court describes DNA in simple terms].

Terry, D. DNA Tests and a Confession Set Three on the Path to Freedom in 1978 Muders, *NY Times* Section 1, page 6, column 1, Vol 145 June 15 1996.

Valentine, P. W. Jailed for murder, Freed by DNA, *Washington Post*, 116, June 29 1993, p. A1, Col. 1.

Watson, J. D., N. H. Hopkins, J. W. Roberts, J. A. Steitz, and A. M. Weiner. *Molecular Biology of the Gene,* 4th ed. Benjamin Cummings, Menlo Park, CA. 1987.

DNA—General

Budowle, B., J. Smith, T Moretti, and J. Dizinno. *DNA Typing Protocols: Molecular Biology and Forensic Analysis.* Biotechniques Books, Eaton, Natick, MA, 2000.

Caetano-Anolles, G. and P. M. Gresshoff (eds.). *DNA Markers: Protocols, Applications, and Overviews.* Wiley-Liss, New York, 1997.

Caplan, Y. H. Current Issues in Forensic Science: DNA Probe Technology in Forensic Serology: Statistics, Quality Control and Interpretation. *Academy News* [American Academy of Forensic Science Newsletter], **18**(6), 1 1988.

Connors, E., T. Lundregan, N. Miller, and T. McEwen, *Convicted by Juries, Exonerated by Science: Case Studies in the Use of DNA Evidence to Establish Innocence After Trial.* U.S. Department of Justice, Office of Justice Programs, National Institute of Justice, June 1996.

Farley, M. A. and J. J. Harrington (eds.) *Forensic DNA Technology.* Lewis, Chelsea, MI, 1991.

Herrin, G. and R. E. Gaensslen, DNA Typing—Criminal and Civil Applications. In *Forensic Sciences*, Chapter 37. Matthew Bender, Albany NY, 2001.

Kaye, D. H. and G. F. Sensabaugh. Reference Guide on DNA Evidence. In *Reference Manual on Scientific Evidence*, 2nd ed. Federal Judicial Center, 2000; also available at http://air.fjc.gov/public/fjcweb.nsf/pages/16.

Kobilinsky, L. Deoxyribonucleic Acid—Structure and Function, A Review. In *Forensic Science Handbook,* Vol. *III.* R. Saferstein (ed.). 287–357, Regents/Prentice Hall, Englewood Cliffs, NJ, 1993, pp. 287–357.

Lincoln, P. J. and J. Thomson (eds.). *Forensic DNA Profiling Protocols.* Methods in Molecular Biology Series, Vol. 98, Humana Press, Totowa, NJ, 1998.

National Institute of Justice. Using DNA to Solve Cold Cases, 2002; available at http://www.ojp.usdoj.gov/nij.

National Research Council. *DNA Technology in Forensic Science.* Committee on DNA Technology in Forensic Science, Board on Biology, Commission on Life Sciences, and the National Research Council. National Academy Press, Washington, D.C., 1992.

Robertson, J., A. M. Ross, and L. A. Burgoyne. *DNA in Forensic Science: Theory, Techniques and Applications.* Ellis Horwood, New York, 1990.

2

Biological Evidence—Science and Criminal Investigation

2.1 CRIME SCENE INVESTIGATION—BIOLOGICAL EVIDENCE

Once a violent crime has been committed and law enforcement personnel have arrived, a crime scene must be officially declared. If a living victim is present, steps must be immediately taken to provide any required medical attention. Responding law enforcement personnel must establish a perimeter and secure the scene to prevent the loss, destruction, or contamination of physical (especially biological) evidence. Each item of evidence has an unknown history, and it is left to investigators and supporting criminalists to determine how that evidence came into being. It is obvious that once the criminalist takes possession of this evidence, it must be protected, documented, and properly labeled so that the chain of custody is preserved. There should be no question raised about possible tampering, mishandling, or contamination of the evidence. Custody of the evidence must be accounted for from the time it is found at a crime scene to the time it is produced in the courtroom.

By analyzing this evidence, it is hoped that the criminalist can reconstruct the events of the crime and shed light on who may have been responsible. Generally, the crime scene is protected by surrounding the area with crime scene tape or barriers, and a guard is posted to prevent entry of unauthorized individuals. The crime scene may be small in area or it may cover a number of square miles. For example, an automobile in which a rape had taken place would be an example of

DNA: Forensic and Legal Applications, by Lawrence Kobilinsky, Thomas F. Liotti, and Jamel Oeser-Sweat
ISBN 0-471-41478-6

the former, whereas a plane that exploded over a vast stretch of land or sea would be an example of the latter.

On the evening of July 17, 1996, a Boeing 747–131 took off from John F. Kennedy International Airport in New York on route to the Charles de Gaulle International Airport in Paris. Approximately 12 min into the flight at an altitude of about 13,000 ft the jet exploded off the shore of Long Island killing 212 passengers and 18 crew members. The crime scene was the Atlantic Ocean stretching for several square miles at a depth of about 100 ft below sea level. Most major parts of the plane were recovered, and the plane was reconstructed to determine what caused the deadly explosion.

On December 21, 1988, Pan Am Flight 103 blew up over Lockerbie Scotland killing 259 on board and 11 on the ground. The crime scene stretch for many miles, but the thorough investigation turned up sufficient evidence to convict a Libyan intelligence agent of planting a bomb on the jet.

Regardless of the size or complexity of a crime scene, the principles of crime scene investigation should be followed. Failure to do so could result in not finding vital evidence and the outcome could be another unsolved case. The following represents what might take place at a typical crime scene. Law enforcement personnel are notified that a crime has taken place and respond by going to the scene.

2.1.1 Help the Victim

The first person on the scene determines if life is at risk and if medical attention is required for the victim. This takes priority over protection of evidence. However, where it becomes necessary to have emergency medical assistance, steps are taken to minimize disturbance of the scene. A route of entry and exit should be established for medical personnel.

2.1.2 Protect the Scene

The area is guarded until trained forensic science personnel arrive to process the scene. Depending on the nature of the crime and scene a sufficient number of experienced, highly trained crime scene investigators should be brought to the scene to start the process of searching, documenting, and packaging. Crime scene investigators should take every precaution not to contaminate or disturb the scene before it is documented. In some cases, investigators will wear masks, head protection, shoe protection, gloves, and protective clothing (see Section 2.1.6). Nothing should be moved until the scene has been fully and properly documented.

2.1.3 Document the Scene

The entire scene is recorded, photographed, and documented in as many ways as practical, including:

1. Schematic diagrams with accurate measurement accompanied by notes
2. Photographing (black and white/color)
3. Videotaping

Records should be maintained detailing photography and videotaping to include not only photographic information (film type, speed, etc.) but also locations within the scene and time of day and date.

2.1.4 Search the Scene

A search of the crime scene is made in such a manner that "no stone is left unturned." By using a geometric pattern, one can be certain that the search is thorough and complete. For example, patterns and search strategies may be described as: area searches in the form of squares or rectangles (see Fig. 2.1*a–c*), concentric circles (see Fig. 2.1*b*), or in a pie configuration (see Fig. 2.1*e*). Failure to search the *entire* scene can result in the case becoming compromised. Exactly what is collected and what is not is difficult to explain since at any one crime scene there may be hundreds of potentially important items of evidence. If crime scene investigators

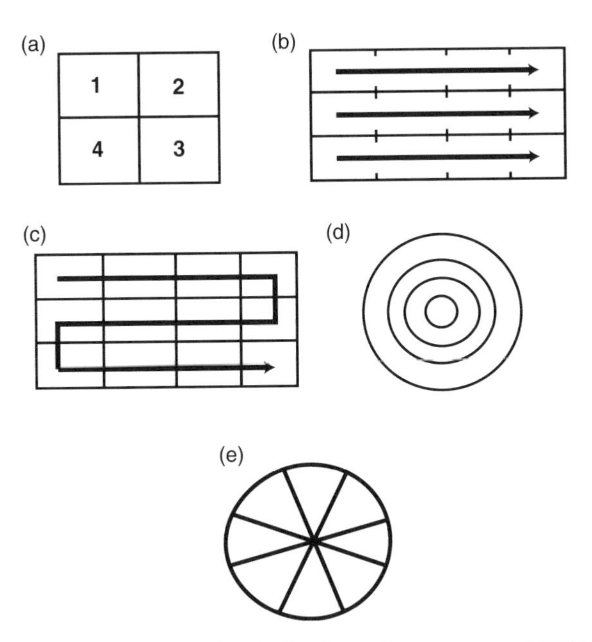

Fig. 2.1 *(a)–(c). Crime scene searches in the form of squares or rectangles, (d) concentric circles, or (e) in a pie configurations.*

submitted dozens of items to the laboratory for each case handled, the lab would quickly be overcome with casework and would cease to function properly. The crime scene investigator must use his or her knowledge and experience in deciding what to retrieve and submit for analysis (Fisher, 1993).

2.1.5 Schematic Drawing Showing Location and Photography of Items of Evidence

Any item of evidence that has been discovered should be photographed in the position in which it was found. Every photograph should be taken with an appropriate ruler or scale placed adjacent to the specimen so that the size and position of items within the photograph can be accurately determined. This is especially important when items are found on top of objects that are not flat but have curved surfaces. Measurements of the precise location of the evidence should be conducted. This can be done by measuring its distance from several nearby more permanent objects or structures. Schematic drawings and videotapes should also be made to supplement written notes. Since a case may not go to trial for many months and in some cases many years after the crime has been committed, the crime scene notes should be thorough and detailed. In this way, time will not distort what had been observed and documented at the scene.

2.1.6 Packaging and Preserving Evidence

Any item that might be relevant to the investigation must be properly packaged so as to protect and preserve the item from damage, loss, or destruction, and to prevent any potential contamination (see Wade, 2003). Crime scene investigators should wear protective garments and nonpowdered gloves (some powders can interfere with DNA analysis) and use tools and instruments that are clean and disinfected. One cannot control contamination of evidence that occurred prior to the arrival of crime scene investigators, but no contamination can be allowed to occur once the evidence is in the possession of crime scene personnel or a laboratory analyst. Gloves should be changed each time a new item of evidence is handled or collected. Biological evidence must be handled in such a manner that there is no contamination of the evidence with bacteria or fungi, both of which are ubiquitous in the environment. In addition, because of the sensitivity of forensic analyses, it is imperative that there be no contamination of biological evidence with (1) other biological evidence (e.g., placing two different items into the same container) or (2) biological material transferred by or from the crime scene investigator or laboratory analyst (Cimino et al., 1990). In general, biological materials are best packaged in "breathable" paper containers so that moisture will not form inside the bag as a result of condensation of water vapor. Until they are brought to the laboratory, they should be stored under cold and dry conditions whenever possible. If evidence is in a liquid or semiliquid state, it should be allowed to air-dry prior to packaging. Any moisture in the evidence container could promote microbial growth with the potential of destroying important biological evidence.

Biological evidence at a crime scene is very important since this category of evidence can be analyzed using the highly sensitive forensic DNA analysis methods that now are routinely used in virtually every crime laboratory in the United States. This analysis can provide genetic information that can be used to reveal the source of this evidencc (Lee, 1982, 2001; Inman and Rudin, 2000). Biological evidence includes blood, semen, saliva, hair, bone, teeth, or any other physiological fluid or tissue (see Fig. 2.2). All of these items contain DNA within nucleated cells. The pattern in which the evidence has been deposited on a substrate may also provide critical information about the events that took place at the time of the crime. Blood spatter on a wall, for example, can sometimes provide information regarding the position of victim and assailant, the type of weapon used, and the force applied (James and Eckert, 1999).

Nonbiological material should also be handled and packaged with caution since such materials may contain traces of biological material, for example, a bullet with a minute amount of human tissue on its surface. Documentation of the packaged evidence might include taping a bar code onto the item, and providing the name (initials), date, time, and the place of collection so that it can be tracked from the moment it is collected until the time it appears in the courtroom. The evidence should be transported as quickly as possible and in an appropriate manner to the laboratory for subsequent analysis. As a general rule, biological evidence is best preserved when maintained frozen (or at least very cold) and dry. Despite these

BIOLOGICAL EVIDENCE

Types & Sources

- **Semen**
  - Sheets, Pillows, Beddings
  - Vaginal Swabs
  - Clothing, Undergarments
  - Condoms
  - Sheets, Pillows Blankets
- **Teeth**
- **Urine**
- **Bone**
- **Blood**
  - Backspatter
  - Stains
  - Whole Blood
  - Autopsy Tissues
- **Saliva**
  - Cigarette butts, Envelope Flaps
- **Hair**
  - Root and Shaft
- **Tissue**

Fig. 2.2 Biological evidence includes blood, semen, saliva, hair, bone, teeth, or any other physiological fluid or tissue.

precautions, DNA is far more stable than proteins. Prior to the DNA era of forensic science, criminalists would primarily study proteins and glycoproteins (proteins containing sugar residues) to determine the origin of biological evidence. DNA can survive many common environmental conditions such as high or low temperatures, low humidity, exposure to certain chemicals, and the like, and it remains in fairly good condition for lengthy periods of time.

2.1.7 Transport to Laboratory

Transport of evidence from the crime scene to the laboratory in a refrigerated container is adequate in most cases. Ice and dry ice are also useful in keeping the evidence cold. In warmer climates, it is essential that biological evidence be preserved in this way. Despite the amazing stability of the DNA molecule described above, transport should not be prolonged since biological evidence such as blood or semen *starts* to decompose or deteriorate the moment it leaves the body. One should not place biological evidence in the trunk of a vehicle on a hot summer day since this would likely cause deterioration of the evidence and decrease the chance that useful information would be obtained by laboratory analysis.

2.1.8 Sexual Assault Evidence

In sexual assault cases, the victim is brought to an emergency room where a physician conducts a physical examination, provides appropriate medical treatment, and collects evidence as indicated by the procedures outlined in the instructions found within a rape kit. The kit contains items that facilitate the collection, documentation, and packaging of important evidence of the assault (Fig. 2.3). Where available, a colposcope is used to collect photographic images of bruising, tearing, laceration, or other trauma to the patient's genitalia. Generally, the physician will collect blood, saliva, fingernail scrapings, oral, anal, and vaginal swabs, scalp and pubic hair, and other trace evidence. Blood is collected in tubes containing EDTA (ethylenediaminetetraacetic acid), an anticoagulant. Blood tubes should be stored refrigerated and not frozen. In addition to the victim's undergarments, any other item (clothing, bed sheets, blankets) that might contain evidence will also be packaged, documented, and brought to the laboratory for analysis. Samples should be separately obtained from the external genitalia and vaginal vault to help determine if penetration had occurred. Swabs are usually supplied in pairs. One of each swab is used to smear a microscope slide that will be stained to enhance contrast under the microscope and studied by a forensic specialist in the laboratory. The swabs are usually the most significant evidence since any foreign substance found on the cotton may provide valuable information leading to the identification of the assailant. Any signs of emotional and physical trauma or lesions must be noted. Some emergency rooms are now equipped with an instrument called a colposcope that can be used to document physical lesions deep inside the vaginal vault or on the cervix of the uterus. In addition to emotional support, the victim should be provided with medication to prevent pregnancy and antibiotic and antiviral medications

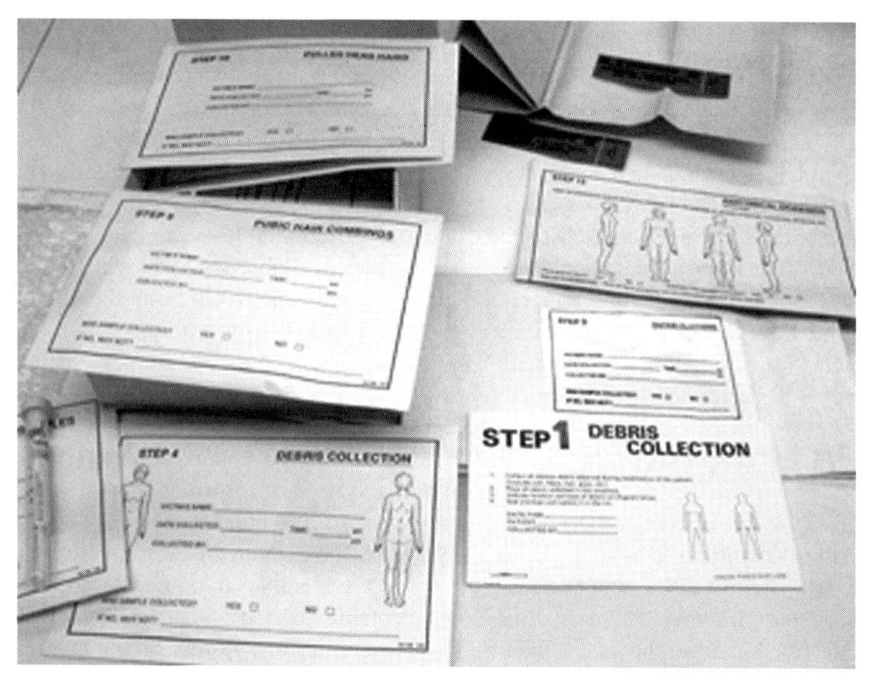

Fig. 2.3 *Rape kit contains items that facilitate the collection, documentation, and packaging of important evidence of assault.*

to prevent infection. The rape kit and its contents are sealed, documented (with name of patient, doctor, time and date, and police officer's name or initials), and brought to the forensic laboratory for analysis.

In the laboratory, the rape kit evidence together with any other physical evidence relevant to the case will be tested. If the assailant ejaculated within the victim's vagina, then the vaginal swab will contain fluid and cellular material from both the victim and assailant. It is important that the victim have the rape kit examination done as soon after the assault as possible because over time semen will drain and valuable evidence will be lost. If the victim had been assaulted by more than one individual, then cells from both assailants may be present on the swab. It is important that the emergency room physician collects these swabs carefully since vaginal penetration must be evident if the assailant(s) is (are) to be charged with the crime of rape. Mixed specimens such as these can pose certain problems for the criminalist since the genetic profiles obtained must be properly interpreted (see Sections 3.3.4.3.27 and 4.13). It is important that there is sufficient DNA from the assailant to develop a genetic profile, and it is important to be able to differentiate the assailant's DNA from that of the victim. The first step in the lab is to snip off a portion of the swab and perform an extraction of cells from the clipping and then a *differential lysis* procedure, which is designed to separate male from female cells. These cells are dramatically different in size and morphology.

They also differ in fragility. Thus, female cells (from the victim) can easily be ruptured resulting in the release of their DNA into the test tube while the more hardy sperm cells remain intact. Once the female DNA has been removed, the membrane surrounding the spermatozoa can be weakened and ruptured by addition of dithiothreitol (DTT) and then male DNA that is released can be collected following centrifugation.

The DNA obtained in each sample following differential lysis can then be quantified using a quantiblot procedure or some other available method (see Sections 3.2.2, 3.3.3, and 3.5.2) and then either concentrated or diluted to bring the volume to the ideal concentration used in the assay (1–2 ng DNA for multiplex kits). Following quantification, the sample is amplified as indicated by the kit manufacturer and/or the laboratory protocol.

Following differential lysis, two separate fractions exist. They are referred to as the "E cell" fraction and the "S cell" fraction, standing for *epithelial cells* (female) and *sperm cells* (male), respectively. Differential lysis is performed to prepare sperm cell DNA free of female DNA. If the procedure is 100% efficient, there will be no mixture of victim and assailant DNA and thus the DNA within the S cell fraction can be compared to DNA obtained from the suspect's exemplar (blood or buccal cell) specimen. However, there is often some contamination of female DNA in the male fraction resulting in a mixture. This means that the results of the analysis must be interpreted. If the mixture can be determined to consist of major and minor peaks on a printout, the laboratory may provide statistics for each profile in the mixture. However, until recently, statistics were not offered in reports since mixture patterns could be subject to different interpretations (Daniel, 1999). If the mixture consists of peaks all of which have approximately the same height, laboratories will generally not offer any statistics in their reports.

When a condom that has been used by an assailant is found at a crime scene, DNA analysis of its contents may produce information that can be used to identify him (Hochmeister et al., 1993).

DNA IN ACTION: GENETIC DESCRIPTION USED TO ISSUE WARRANT AND BEAT STATUTE OF LIMITATIONS

Semen samples taken from a sexual assault victim in Wisconsin were subjected to DNA analysis, and the results were entered into a database for comparison with DNA profiles routinely collected from convicts in the Wisconsin prison system. Milwaukee County Assistant District Attorney Norman Gahn obtained a "John Doe" arrest warrant that used the genetic profile of the defendant. The genetic profile of the assailant had been established after DNA analysis of the evidence in the victim's rape kit. The warrant was obtained just 3 days before the 6-year statute of limitations was to expire. The warrant, which was the twelfth warrant of its kind in the nation, was the first to actually lead to an arrest.

Gahn's rationale for his actions was simple. "We knew who the guy was because we had his genetic code, we just didn't have a name," Gahn reportedly reasoned. "We were up against the statute of limitations and it was very frustrating. We did this to keep the case alive." Bobby Dabney subsequently found himself at the center of a constitutional debate when he submitted a DNA sample after a conviction for armed robbery. His DNA profile matched the DNA profile that was recovered from the sexual assault victim and used to obtain the arrest warrant for "John Doe No. 12." Dabney's attorney, Lynn Hackbarth, argued that a DNA profile was an insufficient description of a suspect and was not enough to obtain an arrest warrant. She also argued that the DNA warrant could not extend the statute of limitations. Circuit Court Judge Jeffrey A. Wagner rejected her reasoning, ruling that the DNA profile met the "reasonable certainty" standard for arrest warrants. "The case is simply a matter of the courts recognizing advances in technology," Gahn reportedly stated.

Source: Peter Page, Warrant Naming DNA Upheld after Wis. Arrest, *National Law Journal* (December 10, 2001).

2.1.9 Evidence Handling in the Laboratory

Once the evidence has been brought to the crime laboratory, it is logged in and cataloged. It may be stored in a freezer/cold room within the laboratory or it may be maintained by a property clerk's office. Most large laboratories are divided into sections or departments, each responsible for analyzing different forms of evidence. When case evidence requires a number of different kinds of analysis, each item is sent to the appropriate division of the lab for analysis. Blood or semen evidence would be sent to the forensic biology section, fibers or hair would be sent to the trace evidence section, shell casings or bullets would be sent to ballistics, and so on. In the case where multiple analyses can be performed on a single item, a decision must be made so that one procedure does not destroy or interfere with the ability to perform a second kind of analysis. For example, a bloody palm print could be photographed to preserve the information derived from studying ridge patterns prior to being sent to the biology section where more destructive testing might be conducted that would alter the pattern or even produce artifacts. In general nondestructive testing should precede any destructive testing and preliminary testing techniques should not interfere with secondary testing.

There is also a legal requirement to preserve a specimen so that the defense counsel can have another expert examine the evidence independently and prepare a report of the findings. Only if there is insufficient material to test and only after the defense counsel has been notified of this fact should the criminalist proceed to test and expend all of the evidence. (The reader should review the legal requirements of discovery in Section 5.2.) Under these circumstances, there is often an agreement made between defense counsel and prosecutor that a scientist obtained by the defense may witness the analysis performed by the crime lab analyst. In this way, each scientist can draw his or her own conclusions from the observed results of testing.

2.1.10 Report Writing

The report is more of a legal requirement than a scientific one. Following the analysis of evidence, the analyst writes a report indicating what was analyzed, how it was analyzed, experimental observations, and conclusions. Where possible, statistics are also included to provide quantitative as well as qualitative information about the findings. The report does not go into the details of evidence handling or chain of custody nor does it attempt to describe how the evidence was produced. It is only meant to be a clear summary statement of the scientific outcomes of the analysis. It may or may not be accompanied by supporting photographs. The report itself becomes evidence that is provided to the court during trial together with the testimony of the analyst (or a designee). Criminalists serve as expert witnesses who may offer opinions about the evidence that they analyzed but only if these opinions are based on their experience and upon accepted scientific principles. Criminalists also assist investigators by using their skills and knowledge to interpret their experimental observations and findings and thereby reconstruct the events leading up to and during the crime. Examples of DNA reports in criminal and civil matters are provided at the end of Chapter 4.

2.2 SEROLOGY

Before individualization of the biological evidence is attempted through forensic DNA analysis, the evidence is tested by the forensic serologist to identify the nature of the material or substance. Techniques to identify body fluids such as blood, semen, and/or saliva are well established. These items are often found at the crime scene or on the victim or suspect following the commission of violent crimes. Serologists use different types of chemical tests to detect and identify these and other physiological substances such as urine, fecal matter, and the like. In most instances, testing is performed in two stages. As a general rule, the smallest amount of evidence that will produce an observable result should be tested. The analyst must always try to preserve as much of the evidence in its original form as possible. Initially, a *presumptive* screening test is used to identify the evidence. Presumptive tests are usually sensitive but not specific, and thus small amounts of the substance can be detected. False-negative results are very uncommon unless the amount of specimen is so minimal that it goes undetected. However, occasionally, false-positive results that could potentially be misleading may be observed. Presumptive tests are useful as preliminary screening procedures that reduce the number of items that would otherwise have to be analyzed. Substances that provide negative presumptive results are not tested further. Presumptive tests that are positive should always be followed by *confirmatory* tests. The latter are less sensitive but more specific and therefore results are more reliable.

2.2.1 Blood

In the case of a suspected bloodstain, in the field or more commonly in the laboratory, a small sample of the evidence is subjected to one of the many presumptive tests that are available such as phenolphthalin, benzidine, leucomalachite green, *o*-tolidine, or luminol (see Fig. 2.4). The first four tests are catalytic, colorimetric tests in which a colorless reagent (chemically reduced) is added to a small amount of questioned substance followed by the addition of hydrogen peroxide. When using any of these four catalytic reagents, if the questioned substance is blood, a fairly rapid color change will take place indicating that blood is probably present. In the case of luminol (3-aminophthalhydrazide), sodium perborate is the preferred oxidizing agent rather than hydrogen peroxide. Rather than a color change occurring in the presence of blood, a faint, relatively longlasting, blue-white light is emitted from

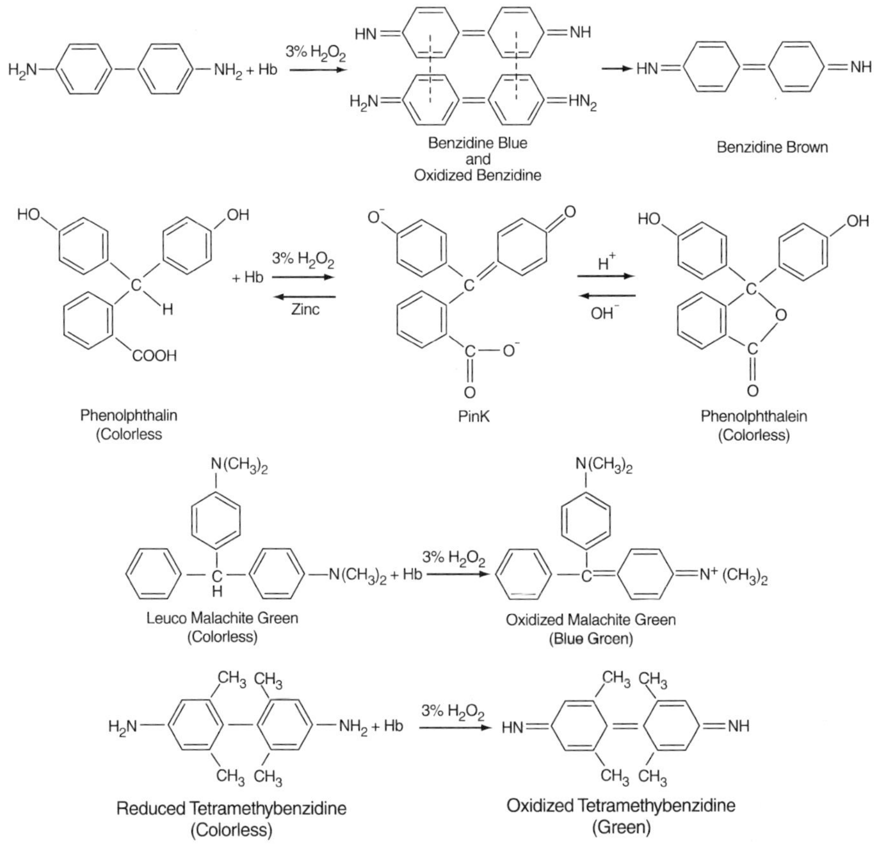

Fig. 2.4 Presumptive tests for the identification of blood include: phenolphthalin, benzidine, leucomalachite green, o-tolidine, or luminol. Saferstein, Richard E., Forensic Science Handbook, 1st Edition, *Copyright 1982. Reproduced by permission of Pearson Education, Inc., Upper Saddle River. N.J.*

the substance. It is necessary to view this reaction in total darkness to visualize the relatively low amount of light that is produced from small specimens. This is referred to as chemiluminescence since the reaction converts chemical energy into light energy. The major benefits of performing a luminol test rather than one of the colorimetric tests is that one can search large areas very rapidly, and the chemical used does not interfere with any of the other presumptive tests that may be performed following luminol testing. If the result of presumptive testing is negative, the analysis is terminated. However, if the result is positive, then a more definitive confirmatory test is performed. The reason for this is that there are a number of substances that can produce false-positive observations including bleach, plant peroxidases, chemical oxidants such as potassium permanganate ($KMNO_4$), copper, brass, lead, zinc, bronze, iron, or cobalt.

There are many confirmatory tests including the hemochromogen (Takayama) crystal test, hematin (Teichmann) crystal test, acetone-chlorhemin (Wagenaar) crystal test, spectrophotometric and immunologic tests (Figs. 2.5*a* and 2.5*b*). If one of these tests produces a positive result, then the analyst should proceed to determine if the blood is human or animal.

The use of specific antisera can reveal if the blood is human or not. Antisera are reagents that are produced in immunized rabbits, sheep, or other animals that can be used by the serologist to determine the species of origin of the questioned bloodstain. Proteins in the evidentiary bloodstain, termed *antigens*, will react with *antibodies* in the antiserum and develop a precipitate that can easily be visualized with the naked eye. The procedure for such a test usually involves either an *Ouchterlony*, double-diffusion plate, or the technique known as crossed-over electrophoresis, but other formats (test tube, capillary tube) can also be used. A positive reaction is one in which the substance in question forms a band of precipitation with only one of several antisera used. In this manner one can easily distinguish blood from humans or various animal species. Many laboratories now combine the confirmatory test for blood and the species test into a single procedure in which the presumptively positive sample is electrophoresed on a plate or microscope slide and then permitted to diffuse and react with a specific antiserum that recognizes human hemoglobin. A positive result indicates the presence of human blood. Another method known as the ABA card has become popular. It too tests for human hemoglobin in the specimen.

Similar approaches are used in the identification of semen, saliva, urine, and the like. The common goal regardless of type of specimen is to identify the substance, which is the foundation for the next stage of tests, which are used to determine who produced the evidence (Culliford, 1971). Any methods employed to identify the substance must not interfere with subsequent typing procedures (Barnett et al., 1992; Shipp et al., 1993).

2.2.2 Semen

Semen contains spermatozoa in a liquid medium known as seminal plasma. This fluid contains a variety of substances but is especially rich in acid phosphatase, phos-

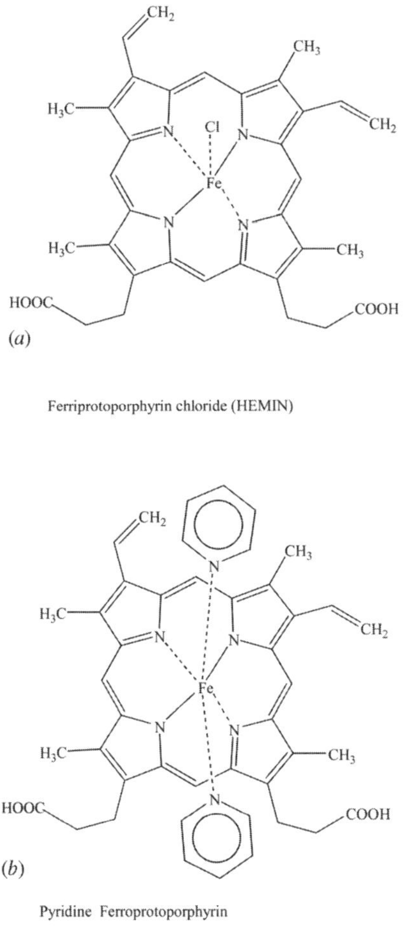
CH2
CH3
H3C
Cl
N
N
CH2
Fe
N
N
H3C
CH3
HOOC
COOH
(a)
Ferriprotoporphyrin chloride (HEMIN)
CH2
CH3
N
H3C
N
N
CH2
Fe
N
N
H3C
CH3
N
HOOC
COOH
(b)
Pyridine Ferroprotoporphyrin

Fig. 2.5 *Confirmatory tests for the presence of blood include (a) the hematin (Teichmann) crystal test, and (b) the hemochromogen (Takayama) crystal test. Courtesy of Dr. Robert Rothchild, John Jay College of Criminal Justice, New York, NY.*

phorylcholine, and spermine. Finding spermatozoa in an evidentiary stain eliminates the need to perform presumptive and confirmatory testing since only semen contains these cells (Fig. 2.6). The evidentiary swab is smeared over a microscope slide, and after fixation the cells are stained with haematoxylin and eosin to enhance contrast and make the sperm cells stand out against the background field. A mature spermatozoon has an oval shaped head 4.0–5.0 μm long and 2.5–3.5 μm wide with a pale anterior half called the acrosome and a darker posterior half. The length-to-width ratio of the head should be 1.50–1.75. The base of the head should be broad and attached to the midpiece, which contains the single spiraled mitochondrion. The midpiece is about 1 μm wide, 7–8 μm long, and is attached to the tail, which is approximately 35 μm long.

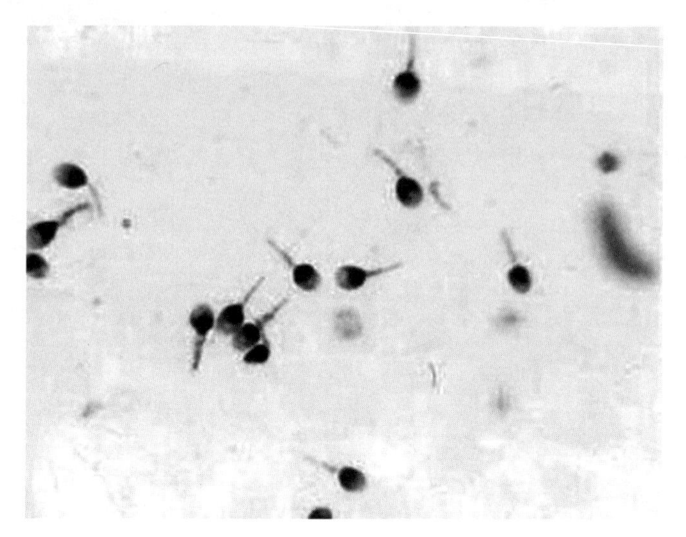

Fig. 2.6 See color insert. Detection of spermatozoa in evidentiary stain eliminates need to perform presumptive or additional confirmatory testing since only semen contains these cells.

Over time, sperm cells will fragment into heads and tails, and it becomes more difficult to identify the cell fragments. Some analysts are conservative and will not identify a sperm cell unless it is intact, while others will identify sperm heads and render a finding anyway.

Because males can be azoospermic or vasectomized, it is necessary to be able to identify semen in the absence of spermatozoa. In the field, dry semen stains can be detected nondestructively using an ultraviolet light source emitting light at wavelengths of 254 and 365 nm since a semen stain fluoresces brightly against a nonfluorescing substrate. In the laboratory, the stain is tested with the light source and then for acid phosphatase activity in a "qualitative" manner. This is one of the most commonly performed presumptive tests for semen. The brentamine reaction for acid phosphatase produces a purple coloration to signal the presence of this substance. Essentially, an enzyme substrate, sodium α-naphthyl phosphate, is converted to sodium phosphate and naphthol by the acid phosphatase enzyme in semen, and a coupled reaction with brentamine fast blue dye takes place, forming a purple color. The intensity of the color produced is graded as either $+$, $++$, $+++$, $++++$, or 0, meaning no color change. As mentioned in the previous section on presumptive blood testing, acid phosphatase presumptive testing can produce false positives because similar enzyme activity can be found in other body fluids, as well as in the presence of fungi, bacteria, and even plants. Fecal stains and vaginal secretions also contain acid phosphatase activity. Other presumptive tests include methods that detect *spermine* or *choline.* These are "crystal" tests. Spermine reacts with the *Barberio* reagent (picric acid) to form specific crystals that can be observed microscopically. Choline reacts with a reagent known as Florence iodine to produce choline periodide, which is seen as characteristic brown needle-shaped crystals. This is known as the *Florence test.* Here too, false-positive results are possible. For this reason, positive presumptive tests must be followed by a confirmatory test: a "quantitative" acid phosphatase test, or the detection of P30, a protein secreted by the prostate gland.

Because acid phosphatase can be found in vaginal fluids as well as in fecal stains but in lower concentration, the finding of a relatively high concentration of acid phosphatase (nitrophenyl phosphate test) has been considered confirmatory for the presence of semen. There remains some question about the critical level of activity required to positively identify a substance as semen. Some forensic experts believe that there is no such critical level and that the test should be used only as a presumptive determinant of semen.

The P30 protein (with a molecular weight of approximately 30,000 daltons) is produced by the prostate gland and therefore is semen specific. It can be detected serologically using antiserum that is specific for the P30 antigen. Thus P30 can be detected using either the Ouchterlony double-diffusion or crossed-over electrophoresis procedures mentioned above with this antiserum. The development of a precipitin band between the P30 antigen and the antibodies within the antiserum indicates the presence of semen. The presence of P30 even in the absence of spermatozoa is confirmatory evidence of the presence of semen.

It is important to determine if a semen stain is human in origin in the same way that it is important to determine if a bloodstain is human. Domestic pets could deposit semen on items that subsequently become crime scene evidence so it is important to be able to differentiate semen from these sources. Semen produced by a number of animal species can produce weak acid phosphatase results. Microscopy can be used to examine the morphology and surface structure of spermatozoa and, thereby, distinguish cat and dog from human.

Prior to the development of forensic DNA technology, individualization was accomplished by grouping seminal stains in the ABO, Lewis, and phosphoglucomutase (PGM) systems. The Lewis system is used to determine the secretor status of the sample donor. Roughly 80% of individuals secrete A, B, and H substance into their body fluids. Which of these substances are secreted depends on their ABO blood type. Thus, testing could determine one's ABO blood group by examination of saliva or semen that had been produced by a secretor. These tests are seldom performed nowadays with the availability of highly discriminating DNA testing methods.

2.2.3 Saliva

Saliva is important when it becomes evidence. Saliva is often found on stamps and envelope flaps, cigarette butts, drinking cups (paper, ceramic) and glasses, and in sexual assault evidence. Saliva is detected chemically by the presence of one of its components *α-amylase*. Other amylase activity exists in semen, vaginal secretions, and in pancreatic secretions and fecal matter. Bacteria are also known to exhibit amylase activity. The enzyme amylase converts starch, a polymer of glucose subunits, to dextrin and ultimately to maltose, which is a disaccharide. The presence of amylase can be determined in two ways. The first is by determining if the specimen breaks down starch, which has been placed either in a test tube (in solution) or into an agarose gel. Regardless of the specific test format, the sample is allowed to react with the starch and in time if amylase is present the starch is converted to maltose. Following the addition of Lugol's iodine solution, which reacts with starch to produce either a blue, red, or yellow color, the presence of starch or its breakdown products can then be determined visually. If the starch has been converted to maltose, no color development will occur at all. Also, because some individuals secrete very little amylase into their saliva obtaining a negative result for amylase does not mean that saliva is definitely absent. Another concern is that this procedure is not specific for saliva. A more specific test for saliva is called the Phadebas reaction. In this reaction the specimen is reacted with the Phadebas reagent and the colored end product is measured spectrophotometrically at 620 nm. If the optical density (related to the concentration and activity of amylase) is sufficiently high, the test is considered to be positive. Prior to the use of DNA to test evidence containing saliva (Walsh et al., 1992), individualization was performed by ABO grouping using a combination of the absorption-inhibition and absorption-elution tests, as well as Lewis antigen testing.

2.2.4 Urine

Although urine stains are less likely to be found at crime scenes than blood, semen, or saliva, urine testing has become a common procedure primarily to test for the presence of illicit drugs. If a positive result is obtained, it sometimes becomes necessary to perform DNA testing to verify that there had not been mislabeling or accidental switching of samples. The serologist detects urine stains based on physical and chemical characteristics. Urine stains can be heated and will produce a characteristic odor as a result of the ammonia present in the stain. Ammonia forms as a result of urea breakdown by bacterial enzymes. The liver produces urea as an end product of protein metabolism. Urea travels through the bloodstream and ends up in urine where it is excreted. In the laboratory, urea can be detected by reaction of the unknown substance with the enzyme urease, thereby producing ammonia and carbon dioxide. In one chemical test for urine, a chemical known as DMAC (*p*-dimethylaminocinnamaldehyde) is reacted with the unknown specimen under highly acidic conditions. In the presence of urea, it rapidly produces a magenta-colored product. This method is also a presumptive one since many substances such as semen, vaginal excretions, sweat, and milk, which contain urea, and amines or amides will produce a positive reaction. Another presumptive test for urine is the creatinine test. Creatinine is another metabolic breakdown product and can be detected in a colorimetric (color-generating) reaction.

2.2.5 Hair

Hair is one of the most common types of evidence found at a crime scene. In crimes of violence, hair may be pulled from the victim or assailant. Furthermore, the average person loses about 100 scalp hairs daily, and therefore, criminals may drop some of their scalp hair during the commission of a crime. The reason for this is that a hair is shed when it reaches the end of its growth cycle (telogen hair) (Fig. 2.7). Once this evidence is recognized, documented, collected, and brought to the lab for analysis, a

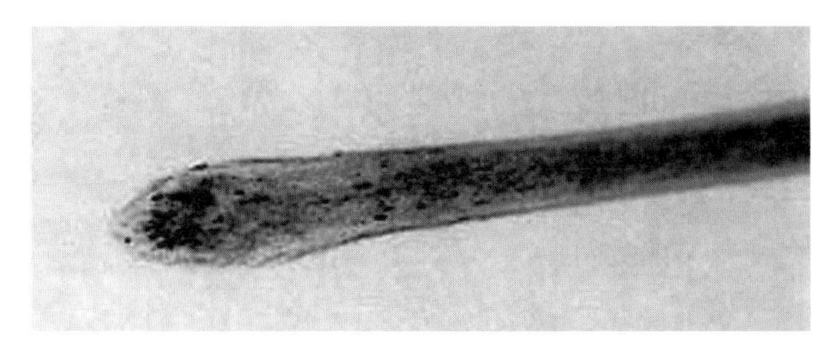

Fig. 2.7 *See color insert. Appearance of shaft and root of telogen hair.*

number of tests can be performed. The hair can be characterized by species, race, and somatic origin (scalp, axillary, chest, pubic). The examination is performed both visually and microscopically. Features such as color, length, and cosmetic treatment, such as dyeing or bleaching, can be seen by the naked eye. However, numerous other features can only be seen with the aid of a microscope. Hair grows from a skin structure known as a hair follicle (Deedrick and Koch, 2004). The hair consists of the root below the skin and the shaft above it (see Fig. 2.8*a*). The root within the follicle consists of many cells, whereas the shaft is mainly proteinaceous and consists of a number of keratin proteins that are cross-linked and form long fibrils. Examination of a typical hair shaft reveals the presence of a thin, transparent cuticle layer on the outside, a cortex layer beneath which contains pigment granules, air spaces, and other granules, and a central portion within the cortex called the medulla (see Fig. 2.8*b*). The cuticle layer appears like overlapping flat cells and has the appearance of roof tiles. Running a hair through one's fingers reveals a smooth surface in one direction but a rough surface in the other. This difference is due to the cuticular cells being arrayed unidirectionally. The medulla may be classified as continuous, discontinuous, or fragmentary depending on its appearance. Microscopically two hairs can be viewed simultaneously and a comparison made. The hair analyst looks at a number of characteristics when comparing head hairs. These include color, length, diameter, shape (curly, wavy, curved, straight), tip (cut, uncut, tapered), root, cross-sectional shape (round, oval, flat), pigment (absent, granular, nongranular, dense), medulla (broken, continuous, absent), cortical fusi, birefringence, cosmetic treatment, cuticle (serrated, narrow, wide, layered, absent), and scales (flat, smooth, level, arched). Known and questioned hairs are compared by examining their gross morphology and microscopic features. In this way the analyst can decide that (1) the questioned and exemplar hairs exhibit different microscopic characteristics and therefore have different origin or that (2) the questioned and exemplar hairs share the same microscopic features and therefore may have the same origin or that (3) no conclusion can be drawn as to any association between the exemplar and unknown hairs. With the development of forensic procedures to analyze nuclear DNA in the root tissue or mitochondrial DNA in the shaft, the microscopic analysis of hair has become more of a screening or preliminary test. Questioned hair that has been excluded by traditional methods need not be further tested, but a questioned hair that is found to match an exemplar hair can be DNA tested owing to the relatively large amount of mitochondrial DNA found within the hair shaft (see Section 3.5).

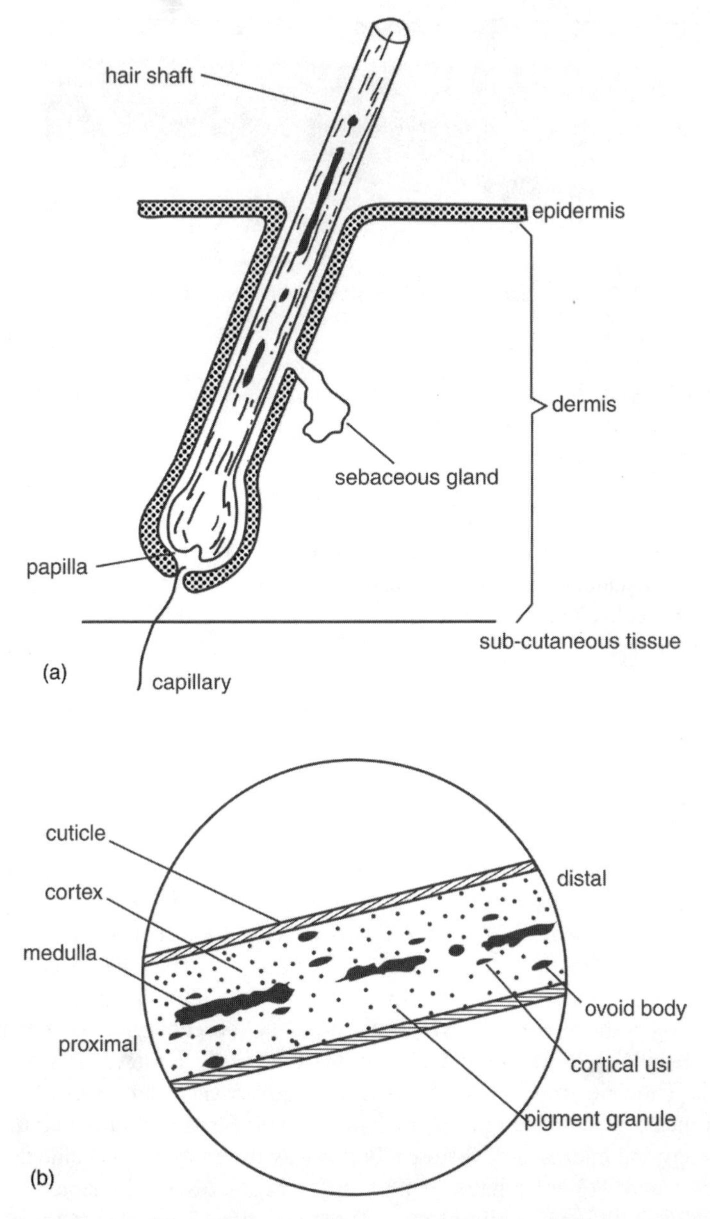

Fig. 2.8 *(a) Hair grows from skin structure known as a hair follicle. (b) Hair consists of the root below the skin and the shaft above.*

2.3 CHAIN OF CUSTODY

From the moment that an item is collected at a crime scene to the moment it is introduced in the courtroom as evidence, a lengthy period of time may elapse. After crime scene investigators locate the evidence, document it, package it, and transport it to the laboratory, that evidence is usually cataloged by a property clerk who labels it and stores it until it is required for testing. Testing in the lab may be conducted by different people and at different times and dates. After each test is completed, the analyst will return the item to the property clerk. Every time the evidence changes hands, the name of the two parties involved, and the date and time of the exchange must be documented. A record of these transfers should be maintained in the evidence file so that it is available for inspection at a later date. It is expected that a number of different individuals will have had possession of the evidence between the time it was collected and the time it reaches the courtroom for consideration of its admissibility. The possession, time and date of transfer, and location of physical evidence from the time it is obtained to the time it is presented in court is called the *chain of custody*.

In cases involving scientific evidence, the chain of custody can be of enormous importance. In cases where weapons and drugs have become evidence, the chain of custody is important because it allows for authentication of the evidence. By tracing the chain of custody, one can determine whether or not the material that is being admitted into evidence is indeed that which was retrieved from the crime scene. In the O.J. Simpson criminal trial, there were issues raised by the defense with respect to the quantity of blood collected from Simpson and the quantity of blood that remained in the tube when it was admitted into evidence. The defense claimed that some of the blood was missing and suggested that it could have been used to plant evidence. The chain of custody documentation became critically important in deciding if the missing blood was "within expected values" or not.

With scientific evidence, factors other than authentication are of concern as well. One concern is the impact of third-party possession on the evidence. When a commercial carrier transports a package containing biological evidence, it is not always obvious how the package was handled and how it was stored. It is known that certain environmental factors can have an adverse effect on unprotected evidence (McNally et al., 1989). Heat and light are two such factors. For instance, a biological sample can break down (degrade) at high temperatures. Mishandled samples can lead to contamination. If, for example, a single 5-inch-long brunette hair is collected at a possible crime scene, packaged, and documented, but upon later examination, two brunette hairs are found in that package, one measuring 1.5 inches while the other measuring 3.5 inches long, one could argue that the evidence had been mishandled or even tampered with, although the likelihood is that the original hair had fragmented into two parts. These concerns are just a few of the factors that make the chain of custody a topic of major importance in a trial involving scientific evidence.

As mentioned above, the location of the evidence is not the only concern when establishing a chain of custody in a case involving DNA. Several other dimensions are also of importance including the storage and handling of the evidence, the temperature of the room or place in which the evidence was stored, whether samples stored in the vicinity of the sample analyzed have been contaminated, the identity and qualifications of those handling, storing, or testing the samples being admitted into evidence, and the frequency with which the samples were accessed or used.

REFERENCES

Barnett, P. D., E. T. Blake, J. Super-Mihalovich, G. Harmor, L. Rawlinson, and B. Wraxall. Discussion of "Effects of Presumptive Test Reagents on the Ability to Obtain Restriction Fragment Length Polymorphism (RFLP) Patterns from Human Blood and Semen Stains." *J. Forensic Sci.*, **37**(2), 69–70 (1992).

Cimino, G. D., K. C. Metchette, J. W. Tessman, J. E. Hearst, and S. T. Issacs. Post-PCR Sterilization: A Method to Control Carryover Contamination for the Polymerase Chain Reaction. *Nucleic Acids Res.*, **19**, 99–107 (1990).

Culliford, B. J. *The Examination and Typing of Blood Stains in the Crime Laboratory.* U.S. Government Printing Office, Washington, D.C., 1971.

Daniel, W. W. *Biostatistics: A Foundation for Analysis in the Health Sciences.* Wiley Series in Probability and Statistics: Applied Probability and Statistics Section, 7th ed. Wiley, New York, 1999.

Deedrick, D. W. and S. L. Koch, Microscopy of Hair Part 1: A Practical Guide and Manual for Human Hairs. *Forensic Sci. Commun.*, **6**(1) (2004). Available at http://www.fbi.gov/hq/lab/fsc/current/research/2004_01_research01b.htm.

Fisher, B. A. J. *Techniques of Crime Scene Investigation*, 5th ed. CRC Press, Boca Raton, FL, 1993.

Hochmeister, M. N., B. Budowle, U. V. Borer, and R. Dirnhofer. Effects of Nonoxinol-9 on the Ability to Obtain DNA Profiles from Post-Coital Vaginal Swabs. *J. Forensic Sci.*, **38**(2), 442–447 (1993).

Inman, K. and N. Rudin. *Principles and Practice of Criminalistics.* CRC Press, Boca Raton, FL, 2000.

James, S. H. and W. G. Eckert. *Interpretation of Bloodstain Evidence at Crime Scenes*, 2nd ed. CRC Series in Practical Aspects of Criminal and Forensic Investigations. CRC Press, Boca Raton, FL, 1999.

Lee, H. C. Identification and Grouping of Bloodstains. In *Forensic Science Handbook.* R. Saferstein (ed.). Prentice-Hall, Englewood Cliffs, NJ, 1982, pp. 267–337.

Lee, H. C., T. M. Palmbach, and M. T. Miller, *Henry Lee's Crime Scene Handbook.* Academic, New York, 2001.

McNally, L., R. C. Shaler, M. Baird, I. Balazs, P. R. DeForest, and Kobilinsky. Evaluation of Deoxyribonucleic Acid (DNA) Isolated from Human Bloodstains Exposed to Ultraviolet Light, Heat, Humidity, and Soil Contamination. J. Forensic Sci., **34** (5), 1059–1069 (1989a).

McNally, L., R. C. Shaler, M. Baird, I. Balazs, L. Kobilinsky, and P. R. DeForest. The Effects of Environment and Substrata on Deoxyribonucleic Acid (DNA): The Use of Casework Samples from New York City. J. Forensic Sci., **34**(5) 1070–1077 (1989b).

Shipp, E., R. Roelofs, E. Togneri, R. Wright, D. Atkinson, and B. Henry, Effects of Argon Laser Light Alternate Source Light, and Cyanoacrylate Fuming on DNA Typing of Human Bloodstains, *J. Forensic Sci.*, **38**(1), 184–191 (1993).

Wade, C. (ed.). *Handbook of Forensic Services.* Available online at www.FBI.Gov. [This handbook was written to provide guidance and procedures for safe and efficient methods of collecting, preserving, packaging, and shipping evidence and to describe the forensic examinations performed by the FBI's Laboratory Division and Investigative Technology Division, Revised 2003]

Walsh, D. J., A. C. Corey, R. W. Cotton, L. Forman, G. L. Herrin, C. J. Word, and D. D. Garner. Isolation of Deoxyribonucleic Acid (DNA) From Saliva and Forensic Science Samples Containing Saliva, *J. Forensic Sci.*, **37**(2), 387–395 (1992).

3

Forensic DNA Analysis Methods

3.1 ASSOCIATIVE EVIDENCE AND POLYMORPHISM

Forensic DNA analysis helps to establish an association between biological evidence and its source, whether it is a suspect, a victim, or a third party. In the analysis, specific genes (alleles) present in a questioned (evidentiary) item are compared with those in a known (exemplar) specimen. Prior to the use of forensic DNA analysis, criminalists employed gene-coded *polymorphic* products to link a suspect to a crime. Polymorphism exists in certain proteins, glycoproteins (proteins linked to sugars), lipoproteins (proteins linked to lipids), and glycolipids (sugars linked to lipids). Because these substances are polymorphic, they can be *typed* using either serological or biochemical methods yielding information about their origin. Glycolipids include the ABO and Lewis antigens, A, B, and H and Le^a and Le^b, respectively. Antigens are substances that react with antibodies found in a specific antiserum to produce some biological or physical effect. For example, blood cells from a type A individual will clump (agglutinate) in the presence of an antiserum that contains antibodies that specifically recognize the A substance. The antibodies that are used to recognize the A and B substances are called hemaglutinins. The A and B substances are called the hemaglutinogens. Antibodies and antigens fit together like the pieces of a puzzle.

The following is a list of some of the polymorphic enzymes (protein catalysts) that forensic serologists used prior to the development of DNA techniques for

DNA: Forensic and Legal Applications, by Lawrence Kobilinsky, Thomas F. Liotti, and Jamel Oeser-Sweat
ISBN 0-471-41478-6

human identification:

PGM	Phosphoglucomutase
EAP	Erythrocyte acid phosphatase
EsD	Esterase D
AK	Adenylate kinase
ADA	Adenosine deaminase
6-PGD	6-Phosphogluconate dehydrogenase
G-6-PD	Glucose-6-phosphate dehydrogenase
Tf	Transferrin

Note: The production of polymorphic enzymes is a result of the existence of polymorphic alleles. Therefore, the analysis of polymorphic proteins is an "indirect" way of analyzing polymorphic DNA.

Let us examine an enzyme that exists in the population in three detectable forms, A, B, and C. All members of the population have this enzyme in one form or another. All three forms of the enzyme have the same functional activity. The particular form that a person has does not manifest itself in any observable physical trait. Nevertheless, these forms are manifestations of one's genetic constitution and they are inherited from one's parents. They are permanent genetic characteristics of an individual. If both the evidence and the suspect were shown to be of the same form (A vs. A), then one could conclude that the two samples may share a common origin. If the two forms were shown to be different (A vs. C), then the suspect could be excluded as the source of the evidentiary specimen. However, because these proteins have relatively poor stability, a low degree of polymorphism, and often are found in insufficient amounts to test, proteins and glycoproteins are not always useful or informative with respect to specifying the source of an evidentiary item. As explained above, polymorphism refers to the existence of multiple alleles at a single locus. Thus, on a particular pair of chromosomes, at a particular locus, let us say that there is a gene that has three common alleles. Thus, in the population under study, three distinct forms of that gene can be detected. At that locus, a person can have either two identical alleles (one on each chromosome of the set) and therefore be *homozygous* or the person can have two different alleles and therefore be *heterozygous.* With two alleles, there will be three types found in the population. With three alleles, there are six possible types that can be found in the population. The type (genotype) is an expression of the two alleles located at the locus that is, "A,C" or "A,A" or "5,7" depending on how alleles are named.

More specifically, in the case of the genetic marker ABO, there are three common alleles, I^A, I^B, and *i.* Because human autosomes (nonsex chromosomes) are arranged in pairs, and because an individual can be homozygous or heterozygous at a locus, you can arrange these three alleles six different ways as described above to form six distinct groups of two. Specifically, a person's genotype (one's genetic make up at a single locus) can be any of the following: $I^A I^A$, $I^B I^B$, $I^A I^B$, $I^A i$, $I^B i$, and *ii.*

However, phenotypically (the way the genotype manifests itself through serological testing), a person may be either ABO type A ($I^A I^A$ or $I^A i$), type B ($I^B I^B$ or $I^B i$), type AB ($I^A I^B$), or type O ($i\ i$). The four different phenotypes are a result of the existence of dominant and recessive alleles. The i allele is recessive to both I^A and I^B. The concept of recessive and dominant genes is important when studying individual phenotypes, where one allele (recessive) can be masked by the presence of the other (dominant). At forensically important DNA (VNTR or STR) loci, *there are no dominant or recessive alleles.* All alleles are expressed equally within a locus. Therefore, if a genetic locus has 3 alleles, and one's genotype can be arranged $3 + 2 + 1 = 6$ ways, then there will be 6 detectable types in the population. Genotype and phenotype become essentially identical. If a locus has 7 known alleles, then the number of possible types would equal $7 + 6 + 5 + 4 + 3 + 2 + 1 = 28$. A locus with 10 alleles can form 55 distinct types (see Fig. 3.1). The higher the level of polymorphism observed at a locus, the more types that can be found in the population. In determining the usefulness of a locus for forensic purposes, one must consider not only the number of alleles that exist but also the frequencies of each. For example, if at a particular locus two or three alleles are very common but most are very rare, then that locus becomes less useful since one will seldom see an individual with a genotype containing one or two of those rare alleles. Of course, if that rare allele should show up in the genotype of the suspect as well as in the evidence, then the value of that observation becomes highly significant.

The ability of DNA technology to associate suspects to crimes has revolutionized the field of criminalistics. Currently, crime labs that conduct DNA testing generally examine 13 (or more) loci. This is especially true if they wish to take advantage of the state and national DNA databases to help solve crimes. The specific loci that are used in DNA testing are chosen because of their high degree of polymorphism at each chromosomal site. There are actually two different kinds of polymorphism in DNA. Alleles can be distinguished either by their relative size (size polymorphism) or by differences in sequence (sequence polymorphism). These are described in Section 3.3.4.1. Higher levels of polymorphism at specific sites within human DNA results in higher levels of discrimination, which means that it becomes easier to identify a person if you test loci that contain a relatively large number of different alleles. Table 3.1 illustrates the 13 CODIS (Combined DNA Index System) (national DNA database) loci. It includes the chromosome number where each of these loci exist, a list of the most common alleles, and the number of possible types that can be detected based on the degree of polymorphism present. Amelogenin is a locus responsible for tooth bud development in the fetus and can be found in two slightly different forms on sex chromosomes X and Y (Roffey et al., 2000) (see Section 3.3.4.3). It therefore can be used to determine the gender of the individual who produced the evidence (Sullivan et al., 1993; Akane et al., 1991, 1992).

All 13 loci and amelogenin can be analyzed using commercially available multiplex kits. PE Applied Biosystems and Promega Corp. manufacture such kits. Profiler Plus is used to genotype the following nine STR loci plus amelogenin: D3S1358, VWA, FGA, D8S1179, D21S11, D18S51, D5S818, D13S317, and

The D5S818 locus has 10 common alleles:

Alleles: 7, 8, 9, 10, 11, 12, 13, 14, 15, 16

10 alleles can be put into 55 groups of 2 as follows:

7,8									
7,9	8,9								
7,10	8,10	9,10							
7,11	8,11	9,11	10,11						
7,12	8,12	9,12	10,12	11,12					
7,13	8,13	9,13	10,13	11,13	12,13				
7,14	8,14	9,14	10,14	11,14	12,14	13,14			
7,15	8,15	9,15	10,15	11,15	12,15	13,15	14,15		
7,16	8,16	9,16	10,16	11,16	12,16	13,16	14,16	15,16	
7,7	8,8	9,9	10,10	11,11	12,12	13,13	14,14	15,15	16,16

Number of possible groups of 2 is given by:
10+9+8+7+6+5+4+3+2+1=55

	CHROMOSOME LOCATION	COMMON ALLELES	POSSIBLE GENOTYPES
D3S1358	3	9, 10, 11, 12, 13, 14, 15, 15.2, 16, 17, 18, 19, 20	91
VWA	12	10, 11, 12, 13, 14, 15, 15.2, 16, 17, 18, 19, 20, 21, 22	105
FGA	4	16, 16.2, 17, 17.2, 18, 18.2, 19, 19.2, 20, 20.2, 21, 21.2, 22, 22.2, 23, 23.2, 24, 24.2, 25, 25.2, 26, 26.2 , 27, 27.2, 28, 28.2, 29, 29.2, 30, 30.2, 31, 31.2	528
D8S1179	8	8, 9, 10, 11, 12, 13, 14, 15, 16, 17, 18, 19	78
D21S11	21	24.2, 25, 26, 27, 28, 28.2, 29, 29.2, 29.3, 30, 30.2, 31, 31.2 32, 32.2, 33, 33.1, 33.2, 34, 34.2, 35, 35.2, 36, 38	300
D18S51	18	9, 10, 10.2, 11, 12, 13, 13.2, 14, 14.2, 15, 16, 17, 18, 19, 20 21, 22, 23, 24, 25, 26	231
D5S818	5	7, 8, 9, 10, 11, 12, 13, 14, 15, 16	55
D13S317	13	5, 8, 9, 10, 11, 12, 13, 14, 15	45
D7S820	7	6, 6.3, 7, 8, 9, 10, 11, 12, 13, 14, 15	66
D16S539	16	5, 8, 9, 10, 11, 12, 13, 14, 15	45
THO1	11	4, 5, 6, 7, 8, 8.3, 9, 9.3, 10, 11	55
TPOX	2	6, 7, 8, 9, 10, 11, 12, 13	36
CSF1PO	5	6, 7, 8, 9, 10, 11, 12, 13, 14, 15	55

Fig. 3.1 *Locus with 10 alleles can have 55 types.*

D7S820 (see Fig. 3.2). COfiler is used to genotype the following six loci plus amelogenin: D3S1358, D16S539, THO1, TPOX, CSF1PO and D7S820.

The combined use of these two kits to analyze evidence produces genotypes for 13 different loci since there is overlap of the typing of D3S1358 and D7S820 loci as well as amelogenin. The redundancy in typing these two loci serves as an additional internal experimental control that demonstrates reproducibility and reliability of the procedure (see Section 4.7 for a discussion about experimental controls).

Short tandem repeat loci are named in two different ways. The newer nomenclature is based on the chromosomal position and the order in which the locus was discovered. Thus D3S1358 is a locus found on chromosome number 3 and was the 1358th sequence to be identified. The older nomenclature named loci based on their position or function, and the following 5 loci are named accordingly:

VWA	von Willebrand Factor (Sajantila et al., 1994)
FGA	*Intron 3 of the human α-fibrinogen gene (Mills et al., 1992)
THO1	*Intron 1 of human tyrosine hydroxylase gene
TPOX	*Introl 10 of human thyroid peroxidase gene (Anker et al., 1992)
CSF1PO	Human c-fms proto-oncogene for CSF-1 receptor gene

TABLE 3.1 CODIS Loci and Their Characteristics

Locus	Chromosome Location	Repeat Pattern	Number Common Alleles (names)	Number Types
CSF1PO	5q33.3-34	TAGA	10 (6–15)	55
D13S317	13q22-q31	TATC	8 (8–15)	36
D16S539	16q22-24	GATA	9 (5–15)	45
D18S51	18q21.3	AGAA	21 (9–26)	231
D21S11	21p11.1	TCTA + TCTG*	22 (24.2–38)	253
D3S1358	3p21	TCTG + TCTA	8 (12–19)	36
D5S818	5q21-q31	AGAT	10 (7–16)	55
D7S820	7q11.21-22	GATA	10 (6–15)	55
D8S1179	8q24.1-24.2	TCTA + TCTG	12 (8–19)	78
FGA	4q28	CTTT	14 (17–30)	105
THO1	11p15-15.5	TCAT	7 (5–10)	28
TPOX	2p23-2pter	GAAT	8 (6–13)	36
VWA	12p12-pter	TCTG + TCTA	11 (11–21)	66
Amelogenin	X, Y			

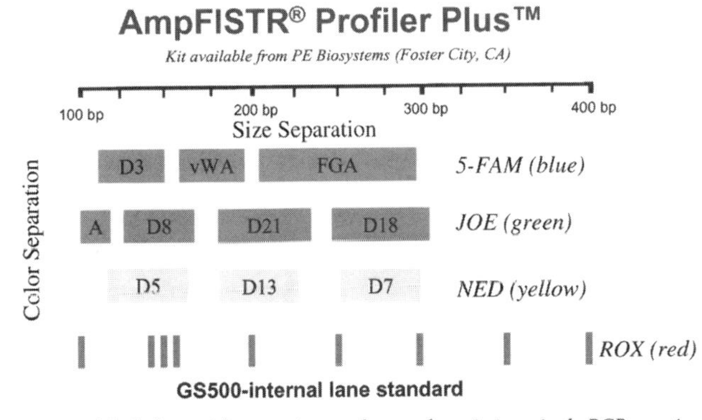

Fig. 3.2 Use of the Profiler Plus Kit produces genotypes for nine different loci and amelogenin. Printed with the permission of Dr. Peter Valone.

*An intron is an intervening, noncoding, sequence within a gene at a particular locus.

The power of discrimination can be calculated to provide an indication of the usefulness of a locus in providing individualizing information (see Section 4.12). For the multiplex AmpFlSTR Profiler Plus (manufactured by PE-Applied Biosystems), this value for Caucasians is approximately 1 in 9.6×10^{10} and for the multiplex COfiler (same manufacturer) this value is approximately 1 in 8.4×10^5. The combined power of discrimination when both kits are utilized is calculated as approximately 1 in 8×10^{16}. The Promega kit PowerPlex 16 can be used to simultaneously amplify and type all 13 CODIS loci—plus amelogenin and two different "penta" loci (penta D and penta E). The penta loci have a repeat core that is five bases in length rather than four and hence they are termed Penta repeats (see Table 3.2). The power of discrimination for the PowerPlex 16 kit is calculated to be 1 in 1.8×10^{17}. One advantage of using Penta STR loci is that there are fewer stutter bands, and therefore the allele patterns can be more easily interpreted. Stutter bands are artifacts created during the amplification process that can sometimes result in genotyping interpretation issues (see Section 3.3.4.3.2.2).

Multiplex Kit	Power of Discrimination*
Profiler Plus	$1{:}9.6 \times 10^{10}$
Cofiler	$1{:}8.4 \times 10^{5}$
PowerPlex 16	$1{:}1.8 \times 10^{17}$
Identifiler	$1{:}2.1 \times 10^{17}$

*Calculation for a Caucasian population.

TABLE 3.2 Variable Number of Pentanucleotide Repeats at a Glance

Advantages	Limitations
1. Most of the advantages of other STRs apply to pentanuclotides.	1. Pentanucleotides generally share the same disadvantages as those described herein for other STRs.
2. Simpler and more precise interpretation of results is possible because the lower percentage of stutter band artifacts allows for cleaner amplification with fewer artifacts.	2. When considered in light of shorter repeats, pentanucleotides are rare in the genome.
3. Some pentanucleotide loci have a high degree of heterozygosity without a significant number of microvariants.	
4. The longer repeat length of pentanuclotides in conjunction with a lack of microvariants allows for more flexibility in separation technique by fostering less extensive separation of amplified products.	
5. Some population genetics research suggests that some pentanucleotide repeats may be capable of assisting scientists in determining the racial origin of the contributor of a particular DNA sample.	

Source: National Commission on the Future of DNA Evidence, *The Future of Forensic DNA Testing Predictions of the Research and Development Working Group* (2000).

3.2 RESTRICTION FRAGMENT-LENGTH POLYMORPHISM

In 1985 Alec Jeffries introduced the technique known as *restriction fragment length polymorphism (RFLP) analysis* to the world of criminal justice. Jeffries worked with the police in the community of Leicester, England, to solve the rape-homicides (Schlesinger, 2004) of two teenaged schoolgirls, one in 1983 and the second in 1986. Following the second murder, an individual confessed but denied any involvement in the first murder. The police felt that he had committed both crimes due to similarities in the manner in which the crimes had occurred. Jeffries performed RFLP testing on the evidence from the two rape kits as well as on the prime suspect. Although he found that the same person had raped both victims, it was not the person who had confessed. The police decided to test every person who could possibly have committed these heinous acts. After analyzing the DNA of 5811 men, Jeffries still had no match. What Jeffries did not know at the time was that a cake decorator by the name of Colin Pitchfork had asked a friend of his to stand in as a substitute for him and provide the police with his blood sample in place of Pitchfork. He told his friend that because he had a prior police record, that the police would try to frame him for the murders. The friend agreed to go to the police and substitute for him. Of course, Pitchfork was cleared after his friend's blood was analyzed. The case would never have been solved if the friend had not talked about the substitution at a bar one day while drinking with friends. A woman at the next table overheard the story and informed the police, who quickly responded by arresting the young man who then explained to police why he had substituted for his friend. RFLP analysis of Pitchfork's blood revealed a match to the rape kit evidence

Jeffries' method is better known as DNA fingerprinting. It is a technique that has had enormous impact on the ability to identify the source of biological evidence (Gill et al., 1985; Hicks, 1988; Jeffreys et al., 1985a, 1985b). RFLP analysis makes use of regions of DNA in which relatively short segments are repeated and arranged tandemly (see Fig. 3.3*a* and 3.3*b*). Figure 3.3*a* shows segments A, B, C, D, and E. Each of these DNA segments contains a repeating core sequence as shown in Figure 3.8. The number of repeats varies significantly between individuals and as a result, the repeats can be used to determine the likelihood that a particular suspect was the source of an evidentiary DNA sample.

The RFLP analysis incorporates a number of procedures. This section will explain the method by examining each biochemical technique that is employed. Initially, DNA is extracted from a questioned biological sample and then purified so that there are no contaminating proteins, lipids, sugars, and so forth.

3.2.1 Isolation of DNA

Biological evidence can be found in many different forms such as blood, semen, and saliva. These substances may be found on a variety of different substrates such as fabric, wood, stone, tile, and skin, in either a liquid or dried state. After the substance has been identified, it is usually removed from the substrate by swabbing, scraping, or cutting. (This collection procedure is different from the chemical extraction described below which is used to isolate DNA from other cellular molecules.) For

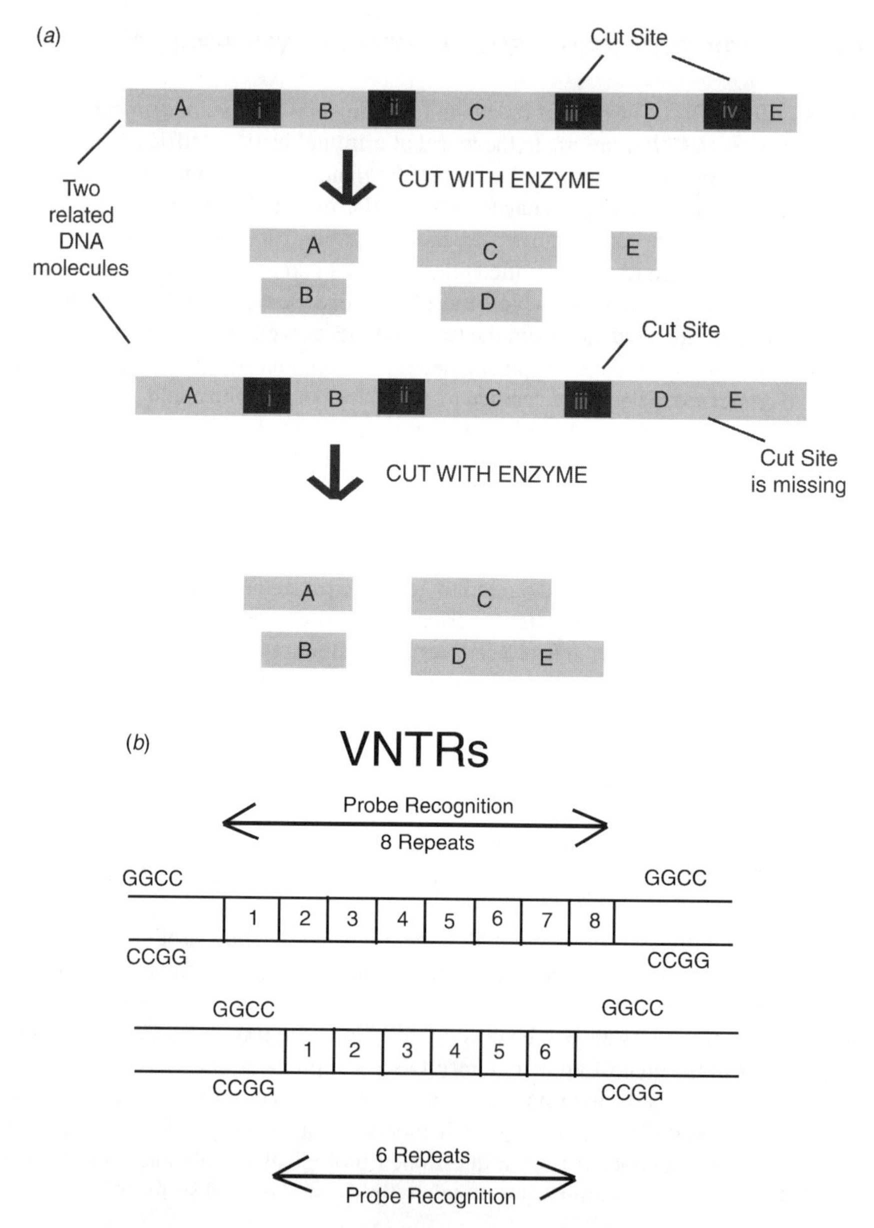

Fig. 3.3 *(a) DNA fingerprinting (RFLP analysis) makes use of regions of DNA in which relatively short segments are repeated and tandemly connected (VNTRs). (b) Variable nucleotide*

example, in the case of a bloodstained fabric, a cutting is made from a stained portion of the fabric (as well as from an unstained nearby area) and each is placed into a tube with an extraction buffer. The tube is gently shaken (vortexed) and then allowed to sit for a period of time at room temperature. (The time depends on the nature of the stain). The shaking tends to remove the biological material from the substrate. Cellular material that has come off the fabric can now be used to obtain DNA. In the cell, almost all the DNA is found in the nucleus, which is surrounded by a host of other molecules including RNA, protein, sugars, lipids, and organic (containing carbon) and inorganic molecules. In order to successfully analyze the nuclear DNA, it must be purified and it must also be in a useful form. There are a number of ways to remove all of these substances from the DNA. There are organic and inorganic isolation methods. Both procedures will produce double-stranded DNA (native DNA), which is essential for RFLP analysis. Regardless of the manner in which it is isolated, the quality and quantity of purified DNA may determine success or failure to obtain results. The nature of the biological evidence will determine just how the DNA is purified. Liquid blood, bloodstains, semen stains, saliva, urine, bone, teeth, hair, and skin are just some of the forms that evidence may take. Different extraction methods are used to isolate DNA from each of these different forms of evidence.

Isolated DNA should be in its native double-stranded form and of high molecular weight, which means that most of the DNA should be larger than 20 kilobases (kb). The reason for this is that following restriction enzyme treatment of the isolated DNA, electrophoresis of the resulting DNA fragments, Southern blotting, and then probe hybridization, the position of the resulting bands reflects fragment sizes up to 10,000 bp in length. Therefore we must use isolation methods that will provide high molecular weight DNA in a form useful for RFLP.

1. *Organic Extraction of Liquid Whole Blood* In this method the red blood cells (which contain no DNA) are lysed by physically rupturing their cell membranes, and then the white cells are centrifuged into a pellet. The white cells are then washed and treated with proteinase K and sodium dodecyl sulfate (SDS) detergent in order to digest proteins and disrupt cell membranes. The DNA is now in solution but mixed with many breakdown chemicals. Phenol–chloroform–isoamyl alcohol extraction is performed to remove proteins and the DNA finds itself in the aqueous phase. (This is a chemical extraction rather than a physical extraction as described above.) It is precipitated with cold ethanol and by addition of sodium acetate (0.2*M*) and then concentrated by centrifugation. The ethanol is removed by use of a Speed-Vac (a centrifuge with a vacuum), and the high-molecular-weight double-stranded DNA is now in a form that can be quantified and processed further. A common approach to purifying DNA after it has been chemically extracted is to filter it through a Microcon 100 filter. This device is manufactured by Millipore and is used to separate high-molecular-weight DNA from smaller molecules, which pass freely through the filter.

2. *Inorganic Methods* Inorganic Methods are based on the use of either sodium chloride or lithium chloride to "salt out" proteins from a mixture of DNA and protein. Liquid whole blood is treated with a lysis buffer containing the detergent Triton X 100 and then centrifuged to remove the supernatant and produce a pellet of white blood cells and cell debris. Proteinase K is then used to break down proteinaceous

material present in the pellet, and salt in the reaction tube causes the proteins to come out of solution (precipitate) while the DNA remains soluble.

3. *Use of FTA Card* Another method of preservation and purification of DNA from reference (exemplar) bloodstains has gained favor recently. Liquid blood is collected and an aliquot placed on an FTA card. The card has been impregnated with a chemical that protects DNA from bacterial enzyme breakdown. Cells within the blood dry and lyse on the paper, but the double-stranded DNA becomes trapped within the fibers of the paper and can remain intact for lengthy periods of time (>10 years) even without refrigeration. When a sample is required for testing, a paper punch is used to remove a small sample. After washing the plug to remove extraneous molecules, the punched paper can be used for any DNA analysis.

4. *Use of Magnetic Resin* Promega Corporation has developed an isolation method that incorporates the use of a magnetic resin to prepare clean DNA samples. To extract stains, the sample is heated in a lysis buffer for 30 min and then centrifuged. The magnetic resin is then added to the solution and the samples are washed using a magnetic stand. After 30 min the genomic DNA is bound to the particles. These samples can be used for either RFLP or for PCR reactions described below and are especially useful when samples are found on the most problematic substrates (denim, leather, and soil). The DNA IQ System permits both DNA isolation from stains or liquid samples as well as providing an approximate quantification of the amount of DNA that has been isolated. This is due to the magnetic resin binding to a specific amount of isolated DNA and therefore excess DNA remains unbound.

3.2.2 Quantification

Once isolated from the rest of the biological sample, the purified DNA is then *quantified* to verify that there is a sufficient amount to test successfully (Budowle et al., 2000b, Waye et al., 1991). Successful RFLP analysis requires 50–100 ng of high-quality, nondegraded, DNA (see Table 3.3).

Several methods can be used to quantify DNA. For RFLP, quantification can be performed by running a "yield" gel to which has been added fragments of DNA of known size to serve as calibration standards. Yield gel electrophoresis is performed by incorporating known amounts of high-molecular-weight DNA ranging from 100 pg up to 200 ng into wells cut within the agarose gel. Isolated "genomic" DNA from questioned and known specimens are placed into other wells. All wells are cut at the top of the gel in the region known as the origin. A voltage difference is placed across the gel with the cathode near the origin and the anode on the opposite end of the gel. Following gel electrophoresis and exposure to ethidium bromide (EB), a fluorescent dye that inserts (intercalates) itself between the grooves of double-stranded DNA, the banding pattern can be visualized as the gel is exposed to ultraviolet light. The series of calibration standards band in an area close to the origin and appear progressively more intense in staining as the quantity of DNA increases. The intensity of band fluorescence for the unknown DNA samples can then be compared to the calibration set, and an estimate of the amount of DNA in

TABLE 3.3 Variable Number of Tandem Repeats (VNTRs) at a Glance

Advantages	Limitations
1. There are a large number of alternative forms (bins), 20–30 at each locus. The heterozygosity is large. This provides a great deal of discriminatory power per locus.	1. The sensitivity of the technique is limited. Generally, 50 ng or more of DNA material is required to obtain clear results. This is a very important disadvantage. Because of their large size, VNTRs cannot be amplified reliably and consistently by polymerase chain reaction (PCR).
2. The techniques are well established, are familiar, and have been widely used.	2. The process is time consuming, taking several days to complete (or even weeks if radioactive probes are used).
3. These tests have been widely accepted by the legal community	3. The number of validated loci are limited.
4. The large number of alleles facilitates mixed-sample analysis.	4. Large RFLP fragments are not suitable for use with degraded DNA samples such as those sometimes found in forensic work.
	5. The necessity of binning introduces statistical complications and sometimes difficulties of interpretation.
	6. Because of the small number of validated loci, VNTRs are of limited value in distinguishing between siblings.

Source: National Commission on the Future of DNA Evidence, *The Future of Forensic DNA Testing Predictions of the Research and Development Working Group* (2000).

these samples can then be made. This method is most accurate when double-stranded DNA is analyzed. Single-stranded DNA will produce much weaker bands with EB staining and the amount will be underestimated. Samples containing more than 1 ng can easily be quantified. DNA that has become fragmented will appear as a long smear starting from the origin and moving anodally toward the opposite end of the gel. A drawback to using this method to quantify isolated DNA is that nonhuman DNA present in the sample cannot be differentiated from human DNA, and therefore the amount of the latter will be overestimated.

Other procedures of DNA quantification include ultraviolet and visible spectrophotometry, fluorometry, and capillary electrophoresis. The quantity of DNA in solution can be determined by measuring its absorbance at 260 and 280 nm and applying a formula.* However, the quality of the isolated DNA cannot be determined in this way. If it is determined that there is a sufficient amount of DNA for RFLP analysis, the DNA must then be processed (or cut) with a specific *restriction enzyme*.

3.2.3 Restriction Enzymes: DNA Scissors

An enzyme is an organic catalyst. Enzymes, *in vivo* or *in vitro*, function to speed up the rate of chemical reactions. Enzymes can be isolated from viruses, bacteria, or

$\frac{O.D_{260}}{O.D_{280}}$ is between 1.8 and 2.0 for pure double-stranded DNA. 50 μg DNA/ml has an optical density of 1.0 at 260 nm in a 1.0 cm path length quartz curvette.

animal and plant cells and can be used in the laboratory for a variety of tasks. Certain enzymes are used to nick or cleave DNA at specific sites within the molecule. These enzymes are called *type II restriction enzymes*, and they play a crucial role in the DNA fingerprinting procedure (Meselson and Yaun, 1968; Roberts, 1983). Unlike type I restriction enzymes that are nonspecific cutters, type II enzymes produce cuts or nicks at specific sites known as restriction sites (Saiki et al., 1985) in the double-stranded DNA molecule.

Restriction enzymes evolved in bacteria as a means of destroying "foreign DNA" introduced by other bacterial and bacteriophage (viral) species. Hundreds of such enzymes with different specificities have been isolated. See Figure 3.4 for a partial list of such enzymes. Restriction enzymes will recognize nonmethylated DNA as foreign and will destroy it by nicking the sugar–phosphate bonds between nucleotide bases on a single strand or on both strands of the double helix. These enzymes recognize short specific pallindromic sequences of bases. The result is that DNA will be fragmented leaving either *blunt ends* (if the cleavage is across both strands) or *sticky ends* if the cleavage is offside (staggered) but involving both strands (see

Name	Source	Activity	
EcoRI	E. coli	5'-G*A A T T C-3' 3'-C T T A A*G-5'	
BamHI	B. amyloliquefaciens	5'-G*GATCC-3' 3'-CCTAG*G-5'	
HaeI#	H. aegyptius	5'-A G G*C C T-3' 3'-T C C*G G A-5'	
HaeII	H. aegyptius	5'-Pu G C G C*Py-3' 3'-Py*C G C G Pu-5'	(Pu=Purine A or G (Py=Pyrimidine C or T)
HindIII	H. influenza	5'-A*A G C T T-3' 3'-T T C G A*A-5'	
PstI	P. stuartii	5'-C T G C A*G-3' 3'-G*A C G T C-5'	
MstII	Microcoleus sp.	5'-C*C T N A G G-3' 3'-G G A N T C* C-5'	(N=Purine or Pyrimidine)
TaqI	T. aquaticus	5'-T*C G A-3' 3'-A G C* T-5'	
AluI#	A. luteus	5'-AG*CT-3' 3'-TC*GA-5'	

#This restriction enzyme produces blunt ends.

Fig. 3.4 *Restriction enzymes evolved in bacteria as a means of destroying "foreign DNA" introduced by other bacterial and bacteriophage (viral) species.*

5'...............TC*GA..................3'
3'...............AG*CT..................5'

The restriction enzyme nicks both upper and lower strands symmetrically forming two molecules with blunt ends:

(a)

5'...............TC3' 5'GA..................3'
3'...............AG5' 3'CT..................5'

19b: Sticky Ends (HindIII Restriction Enzyme)

5'..............A*AGCTT..................3'
3'..............TTCGA*A..................5'

The restriction enzyme nicks both strands symmetrically but around a central site Resulting in two molecules with "sticky" ends:

(b)

5'..............A AGCTT..................3'
3'..............TTCGA A..................5'

Fig. 3.5 *Restriction causes DNA to be fragmented leaving either (a) blunt ends (if the cleavage is across both strands) or (b) sticky ends if the cleavage is offside (staggered) but involving both strands.*

Fig. 3.5). Like other enzymes, restriction enzymes work best at a specific temperature and pH. Hundreds of restriction enzymes have been identified and isolated and each is very specific in its action. These enzymes act like molecular scissors, recognizing certain base sequences and cutting DNA only at those sites.

Scientists have been able to use these enzymes for their own purposes. Genomic DNA can be cut by restriction enzymes, and the fragments produced can then be separated by a technique known as electrophoresis, which is a method that separates molecules based on mass (fragment size) and net charge. The separation of DNA fragments is based only on fragment size since all fragments have the same mass-to-charge ratio. Electrophoresis of DNA fragments will be described in the next section.

One of the most commonly used restriction enzymes is Hae III, which recognizes the sequence 5′-GGCC-3′ and cuts the bond in the center of this sequence where G is linked to C. This happens on both of the strands that make up DNA. During RFLP, after the DNA is isolated, purified, and then restricted, the resulting DNA fragments are then subjected to the separation procedure called electrophoresis.

3.2.4 Gel Electrophoresis

In the *gel electrophoresis* method, fragments of DNA are separated based on their size differences (numbers of base pairs). See Figure 3.6. DNA molecules, regardless of size, have a net negative charge when placed into a neutral solution. During gel electrophoresis, 50–100 ng of the purified DNA is pipetted into each of the small

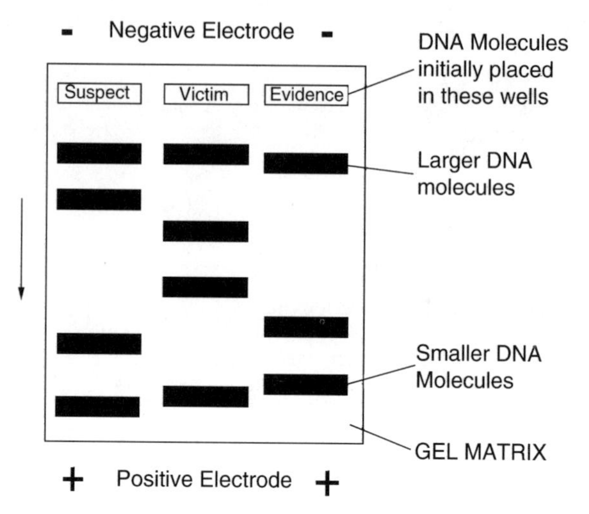

Fig. 3.6 *In gel electrophoresis method, fragments of DNA are separated based on their size (numbers of base pairs) rather than simply ratio of charge to mass.*

wells or slots that have been formed in the gel. The gel is prepared from either agarose or acrylamide. The solution is poured onto a glass plate so that the gel will form in a thin slab format. A comb is inserted at one end so that after the gel hardens and the comb is removed, small wells are present on one end of the slab gel. The wells are located at the origin near the position of the negative electrode (cathode). A positive electrode (anode) is placed at the opposite end of the gel. When the power is turned on, there is a voltage difference produced across the gel. On a microscopic level the gel is internally porous and will act like a sieve during the electrophoresis procedure. Gels can be prepared so that the optimum pore size can be produced. The negatively charged DNA molecules migrate within the gel away from the cathode and toward the anode. The greater the voltage difference across the gel, the faster the DNA molecules will migrate. The larger the DNA molecule the more difficulty it will have moving through the gel. Consequently, over a fixed period of time, the smaller DNA molecules will have moved further in the direction of the positive electrode than the larger molecules. At this stage the bending pattern cannot be seen.

3.2.5 Southern Blotting

After the DNA molecules have been separated by size, the gel that contains them is subjected to a technique called *Southern blotting*. Southern blotting is named for Edward Southern, who invented the procedure at Edinburgh University in the 1970s. During the Southern blotting procedure, the DNA fragments within the agarose gel are first denatured into their single-stranded constituent molecules. The single-stranded DNA molecules are then transferred to a more sturdy, nylon membrane, where they become covalently bound. Figure 3.7 illustrates an agarose gel containing DNA fragments that have been separated based on differences in their size. A nitrocellulose membrane is placed over the gel and a stack of paper towels

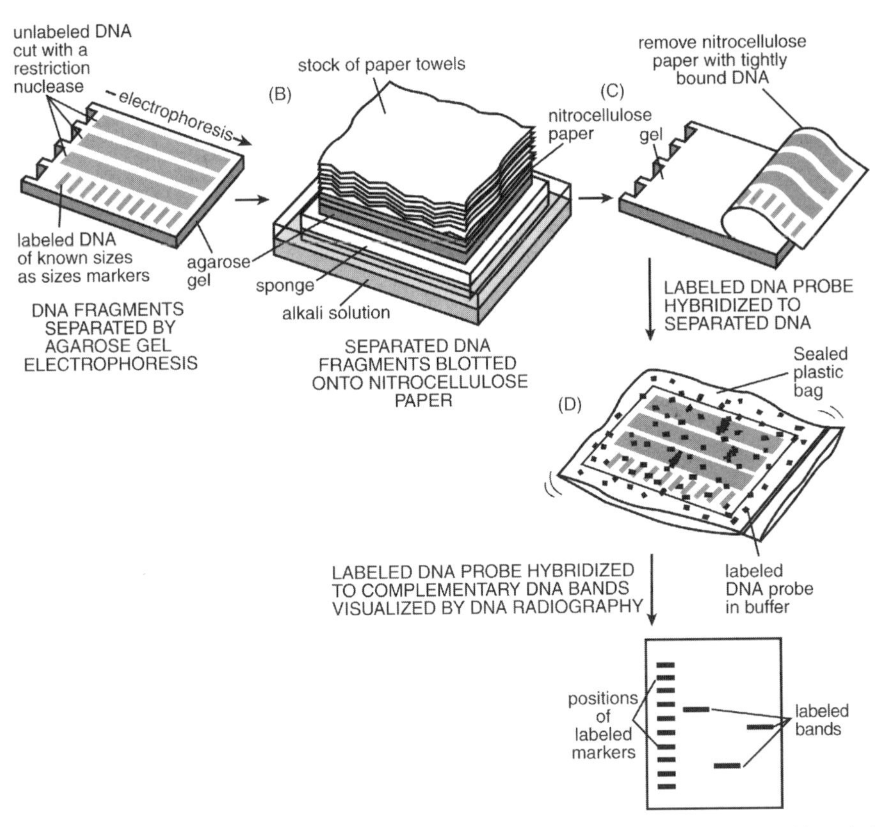

Fig. 3.7 During the Southern blotting procedure, DNA fragments are first denatured into their single-stranded constituent molecules. Single-stranded DNA molecules are then transferred to a more sturdy, nylon membrane, where they become covalently bound. Essential Cell Biology, *by Bruce Alberts, et al. Copyright 1997. Reproduced by permission of Garland Science/Taylor & Francis Books, Inc.*

is placed on top of that. Beneath the gel is a sponge sitting in an alkali solution (0.5*M* NaOH), which will denature the fragments into single-stranded DNA. As the fluid moves upward through the stack by capillary action, the single-stranded fragments are transferred to the membrane over a period of several hours. A vacuum blotting apparatus can also be used to speed up the transfer. The fragments become bound in the same positions on the membrane as they were in on the gel from which they were transferred. The single-stranded DNA on the membrane is now treated with ultraviolet light causing it to become cross-linked (covalently bound) to the membrane. It is now ready for hybridization. This transfer of DNA from agarose to either a nylon or nitrocellulose membrane is important because gels are very hard to manipulate and preserve. Nylon membranes have about five times the DNA binding capacity as nitrocellulose membranes. The nylon membrane also makes it easier to do further testing on the separated DNA molecules since it is less fragile than the nitrocellulose membrane. For P^{32}-labeled probes, a positively charged membrane (Biodyne B) is used. For chemiluminescent detection, an amphoteric (charged both positively and negatively) membrane (Biodyne A) is used. Chemiluminescent RFLP analysis has a number of advantages over the use of radioisotopes (Budowle et al., 1995b).

3.2.6 Hybridization

Prior to *hybridization* with a labeled probe, the membrane is "prehybridized" with nonspecific single-stranded DNA. This tends to prevent the labeled probe from binding nonspecifically to DNA fragments on the membrane. The membrane is then bathed in a solution containing a labeled probe, which is a short segment of single-stranded DNA chosen to be complementary to a specific DNA sequence. The probe will recognize and hybridize to the target portion of the single-stranded DNA molecule (Webb et al., 1993). Short probes are generally more specific. The specificity of probe binding is determined by the degree of stringency used. Stringency is determined by the concentration of formamide (or urea), salt, and detergent. The higher the stringency, the higher the degree of sequence-specific binding by the probe. Typically, one can detect as little as 0.1 pg of target DNA.

The target DNA consists of a core of repeated sequences called *variable number tandem repeats*, or VNTRs (Fig. 3.8). These are DNA sequences composed of repeating units, each of which has from 8 to 80 bp and which are connected tandemly or adjacent to each other. The length of the complete VNTR segment, including the regions that flank the core repeats, defines the different alleles within the locus. The probe, which is single-stranded DNA that corresponds to the VNTR, can either be labeled with a radioisotope or with a substance that renders the probe–VNTR complex chemiluminescent.

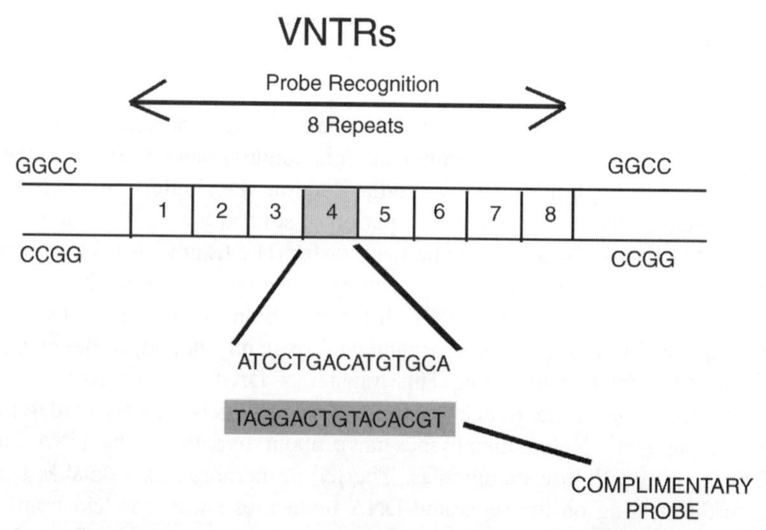

Fig. 3.8 *Target DNA consists of core of repeated sequences called variable number tandem repeats, or VNTRs.*

A radiolabeled probe allows the technician to locate the position of a particular DNA sequence within the single-stranded DNA on the membrane. The probe will only attach to DNA molecules that contain the complementary sequence, that is, the repeated VNTR sequence. The analyst places an intensifying screen inside of a film cassette, the hybridization membrane that has been wrapped in cellophane, and a sheet of high speed X-ray film. The film cassette is closed, wrapped in foil (to prevent light leaks), and allowed to sit in a $-70°C$ freezer. A chemiluminescent probe will cause the DNA fragment to emit light when subjected to a certain stimulus. After a period of exposure, the X-ray film is removed and developed and the resulting banding pattern can be visualized and analyzed.

To avoid the use of radioisotopes, a chemiluminescent method was developed. The single-stranded DNA probe is labeled with horseradish peroxidase (HRP). Stringency of the hybridization to the membrane-bound DNA fragments is controlled by altering the salt concentration. The membranes are washed free of excess nonhybridized probe. In a series of coupled reactions luminol is oxidized, thereby producing emission of a blue light. The membrane is placed next to a sheet of X-ray film, which is then placed into a film cassette. In a couple of hours the film has been sufficiently exposed and is ready for development.

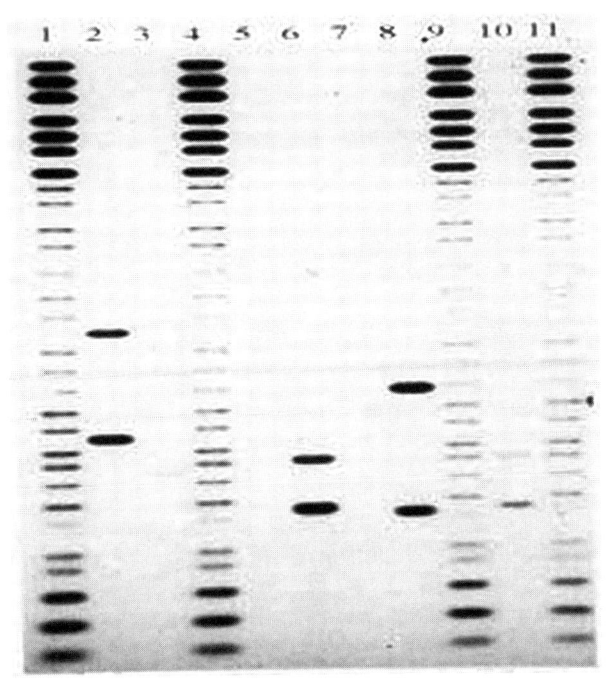

Fig. 3.9 Banding pattern on autoradiograph resembles a supermarket barcode.

3.2.7 Autoradiography and Visualization of DNA Banding Pattern

Use of a radiolabeled probe and exposure to X-ray film for the appropriate period of time will allow visualization of a banding pattern. This process, in which the radioactive DNA is allowed to "take its own picture" on a sheet of photographic film is commonly referred to as *autoradiography* (Duewer et al., 1995). The banding pattern on the developed film, which is called an *autoradiograph,* tends to resemble a supermarket barcode (Fig. 3.9). Alternatively, the use of a chemiluminescent label allows for the visualization of the banding pattern without the use of hazardous radioisotopes. The developed film is called a *lumigraph.* The banding patterns of DNA of known origin or "knowns" (e.g., a suspect's DNA) and DNA obtained from a crime scene ("questioned") can then be compared first by eye and then by computer scanning (see Fig. 3.10).

3.2.8 Analysis of RFLP Results

Once the above procedures have been performed, analysis and interpretation of the banding patterns follows. Visual observation is too subjective and unreliable. Some lanes contain bands produced by the DNA of a suspect. Other lanes contain DNA produced from evidence collected at a crime scene. Several lanes will contain a number of samples that serve as *positive* and *negative controls.* The positive control usually consists of DNA from a known cell line that produces a banding pattern consisting of fragments of known size at a specific locus. In addition, a number of calibration samples are run on the gel. They appear at strategic locations on

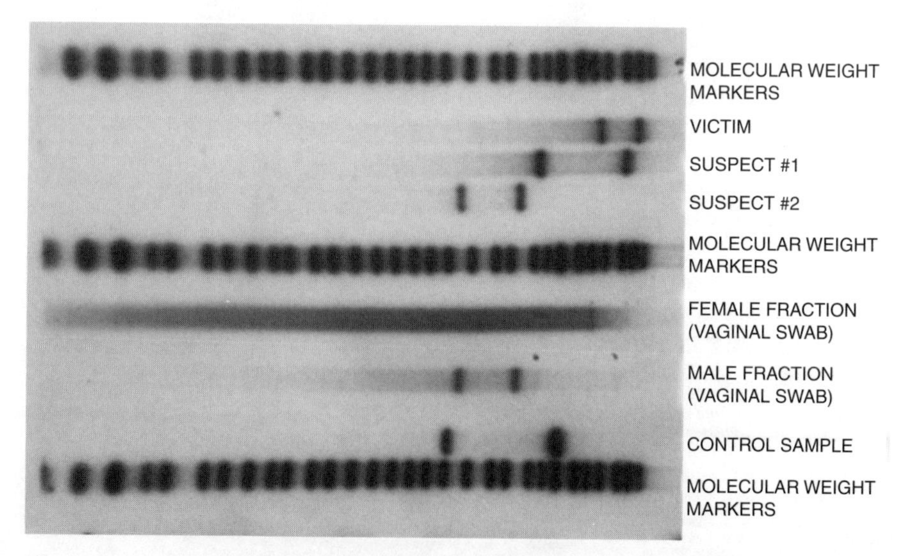

Fig. 3.10 *Banding patterns of known (exemplar) DNA and questioned DNA (from a crime scene) can be compared. Top of figure is to the right.*

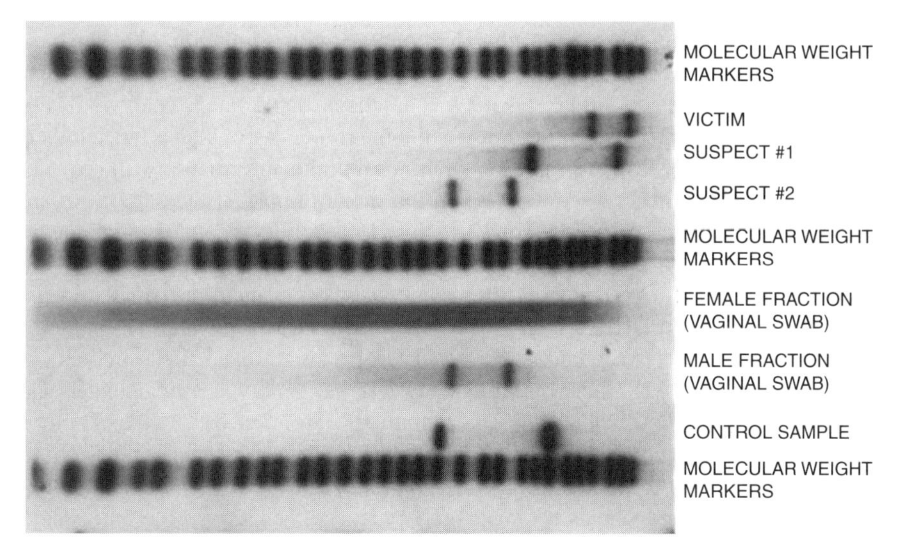

Fig. 3.11 *Molecular weight markers are placed at strategic locations on gel, usually on the ends and in the middle. Top of figure is to the right.*

the gel, usually on the ends and in the middle (Fig. 3.11). These samples consist of mixtures of DNA fragments of various molecular weights and are called *molecular weight ladders.* The bands present in the ladders allow a computer to interpolate molecular weights of unknown sample bands by comparing their positions to those of the bands within the ladders. Homozygotes at any locus will display only one band. Heterozygotes will display two bands. The observation of three or more bands generally means that there is a mixture of DNA from more than one person. Based upon certain preestablished standards, the bands of the questioned and known samples are either declared a match or a nonmatch. A nonmatch indicates an exclusion of a suspect as the origin of the biological evidence. A match indicates that the suspect *may* have been the source of the questioned specimen. Occasionally a three-band pattern will be produced in a specimen obtained from a single individual (Waye et al. 1994). The simplest explanation for this observation is that an additional restriction site occurs within the VNTR region on one of the two homologous chromosomes of a pair. In this case, the restriction enzyme will cut at the ends and within the VNTR region, thereby creating two DNA fragments from one chromosome and a third fragment from the other chromosome. Rather than causing confusion, the presence of a three-banded pattern in both known and questioned specimens reinforces the match between the two.

If a match is declared, then the final stage of the procedure is to determine how rare the banding pattern is in the relevant population. The relevant population includes the population in the relevant geographical area. It can also mean in the relevant age group. In the case of rape where a semen specimen is analyzed, women and young children would not be part of the relevant population.

What is the significance of each band on the autoradiograph? Each band, represents an allele (gene), which has a frequency of occurrence in the relevant population. Once the individual's genotype has been determined, analysis of an established database provides information about the frequency of alleles at a number of loci (Smith et al., 1990). A simple calculation using this information will provide the frequency of the genotype in the population. For example, if the pattern for an individual consists of two bands (corresponding to two alleles) and the first has a frequency of 1 in 10 ($p = 0.1$) while the second has a frequency of 1 in 20 ($q = 0.05$), then the frequency of the genotype (both bands) is calculated as $2pq = 2\ (0.1)(0.05)$ or 0.01 which corresponds to 1 in 100 people. The pattern for an individual who is a homozygote consists of a single band. If its frequency is 1 in 15 ($p = 0.067$), then the frequency of the genotype is calculated as $p^2 = (0.067)(0.067) = 0.0045$. These calculations are based on the Hardy-Weinberg formula, which is the foundation of population genetic theory (see Section 4.2). If the population is found to be in *Hardy–Weinberg equilibrium* we can use the database of allele frequencies to make genotype and genetic profile frequency calculations.

3.2.9 Probe Stripping from Membrane

After a pattern is recorded for one locus, the probe is stripped from the nylon membrane by treating it with a solution containing formamide (to denature DNA) and SDS (ionic detergent). At this point a second probe can be applied to the membrane. This probe recognizes a VNTR sequence at another locus. This banding pattern is different from the first, reflecting different DNA fragments that combine with the labeled probe. The new pattern is recorded and statistical calculations are made to determine the frequency or rarity of this pattern (genotype). Genotype frequencies can be calculated assuming that the gene frequencies are in Hardy–Weinberg equilibrium (see Sections 3.2.11, 3.3.4.2.1, 3.6.5, 4.1.1, and 4.1.3.) The stripping procedure can be performed repeatedly as long as the original fragments remain bonded to the nylon. Despite the covalent binding, there are limitations to how many times a membrane can be repeatedly stripped and probed because stripping tends to deplete some of the membrane bound DNA fragments. The statistics produced for each genotype can be multiplied with each other to determine the frequency of the entire genetic profile. The assumption here is that genes at different loci are in linkage equilibrium (see Section 4.2.1). Multiplication of individual frequencies to determine overall frequencies is legitimate according to the laws of probability and the product rule. The concept is that if you know the probability of one event (such as a coin flip producing a head), p_1, and if you know the probability of a second event, p_2, then the probability of both events occurring simultaneously—assuming that they are independent of each other—(flip two coins at the same time and get two heads) is the product of the two individual probabilities, $p_1 \times p_2$. In RFLP analysis, the product rule can be applied repeatedly for each locus bringing a match statistic into the range of one in hundreds of millions or billions or even trillions, depending on how many loci are tested.

3.2.10 Match Criteria

At a particular locus, VNTR alleles may differ by only a single repeat. Electrophoresis is a good analytical tool but insofar as reproducibility is concerned, there is a small level of variation that occurs. One cannot always differentiate alleles that differ by only one repeat. VNTRs are not discrete alleles but rather fall into a continuum. The range of this variation is important since we cannot easily determine the true size of an unknown fragment with this technique. Even more important, the analyst must decide if two DNA fragments match in length, that is, that the fragments derived from the evidentiary and exemplar specimens are identical in size. One way to address the issue experimentally is to repeatedly analyze the sample by electrophoresis to determine the size of a particular fragment. Assume that we have performed 100 RFLP analyses of a single fragment and have obtained a mean score of 1100 bp with values ranging from 1070 to 1130 bp. Thus when we make any measurement, the true value lies someplace within 1100 ± 30 bp, which is approximately $\pm 2.5\%$ of 1100 bp. A match window can then be set up around the value of 1100 such that an observation of 1070 (-2.5%) or 1130 ($+2.5\%$), or anything in between is accepted as a match. We can now compare two fragments, each in a different lane of the gel, to determine if there is a match or an exclusion. If one fragment is evaluated as 1120 bp while the second is 1080 bp, the two fragments would be declared a match since $1120 \pm 2.5\%$ overlaps $1080 \pm 2.5\%$ (see Fig. 3.12). On the other hand if the determined sizes are 1130 and 1070 bp, the respective windows ($1130 \pm 2.5\%$ and $1070 \pm 2.5\%$) will not overlap, and we can conclude that the two fragments have different origins (see Fig. 3.12).

Some laboratories may decide to use different match criteria, for examples, values of $\pm 1.5\%$ or $\pm 1.8\%$ depending on their empirical laboratory studies. The larger the "window" the more likely that two fragments will be found to "match." The smaller the window the less likely that they will match even if

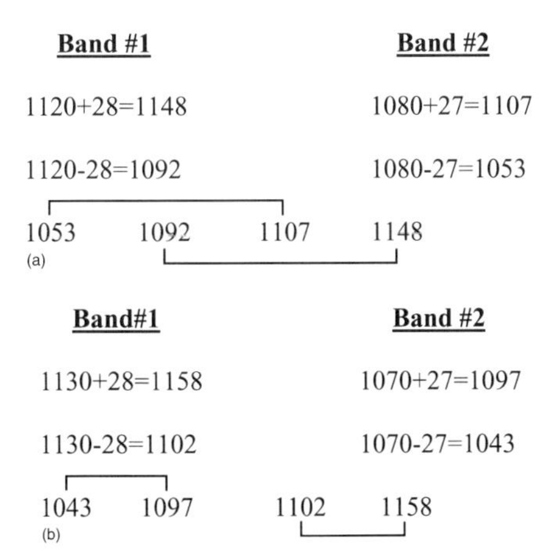

Fig. 3.12 (a) Bands 1 and 2 are each found in a bin that overlaps each other. (b) Bands 1 and 2 are each found in bins that do not overlap.

they are in fact identical. This window concept is used to "bin" alleles. That is, at a particular locus, a number of alleles can be assigned to specific bins. Instead of considering the hundreds of continuous alleles that can exist at a VNTR locus, you are now considering 20–30 different alleles (bins). In essence we are converting the continuum of alleles into a set of discrete alleles (Budowle et al., 1991) and therefore all the alleles within a single bin are assigned the frequency corresponding to the entire group of alleles. You can certainly have many alleles contained within each bin.

Prior to using RFLP in casework, it is important to perform a series of validation tests to determine that it produces reliable and reproducible results regardless of how the evidence has been treated (Laber et al., 1994; TWGDAM, 1989, 1990).

3.2.11 Statistics and the Product Rule

If it is found that a match does not exist, then the suspect has been excluded. The suspect may match at a number of loci, but if he is excluded at one locus, he is excluded as the source of the evidence. On the other hand, once it is apparent that a match exists in fragment patterns and therefore in the size of alleles for a number of loci, a statistic must be calculated so that the significance of the match can be ascertained (Risch and Devlin, 1992). If the profile is very common, then the finding of a match is of limited significance to a jury. Once again the concept of binning must be applied. With the method of electrophoresis, VNTRs cannot be precisely measured. Rather than existing as discrete alleles, they form a continuum of alleles. VNTRs have a range of sizes, and therefore it becomes impossible to determine their precise allelic frequencies in the relevant populations. For the reasons described above it is important to be able to look at the VNTR alleles as if they are discrete. In this way we can determine allele frequencies and then apply the product rule. For each VNTR locus a database must be constructed for each of the major racial/ethnic groups (Budowle et al., 1994a, 1994b). For example, a database consisting of 500 individuals, all within the same ethnic/racial group, will provide information on the frequency of (most) alleles that exist within that locus based on an analysis of 1000 alleles. Even rare alleles can be observed given 1000 alleles in the database. If we assume that the allele sizes for this locus range from 0 to 12,000 bp, we can now divide the 12,000 bp by 20 equally sized bins. The first bin will run from 0 to 600, the second will run from 601 to 1200, the third from 1201 to 1800, and so on. The last bin will include alleles in the range of 11,401 to 12,000 bp (see Fig. 3.13). This is called the fixed-bin method (Budowle et al., 1991, Budowle and Giusti, 1995). Its divisions are arbitrary. Now we determine how many of the 1000 alleles fall into each bin, and this determines that bin's frequency.

When we are finished, we will have frequencies for alleles that fall into each of the 20 bins. Of course, in some instances a bin may contain no alleles or only a few. The FBI has established guidelines on how to calculate statistics given this kind of situation. Another common situation is when we are dealing with an allele found to be, say, 2400 bp but whose real size is between 2300 and 2500; the range of values may not be found in one bin but rather in two adjacent bins (see Fig. 3.14). In such a

Binning and Databasing (n=500 individuals)

Bin Number	Size Range of Alleles (bp)	Number of Alleles	Allele Frequency
1	0-600	5	0.0050
2	601-1200	15	0.0150
3	1201-1800	24	0.0024
4	1801-2400	12	0.0012
5	2401-3000	32	0.0032
6	3001-3600	30	0.0030
7	3601-4200	46	0.0046
8	4201-4800	65	0.0065
9	4801-5400	62	0.0062
10	5401-6000	88	0.0088
11	6001-6600	35	0.0035
12	6601-7200	85	0.0085
13	7201-7800	120	0.0120
14	7801-8400	95	0.0095
15	8401-9000	86	0.0086
16	9001-9600	54	0.0054
17	9601-10,200	36	0.0036
18	10,201-10,800	46	0.0046
19	10,801-11,400	54	0.0054
20	11,401-12,000	10	0.0010
		Total: 1000	1.0000

Fig. 3.13 *Statistical method of binning helps to determine if two alleles have the same size or are significantly different.*

Binning and Databasing (n=500 individuals)

Bin Number	Size Range of Alleles (bp)	Number of Alleles	Allele Frequency
1	0-600	5	0.0050
2	601-1200	15	0.0150
3	1201-1800	24	0.0024
4	1801-2400	12	0.0012
5	2401-3000	32	0.0032
6	3001-3600	30	0.0030
7	3601-4200	46	0.0046
8	4201-4800	65	0.0065
9	4801-5400	62	0.0062
10	5401-6000	88	0.0088
11	6001-6600	35	0.0035
12	6601-7200	85	0.0085
13	7201-7800	120	0.0120
14	7801-8400	95	0.0095
15	8401-9000	86	0.0086
16	9001-9600	54	0.0054
17	9601-10,200	36	0.0036
18	10,201-10,800	46	0.0046
19	10,801-11,400	54	0.0054
20	11,401-12,000	10	0.0010
		Total: 1000	1.0000

Fig. 3.14 *Allelic values may not be found in one bin but rather in two adjacent bins.*

case, the frequency assigned to that allele is the larger of the two bin frequencies. This is a very conservative way of estimating frequencies and produces a number favorable to the defendant. Another approach to determining match probabilities is to use the floating-bin method. In this method a window around a fragment size is established, for example, 2400 $\pm$ 1.8%, and the database of allele frequencies is examined using the same size window ($\pm$1.8%). The frequencies for all alleles within this window are combined to approximate the frequency of the allele being examined.

Now let us assume that we have tested samples at five VNTR loci, and four have resulted in heterozygous patterns, while the fifth produced a homozygous banding pattern. We can check our databases for these alleles using the appropriate ethnic group (relevant population). We must determine genotype frequencies at each locus and then use the product rule to determine the frequency of occurrence across all five loci. For heterozygous loci we can use the formula $2pq$, which is derived from the Hardy-Weinberg equilibrium formula ($p^2 + 2pq + q^2 = 1$). In this formula p refers to the bin frequency of the first allele and q refers to the bin frequency of the second allele. Therefore if $p = 0.01$ and $q = 0.02$ at locus 1, then $2(p)(q) = 2 \times 0.01 \times 0.02 = 0.0004$, which is 4, in 10,000 (or 1 in 2500). For homozygous loci the genotype frequency is calculated as $2p$, where p is the frequency of the single detectable allele. Therefore if $p = 0.015$ at locus 2, then $2(p) = 2 \times 0.015 = 0.03$ or 3 in 100. Had we calculated the genotype frequency of the homozygote as $(p)^2$, which is what the Hardy–Weinberg formula calls for, it would have resulted in a frequency statistic that could underestimate the number of homozygotes in the subpopulation ($p^2 = 0.015^2 = 2.25 \times 10^{-4}$). Therefore, we replace the p^2 calculation with the $2p$ calculation. If we err in the calculation, it is an error in favor of the defendant. Use of the product rule *across* these two loci (locus 1 $\times$ locus 2) produces the statistic 0.0004×0.03, which is 12×10^{-6} or 1.2×10^{-5} or roughly one in 100,000 people. Using the product rule for the additional three loci will bring the final statistic to one in greater than hundreds of millions or billions.

It should be noted that when examining gene frequencies following RFLP analysis, one *may* note an excess of homozygotes. That is, we may find more single-banded patterns than expected. There are a number of explanations for this observation, some having to do with the detection method employed (small alleles migrating off of the electrophoretic gel). However, another explanation is the existence of substructure within the larger population. This observation is referred to as the *Wahlund effect.* Typically one observes more homozygotes than expected if the population were in Hardy–Weinberg equilibrium. Substructure can also result in the observation of linkage disequilibrium (see Section 4.2.1).

In preparing a written report, it is appropriate to indicate the product rule statistics for each of the relevant major ethnic groups. One should not present the statistics for the ethnic group of the defendant alone, since we cannot say with scientific certainty who was the source of the evidence. Also since the product rule only provides an approximation of the true value, it should be stated that the correct frequency lies someplace between 10-fold above and 10-fold below the calculated figure. It is also informative to provide in the report frequency statistics that would be expected for siblings and close relatives (Brookfield, 1995) and to provide a correction factor (θ) in statistical calculations to compensate for the existence of subpopulations (Schneider et al., 1991; Sharma et al., 1995; Brookfield, 1995; Budowle et al., 1994c). This correction factor should be empirically determined by each testing lab since the location of the lab with respect to surrounding populations is important. This very conservative approach to reporting statistics is important because it shows no bias against the defendant while at the same time the statistics calculated for an inclusion are so rare that they carry great significance for the jury to consider.

In 1989, the Technical Working Group on DNA Analysis Methods (TWGDAM) published guidelines for a quality assurance program for RFLP analysis. The next year it published guidelines for a proficiency testing program for laboratories performing RFLP (TWGDAM, 1989, 1990)

CLOSER LOOK AT ELECTROPHORESIS

Electrophoresis is a method of separating proteins, nucleic acids, and other chemical substances based on differences in charge and mass. There are two common forms of electrophoresis used in forensic laboratories—gel and capillary electrophoresis. Capillary electrophoresis will be discussed below (see Section 3.3.4.2.1). Agarose gel electrophoresis is the method of choice for RFLP analysis. The pore size can be changed by changing the concentration of the agarose–buffer solution. Gels should be made up with the appropriate type of agarose. Some agarose preparations (mid to high EEO) are charged thereby creating electroendosmotic problems when performing electrophoresis. These charged molecules resulting from contaminants in the agarose preparation can cause water to move in the direction opposite to the moving protein or DNA fragment, thus changing the migration rate and resulting in drying on one side of the gel. It is important that low-EEO agarose is used for electrophoretic separation of proteins and DNA molecules, although there are some procedures that call for high-EEO agarose. Polyacrylamide gels provide higher resolution than agarose gels. The pore size can easily be altered by changing the ratio of the monomer acrylamide and the cross-linking monomer (methylene bis-acrylamide). The monomers are cross-linked and polymerized by ammonium persulfate and TEMED (N,N,N′,N′-Tetramethylethylene diamine). As a result of the polymerization, the solution hardens and forms a gel. The gel is made up with Tris–borate–EDTA buffer, which is also used in the buffer chambers of the electrophoresis apparatus as well. As is the case for agarose gels, the DNA fragments are separated in acrylamide gels based on their size, and so the gel works like a sieve allowing smaller molecules to move faster than larger ones. Polyacrylamide is not charged and therefore will not interact with DNA. The monomer is neurotoxic and must be used with caution.

The gel is formed on one side of a glass plate. After the liquid has hardened, wells are cut at one end and samples of DNA to be separated are pipetted into each well. The gel can be used to observe the quality and quantity of DNA that had been previously isolated. Such a gel is called a *yield gel.* The yield gel is usually a short slab gel in which isolated DNA is run together with a set of calibrating standards consisting of high-molecular-weight double-stranded DNA ranging from picogram to nanograms quantities. After staining with ethidium bromide, the analyst can tell if the DNA is intact (band remains close to origin) or fragmented (bands form a continuous smear from top to bottom) as well as estimate the quantity of the DNA sample that was run on the gel by comparing the intensity of staining for the sample with the known set of calibrator standards (Waye and Fourney, 1990). The *analytical gel*, on the other hand, is used to compare known and questioned specimens and to determine the approximate size of each fragment. In this kind of gel, wells are loaded with questioned and known specimens as well as control specimens. One of the positive controls includes DNA obtained from a known cell line. In addition, several wells are loaded with a mixture of molecular weight markers. The known cell line produces a banding pattern containing bands of known size. The molecular weight marker is used to calibrate the gel, and thereby bands from all other samples can be objectively sized by comparing band positions. The fragment size is related to its band position. Once the wells are filled, an electric field is placed across the gel, such that the negatively charged DNA fragments within each well will migrate away from the cathode and toward the anode, the positively charged terminal. In this mode the larger fragments remain closer to the origin and the smaller fragments migrate through the pores of the gel more rapidly toward the anode. Thus depending on the size of the fragment, the migration will occur at a different rate and at any moment during the process, fragments of different size will occupy different positions.

3.3 POLYMERASE CHAIN REACTION

3.3.1 Development and Theory

The polymerase chain reaction was developed by Kary Mullis who was awarded the Nobel Prize as well as a number of other prestigious awards for his research efforts. In 1985 Mullis had visualized a way of replicating DNA *in vitro* much the same way as DNA is replicated in every cell of our bodies. This technique has had profound impact on medical diagnostics and therapeutics as well as for forensic science (Sensabaugh and von Beroldingen, 1991).

Polymerase chain reaction is a simple and elegant procedure. As noted in an earlier chapter, the four deoxyribonucleoside triphosphate (dNTP) subunits that are used to construct a molecule of DNA have molecular structures that allow complementary base pairing within the DNA molecule. The molecule is constructed so that an A on one strand always pairs with a T on the other strand. Similarly, the C on one strand always pairs with the G on the other strand. These letters represent the four nitrogenous bases, adenine (A), cytosine (C), guanine (G), and thymine (T), which were described in Chapter 1. When cells are ready to divide and DNA must be replicated, they employ enzymes called *DNA polymerases.* There are actually several different polymerase enzymes used for DNA replication as well as to repair damaged DNA. Only one polymerase enzyme is used to synthesize DNA in a test tube. The method used to replicate DNA *in vitro* is known as the polymerase chain reaction (PCR). In addition to a polymerase enzyme, PCR requires the four nucleotide building blocks, inorganic chemicals, as well as a device that can rapidly and accurately change and maintain temperature for short periods of time, the *thermal cycler.* PCR is an extremely important technique for criminalists because when evidentiary DNA is in small quantity and/or when the specimen is old and the DNA may be fragmented, nicked, or otherwise damaged, the process can produce many copies of the original molecule thereby making detection and analysis possible (Erlich, 1989; Erlich et al., 1991; Haff et al., 1991).

To replicate DNA *in vitro*, it is necessary to unwind and separate the two strands of the double helix. This is called *denaturation.* Each of the original strands in the sample DNA is used as a *template* for the synthesis of a new strand. A template strand exhibits a sequence of nitrogenous bases that can be used to synthesize a complementary sequence on a new strand. The template determines how the new strand will look. Because the strands of DNA are held together by the weak hydrogen bonds that form between the purine and pyrimidine bases, it is quite easy to cause the strands to separate (denature) simply by subjecting native (double-stranded) DNA to elevated temperatures (95°C) for approximately 1 min. Double-stranded DNA sequences rich in G-C content will be held together more tightly than sequences that are rich in A-T since there are three hydrogen bonds holding the G-C bases together while there are only two hydrogen bonds that hold the A-T bases together.

Once the strands have become separated, the temperature is lowered to approximately 59°C to allow the primers to anneal (hybridize) to each of the two strands (see Fig. 3.15). The *primers* are short fragments of single-stranded DNA (15–30 bases long) that are synthesized so that they have a complementary sequence to sites on the template strands (see Fig. 3.16). The sites at which these primers attach are very important because these sites mark the boundaries of the areas to be amplified. The primer set determines the specific region for the new DNA synthesis to begin, and they also "prime" the reaction. The two different primers determine the boundaries for the portion of the DNA molecule that is to be replicated (see Fig. 3.17).

The primer binding sites flank the region of interest that is to become amplified. The polymerase enzyme uses the target DNA strand as the template. The enzyme recognizes the site where the primer has become annealed and starts to synthesize the new DNA strand in the 5′ to 3′ direction by adding the dNTPs, one at a time,

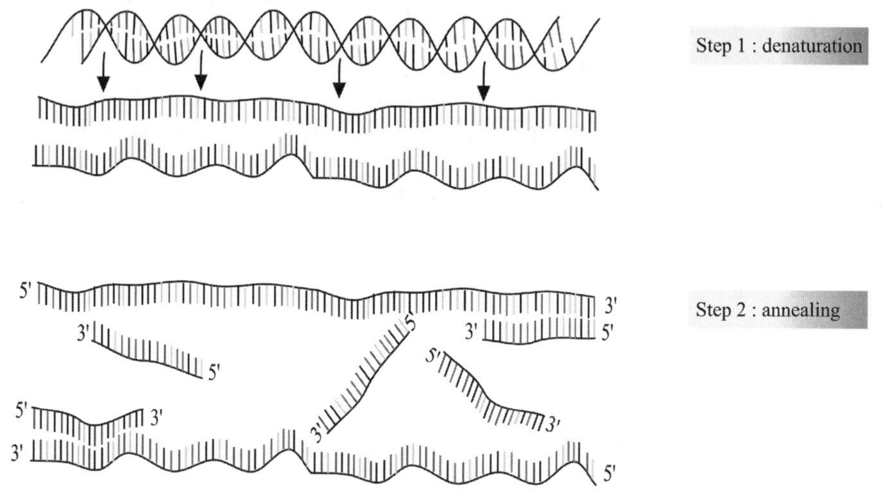

Fig. 3.15 After the template DNA is denatured as a result of heating, a primer now binds (anneals) to each of the two separated strands based on a complementary base pairing.

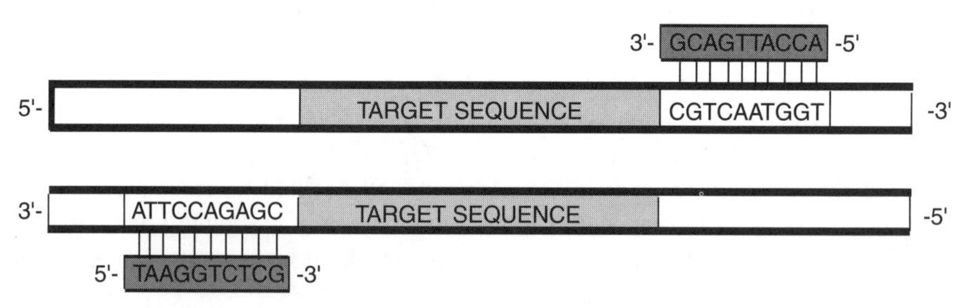

Fig. 3.16 Primers are synthesized so that they bind to the template strand by complimentary base pairing.

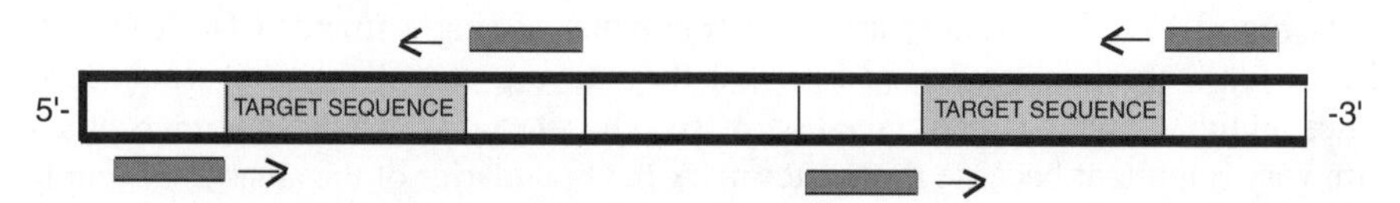

Fig. 3.17 Primers set boundaries for portion of DNA molecule that is to be replicated. Reproduced with the permission of Andy Vierstraete.

to the 3′ end of the primer. (The N in dNTP can refer to any one of the four nitrogenous bases.) The new dNTPs are assembled into a complementary sequence that is determined by the template DNA.

It is very important that primers have optimal configuration including: (1) *Melting temperature* in the range of 55–72°C, (2) G-C content in the range of 40–60%, (3) a unique base sequence to avoid binding to nonspecific sites, (4) a sequence that will not permit a primer to form a hair-pin loop on itself due to internal complementary binding, or (5) to form dimers by binding with other primer molecules (*primer-dimers*) rather than annealing with template DNA. Melting temperature is the point at which the hydrogen bonds between the primer and template DNA start to break, freeing the primer from its target sequence. There are a number of commercial software programs available to help in the design of efficient primers.

Polymerase chain reaction requires template DNA, the four dNTP building blocks, two specific primers, and a DNA polymerase that needs magnesium (Mg^{2+}) to function properly. *Taq polymerase* is the enzyme most often used in forensic work. It is a heat-stable enzyme that is obtained from a bacterium called *Thermus aquaticus*. This organism lives in hot springs where the temperature of the water is close to boiling. Before this temperature-stable enzyme was discovered, PCR had to be conducted manually, and temperatures had to be raised and lowered in sequential steps using water baths. Because of the effect of high temperature on enzyme activity, the polymerase enzyme had to be replaced often throughout the process of amplification. The heat stability and extraordinary fidelity of the Taq enzyme allowed for many cycles of temperature ramping and faithful amplification (Tindall and Kunkel, 1988; Kunkel, 1992; Krawchak et al., 1989) without loss of the DNA synthesizing enzyme due to the elevated temperature.

The PCR reaction also requires Mg^{2+} as a cofactor for enzyme activity and a buffer to maintain a neutral pH. It is essential that sterile technique is used to load the PCR tube with the necessary reagents because any contamination that is accidentally introduced can adversely effect PCR results. As a result of the amplification process, two new strands are created, each attached to one original strand (see Fig. 3.18). This mechanism of synthesis is called semiconservative replication. PCR takes place over 30–33 cycles each of which consists of 3 steps (see Fig. 3.19). Each step takes place at a different temperature.

Typically, there is an initial incubation period at 95°C for 11 min followed by 30 cycles consisting of denaturation at 94°C for 1 min, annealing of primers at 65°C for 1 min and extension of the primers on the template at 72°C for 1 min. After the cycles are complete, a final extension step is performed at 60°C for 45 min (see Section 3.3.4.3.2.1) and the specimen is then left to soak at 25°C until it is removed for the next stage of its analysis. (Note: Optimum temperatures and length of time for each cycle depends on the specific primers and template DNA to be amplified and should be experimentally determined.)

In the first cycle, the two strands of the DNA double helix are denatured by increasing the temperature to 94–95°C for 30 to 60 sec. Once the DNA is denatured, the test tube is cooled to about 65°C for another minute. At this temperature the two primers will bind to their complementary sequences on each of the two DNA strands in a process known as *annealing*. The temperature for annealing may vary from 45 to 65°C and depends, in part, on the sequence of the primers. The optimal temperature must be determined experimentally. The annealing process takes from 20 to 50 sec. During the third step, the temperature of the test tube is raised to approximately 72°C for one min and the Taq polymerase synthesizes the new strand of DNA complementary to the template strand. The Taq polymerase begins adding nucleotides to the primer, one at a time, and continues to copy the template DNA until the new strand is complete. The incoming nucleotides are added onto the 3′ end of the primer so that they correspond in a complementary fashion with the sequence on the template strand (Fig. 3.20). The new strand lengthens in the 5′ to 3′ direction. If the template sequence of bases is ordered T-A-G-A-G-C-T, then the order of the new strand will be A-T-C-T-C-G-A. This step is referred to as *extension*. This extension process is followed by *ligation* or bonding of the newly incorporated bases to each other. It is the Taq polymerase that is responsible for extension and ligation.

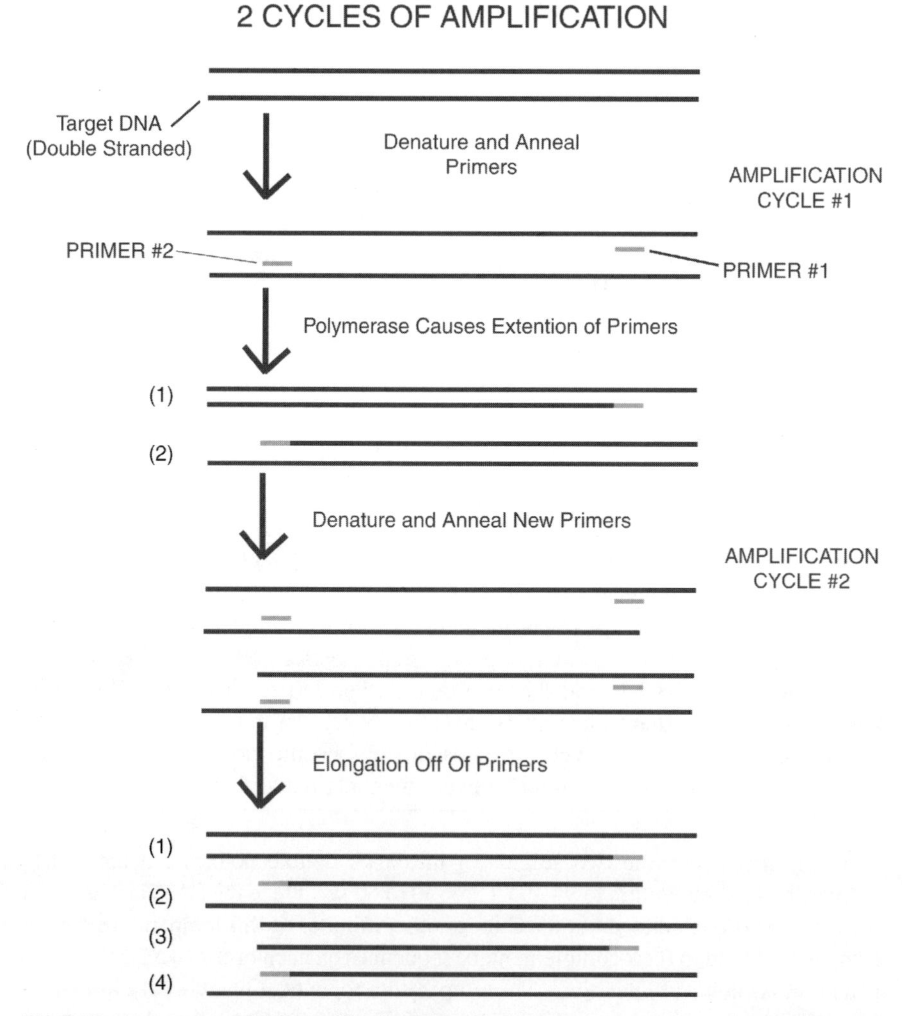

Fig. 3.18 *In process of PCR amplification, two new strands are created, each attached to one original strand.*

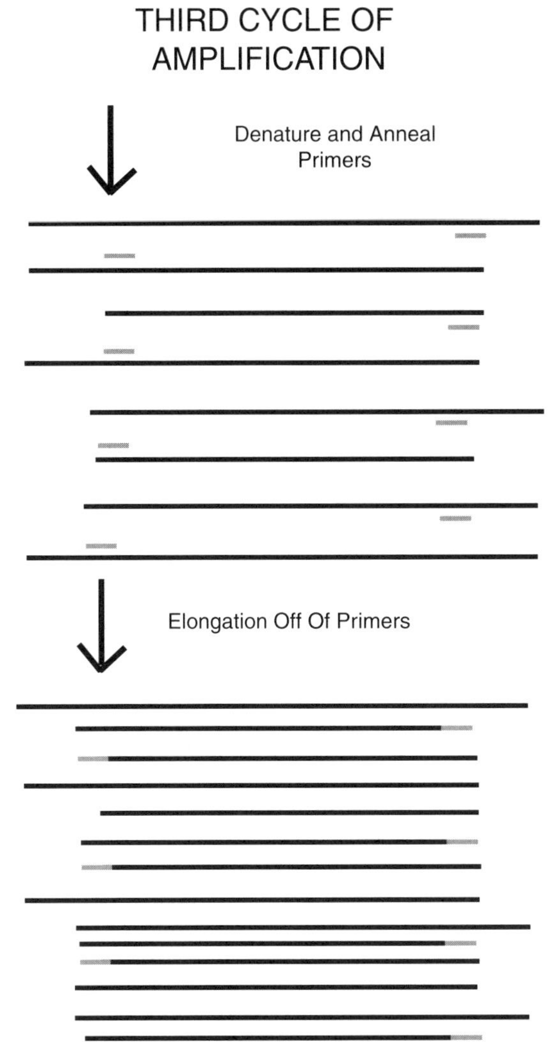

Fig. 3.18 *Continued.*

Thus by the end of the first cycle the target region has been duplicated. In the second round of amplification, both of the two new double-stranded molecules are denatured, primers anneal, one to each strand, and the polymerase enzyme again creates a new complementary strand. Thus by the end of the second cycle there are four double-stranded DNA molecules. Four new strands have been created, each attached to a template strand. In the third cycle, amplification takes place as described above resulting in eight molecules of DNA, each containing the target region, that is, the locus of interest. The products of amplification are not all the same size. The Taq

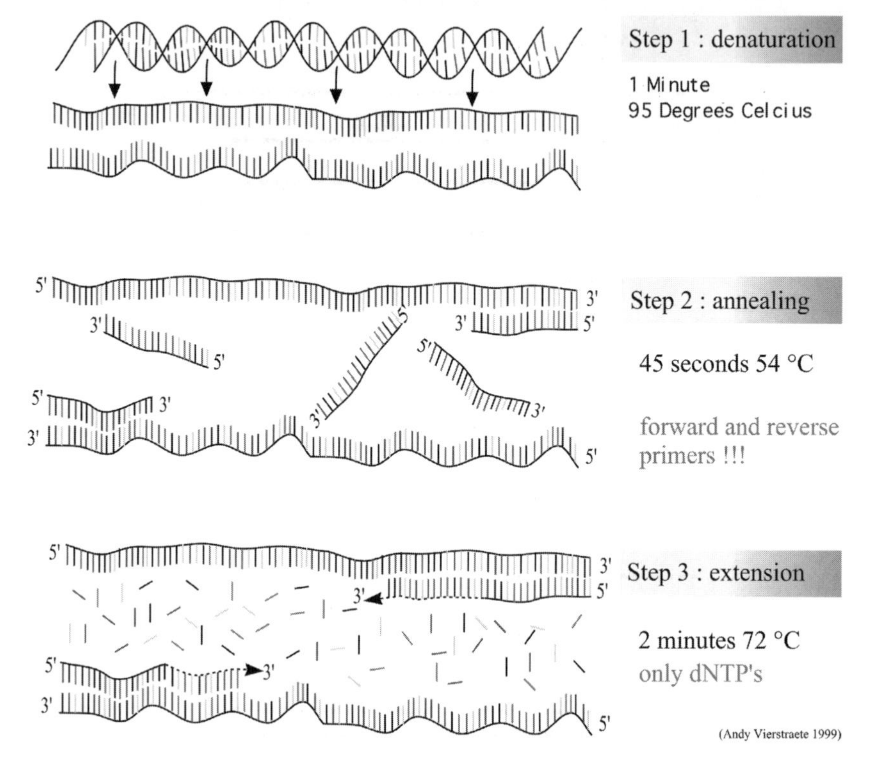

Fig. 3.19 PCR takes place in a number of cycles each of which consists of three steps. Reproduced with the permission of Andy Vierstraete.

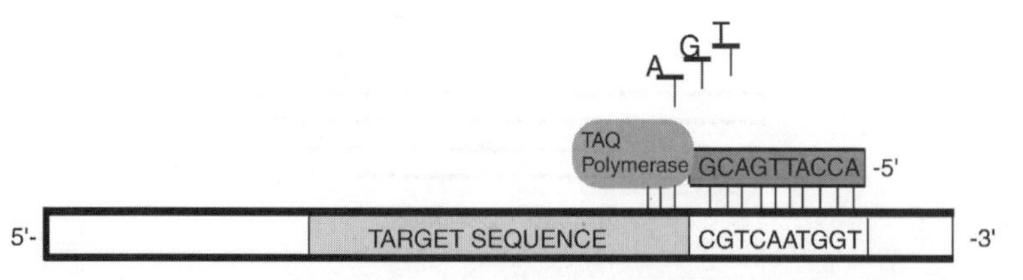

Fig. 3.20 In amplification phase of cycle, the Taq polymerase extends primer strand one nucleotide at a time (onto the 5′) so that new base corresponds with complementary base on DNA template strand.

polymerase processes the smaller molecules more efficiently than the longer ones. As a result after the third cycle of amplification, virtually all amplified product is identical in size and sequence with its boundaries delimited by the primers (see Fig. 3.21). In this way DNA is amplified from one molecule to two, and from two to four, and from four to eight molecules, resulting in exponential amplification.

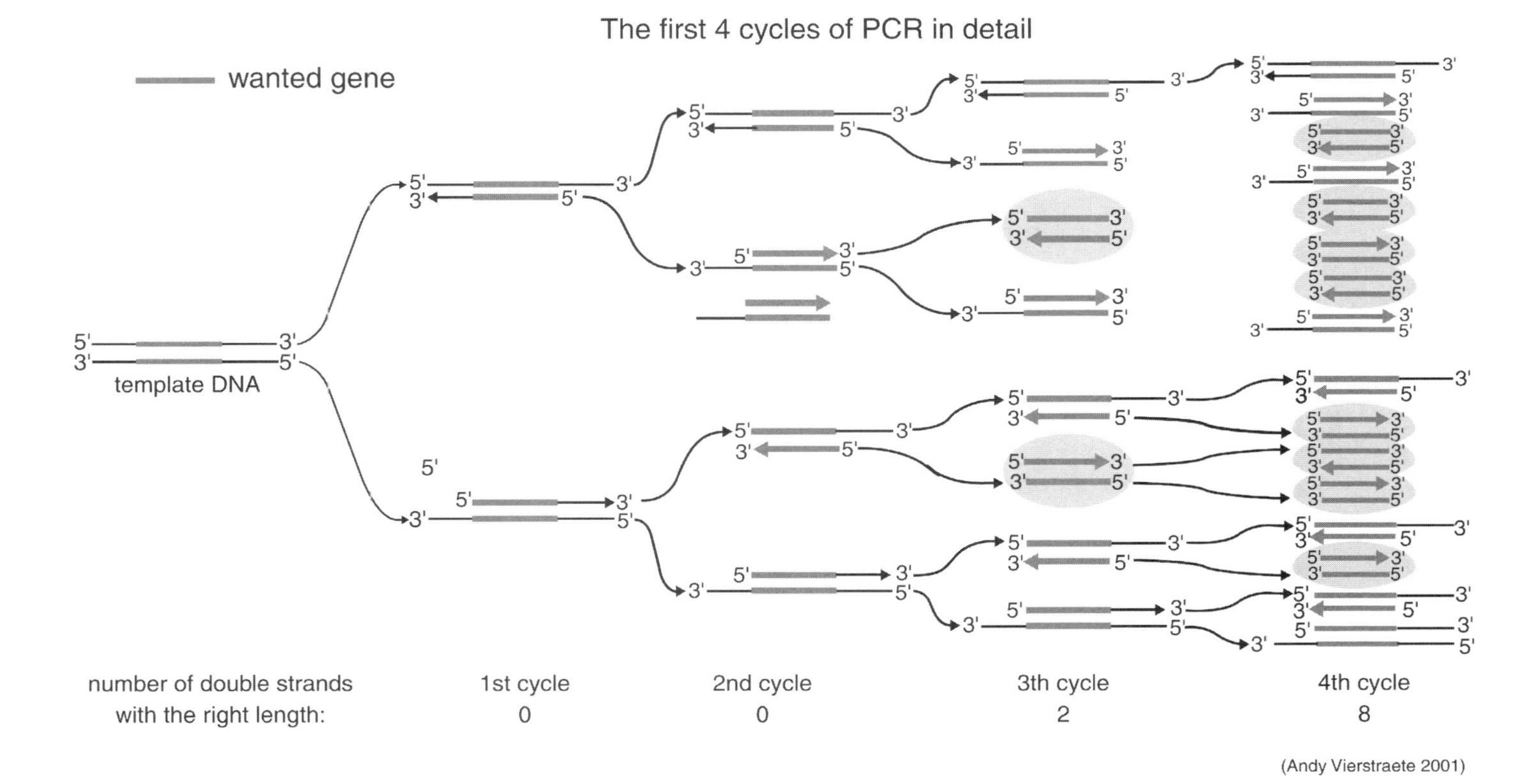

Fig. 3.21 *The Taq polymerase processes smaller molecules more efficiently than longer ones. As a result, after third cycle of amplification, virtually all amplified product is identical in size and sequence with their boundaries delimited by primers. Reproduced with the permission of Andy Vierstraete.*

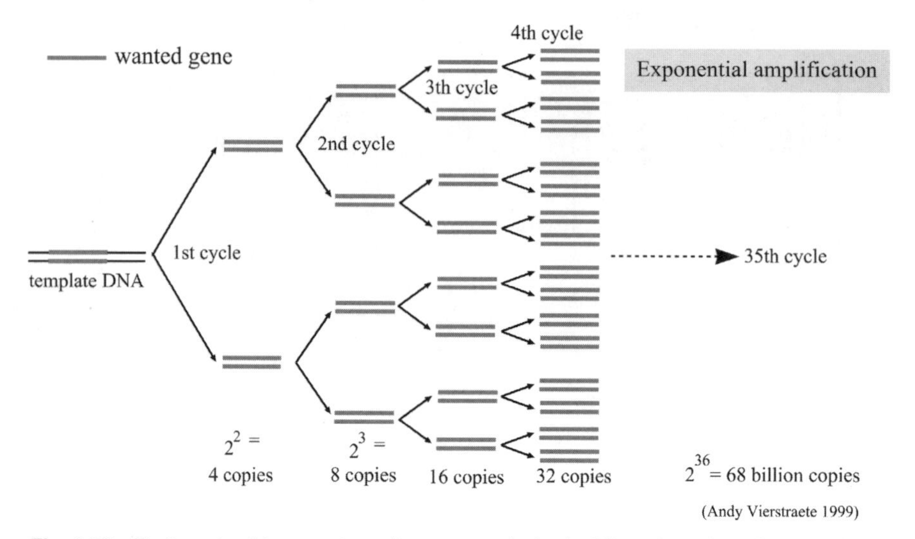

Fig. 3.22 Each cycle of temperature changes results in doubling of number of targeted DNA molecules. Reproduced with the permission of Andy Vierstraete.

Each cycle of temperature changes results in a doubling of the number of targeted DNA molecules (see Fig. 3.22). In about 2 hr the original template has been replicated to approximately one billion copies.

Many laboratories have now begun to use AmpliTaq Gold DNA polymerase (PE Biosystems) rather than the standard Amplitaq preparation because the former enzyme is inactive until it is exposed to elevated temperature, thus making it more efficient and increasing the yield of amplified DNA. It starts to become active after it has been maintained at 95°C for 2 min and reaches its maximum activity after approximately 16 min. It should be noted that the PCR process is covered by patents owned by Roche Molecular Systems, Inc., and F. Hoffmann-La Roche Ltd.

3.3.2 Isolation of DNA

Unlike RFLP, which requires the isolation and restriction of double-stranded DNA, PCR reactions require single-stranded DNA to serve as template. Therefore any of the extraction methods that can produce double-stranded DNA (see Section 3.2.3.2.1) can be used for PCR. The most often used method employs the use of a cation exchange resin known as Chelex 100. The sample is treated with 5% Chelex 100 at 56°C for 30 min and then boiled for 8 min. Centrifugation will leave DNA in a soluble form while other cellular constituents will be found in the pellet. Boiling in Chelex 100 results in purified single-stranded DNA that can then be quantified and amplified. Failure to completely isolate and purify DNA can result in contaminants and PCR inhibitors (such as hematin) being included in the reaction mixture. It may be possible to minimize the inhibition caused by a contaminant by adding bovine serum albumin (BSA) to the reaction mixture. This protein may stabilize the polymerase, thereby protecting its activity. PE Biosystems now includes BSA in its AmpF*l*STR multiplex kits.

3.3.3 Quantification

For successful PCR analysis only 1–2.5 ng of purified DNA is needed to serve as template. A human diploid cell contains approximately 6.1 pg of DNA. Therefore, approximately 164 cells are needed to obtain sufficient DNA for amplification. Furthermore unlike the DNA that is required for RFLP, the template DNA for PCR may be somewhat degraded, nicked, fragmented, and it may also be partially denatured or completely single stranded. The amplification process will still succeed. Prior to PCR amplification, DNA should be quantified, and this can be done in a number of ways (Budowle et al., 1995b). For example, isolated DNA can be quantified using a slot-blot device and a human-specific probe. The benefit of using a human-specific probe is that the analyst can determine the amount of human DNA present in a sample that may also contain contaminating bacterial or fungal DNA. Yield gel analysis of a specimen that is heavily contaminated with bacterial DNA will result in an overestimate of the amount of useful DNA in the sample, and as a result the specimen will be diluted resulting in an insufficient amount of human DNA to produce a result. The Quantiblot Human Identification Kit is a commercial kit manufactured by Applied Biosystems and is based on the hybridization of a human-specific probe (D17Z1) to purified DNA samples (see Fig. 3.23). Because the probe has been biotinylated, detection of probe-bound DNA can be performed either colorimetrically or by chemiluminescence. Test samples are visually compared to a set of calibrated standards. The kit can quantify either single- or double-stranded DNA in the range of 0.15–10 ng. The probe is known to react with primate DNA as well as human DNA. This analysis provides a good indication

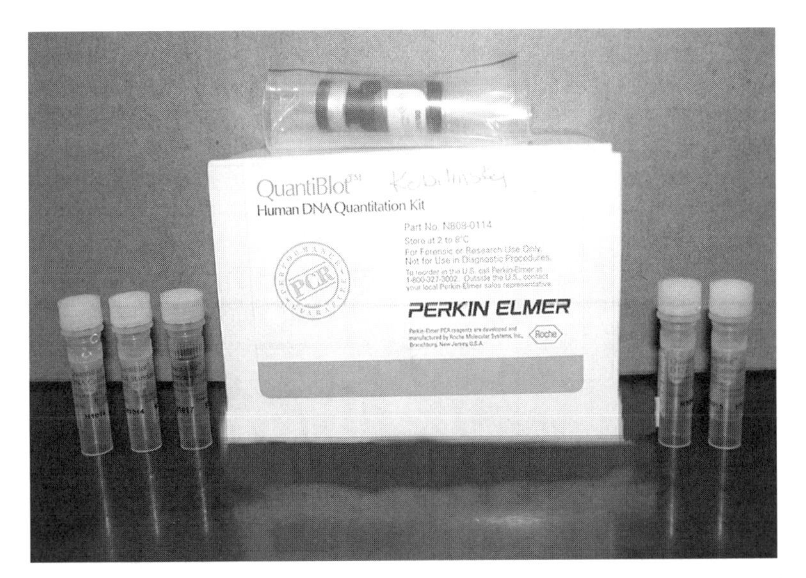

Fig. 3.23 *Quantiblot Human Identification Kit is commercial kit manufactured by Applied Biosystems and is based on hybridization of human-specific probe (D17Z1) to purified DNA samples.*

of the quantity of human DNA present regardless of the presence of bacterial or fungal DNA. Another method to quantify DNA prior to PCR amplification is known as the AluQuant Human DNA Quantitation System developed by Promega Corp. Human-specific probes are used to quantitate DNA in the range of 0.1–20 ng in a solution-based assay. The probe recognizes the repetitious DNA known as the *Alu inserts* (see Section 1.2.3). It should be noted that this probe like the D1721 probe will react with DNA obtained from higher primates. This method requires no gels or blotting equipment and works with partially degraded DNA. It uses a luciferin-luciferase reaction to produce an amount of light that can be detected and measured in a luminometer. The amount of emitted light is related to the amount of human genomic DNA in solution (see Fig. 3.24). Applied Biosystems has recently developed another method to quantify DNA using real-time PCR amplification with the Quantifiler Human DNA Quantification Kit. It is reported to be able to quantify samples with DNA concentrations in the range of 0.023 to >50 ng/μL. As the real-time PCR assay progresses, each amplification cycle produces more fluorescently labeled product, which is recorded by the instrument along with cycle number. The software calculates the cycle number at which the amount of product crosses the "noise" threshold for each reaction. A number of DNA standards of varying concentration are treated in the same manner and used to develop a standard curve. Thus the cycle number of each of the samples can be compared to the cycle number of the standards and an estimate of the DNA concentration can be made.

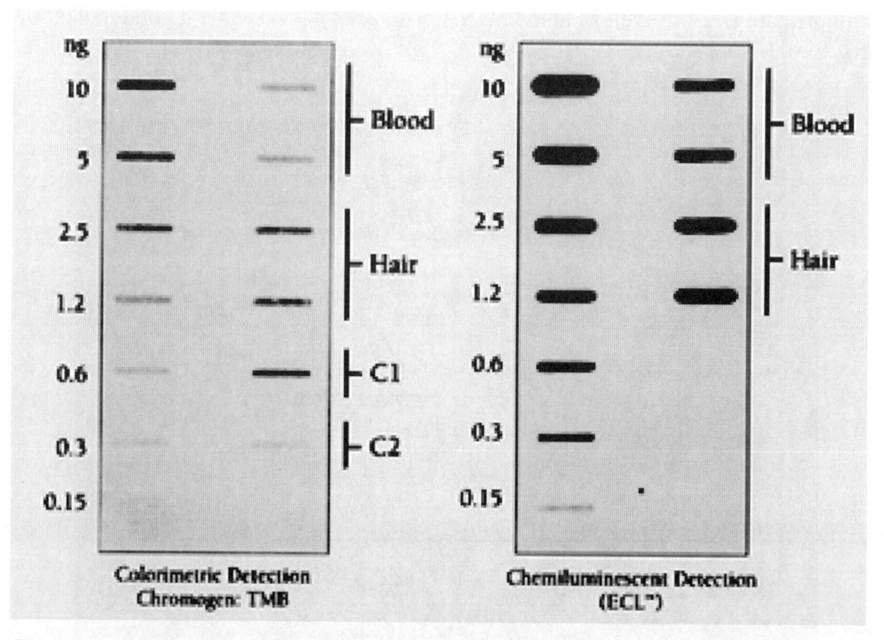

Fig. 3.24 DNA. It uses a luciferin-luciferase reaction to produce an amount of light that can be detected and measured in a luminometer. Amount of emitted light is related to amount of human genomic DNA in solution. Courtesy of Applied Biosystems Inc., Foster City, CA.

3.3.4 Techniques

3.3.4.1 Typing of HLA-DQA1 and AmpliType PM There are two kinds of polymorphism within DNA. Sequence polymorphisms are variations in the base sequence of an allele. The second is length polymorphism where there is variation in the lengths of alleles within a locus. In the late 1980s PCR amplification was performed on loci that were neither VNTRs nor STRs, but rather they contained alleles that existed as sequence variants. The tests were designed to identify sequence differences in several well-characterized loci (DNA Commission Recommendations, 1992). The first commercial DNA typing kit to be used by forensic analysts was the *AmpliType HLA-DQα* kit (Allen et al., 1993; Cetus Corporation, 1990). The HLA DQα locus contains eight common alleles, consisting of allele 1 (three subgroups), 2, 3, and 4 (with three subgroups). Many lab directors had decided not to perform RFLP testing because of the use of radioactivity and hazardous waste as well as the fact that RFLP is a labor-intensive technique that sometimes requires many weeks before results become available. Some wanted to become involved in PCR-based allele detection. They welcomed the production by Perkin-Elmer of HLA-DQα kits. These kits could differentiate alleles 1.1, 1.2, 1.3, 2, 3, and 4. The HLA-DQα kit has been characterized for forensic purposes and allele and genotype frequencies determined for many populations (Sajantila et al., 1991; Sullivan et al., 1992; Comey et al., 1993). It has also been validated for use with casework (Wilson et al., 1994; Presley et al., 1993). The more advanced *Amplitype HLA-DQA1* kit allowed for the additional separation and detection of the 4.1 from 4.2/4.3 alleles. Although this second-generation kit improved the resolution of the common HLA DQ alleles, it still could not differentiate the 4.2 allele from the 4.3 allele. Because of this, the interpretation of the results is very important if both known and questioned samples come up with matching profiles such as 2,4.2/4.3. In this case the former could be 2,4.3 and the latter could be 2/4.2. What seems to be an inclusion may, in fact, be an exclusion (see Section 3.5.4).

In the procedure, amplified products would be applied to typing strips that contain a number of colorless circular dots arranged in a linear fashion. Each "dot" is prepared with a specific probe that is covalently bound to the nylon substrate of the typing strip. This is called the *reverse dot blot format* of testing. Tests in which the probes are added to amplified DNA that is fixed to a substrate are known as dot blot analyses (see Fig. 3.25). In the case of HLA DQA1, a total of 11 sequence-specific oligonucleotide (SSO) probes were bonded onto the typing strip, allowing the analyst to detect seven different sequences (alleles) within the locus. If the amplified product consisted of DNA that complemented, base for base, one of the membrane-bound probes, a chemical reaction would result in a strong blue color. The basis for this colorimetric reaction is that the primers used to amplify the HLA-DQA1 alleles are labeled with biotin at their 5′ ends. Thus every amplified product contains this biotin label. After allowing the amplified product to hybridize to the membrane-bound probes, and then washing away any unbound amplified product, a reagent known as streptavidin–horse radish peroxidase (SA-HRP) is added. Streptavidin binds tightly to biotin. As a result, the HRP

Dot Blot Format

The amplified product is dotted onto the typing strip and the labeled allele specific probe is added to the strip. Where the probe recognizes the corresponding allele, a blue color will develop.

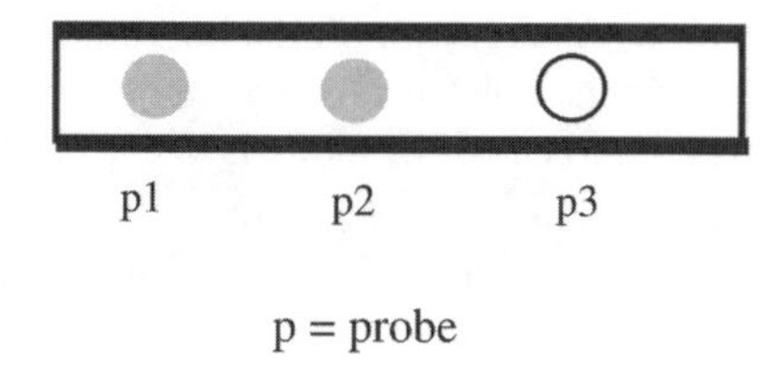

Reverse Dot Blot Format

Oligonucleotide probes that are specific for each allele are covalently bound to the typing strip. The labeled amplified product is added to the strip. If the probe hybridizes to the amplified alleles, a blue color will develop.

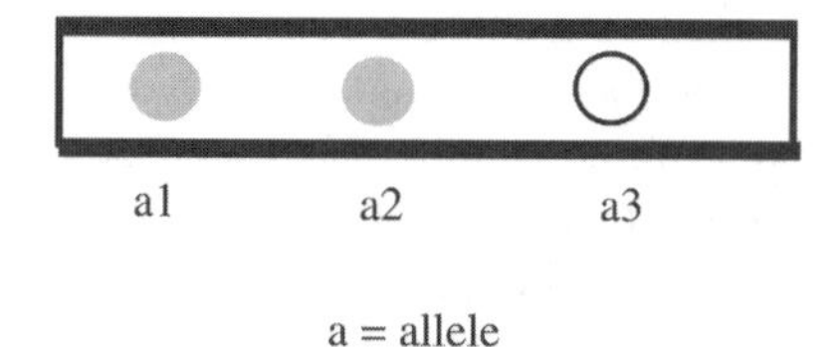

Fig. 3.25 *See color insert. Reverse dot blot format of testing.*

becomes bound to the amplified product. This enzyme oxidizes and converts a colorless soluble reagent (tetramethyl benzidine) to a blue precipitate, and the dot containing the appropriate probe turns blue (Fig. 3.26*a* and *b*). An examination of the pattern of blue dots reveals the genotype of the source of the DNA specimen (see Fig. 3.27*a* and *b*). The information obtained from a single locus is somewhat limited but can be supplemented by amplifying and analyzing additional loci. Ordinarily, amplification of two or more loci would require two or more separate reactions using two or more samples taken from the evidentiary specimen.

Multiplexing refers to the simultaneous amplification of more than one locus using a single specimen. A product known as *AmpliType PM* or Polymarker was developed by the same company. It was a multiplex kit capable of simultaneously amplifying five loci (see Table 3.4). Polymarker loci consist of low-density lipoprotein receptor (LDLR), glycophorin A (GYPA), hemoglobin gamma globin (HBGG), a locus on chromosome 7 (D7S8), and group-specific component (Gc).

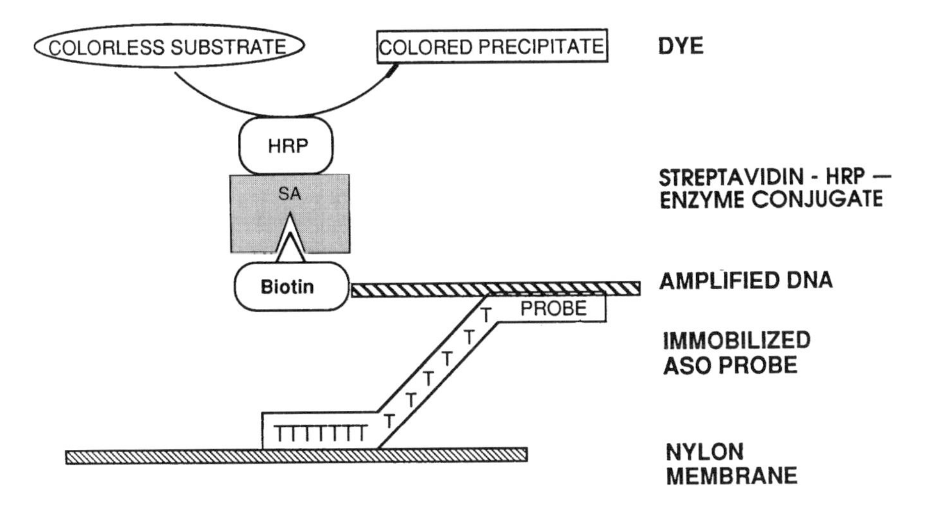

Examples:

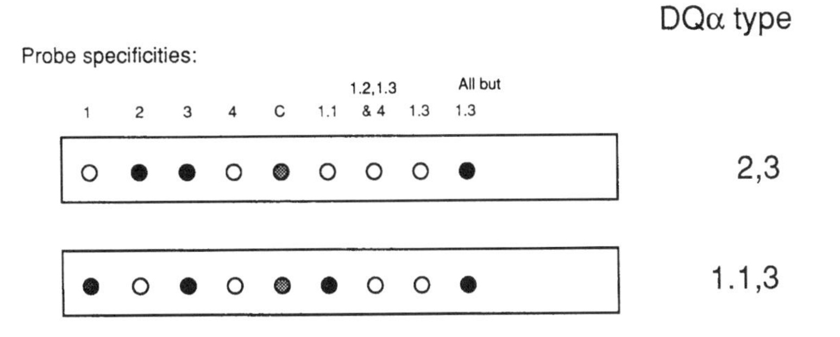

(a) *Allele-Specific Oligonucleotide

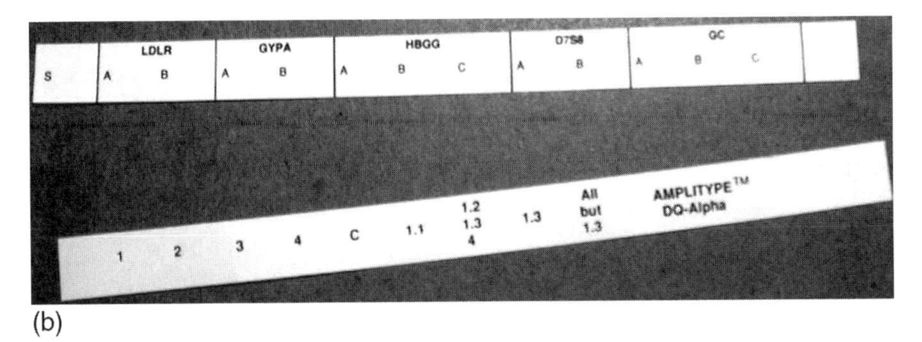

(b)

Fig. 3.26 *(a) Colorless soluble reagent (tetramethyl benzidine) is converted to blue precipitate and dot containing the appropriate probe turns white to blue. (b) Typing strip prior to color development. Reprinted with the permission of Roche Molecular Systems, Alameda, California. Cetus Amplitype User Guide-Laboratory Setup, Copyright 1990.*

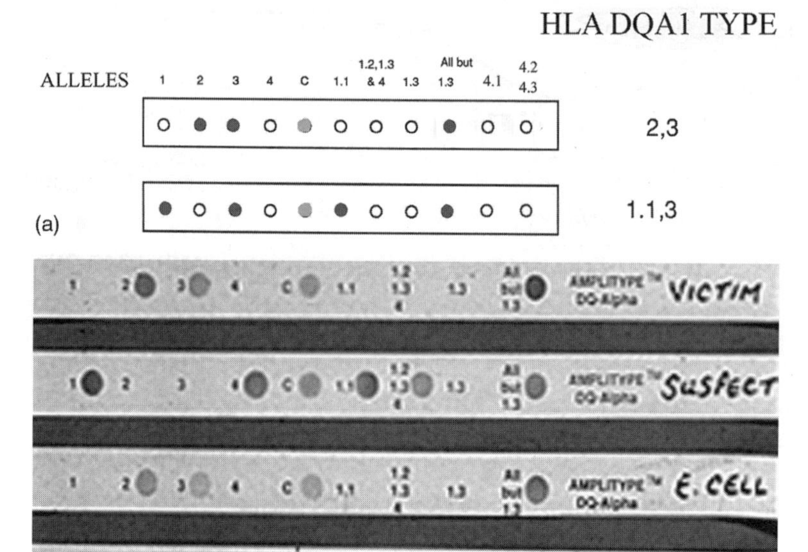

(b)

Fig. 3.27 *(a) Examination of pattern of blue dots reveals phenotype (and genotype) of source of DNA specimen. (b) See color insert. Comparison of patterns can be made to determine if suspect can be excluded.*

A typing strip containing membrane-bound probes for the five AmpliType PM loci is shown in Figure 3.28.

Following amplification, a test gel is run to verify that all loci have been successfully amplified. The bands appear in the region between 138 bp (Gc is the smallest) and 242 bp (DQA1 is the largest) on the test gel. The remaining four loci (LDLR,

TABLE 3.4 AmpliType Polymarker (PM) Kit at a Glance

Advantages	Limitations
1. PM is a quick and simple and the results can be visually interpreted.	1. PM lacks the discriminating power of VNTRs and STRs because the current number of alleles and developed loci are small.
2. PM is widely used and has gained recognition.	2. The limited number of alleles and loci per locus makes it harder to identify the components of mixtures than it is to do so when using VNTR and STR loci.
3. PM incorporates PCR techniques, so it too can be used to analyze small or degraded samples.	

Source: National Commission on the Future of DNA Evidence, *The Future of Forensic DNA Testing Predictions of the Research and Development Working Group* (2000).

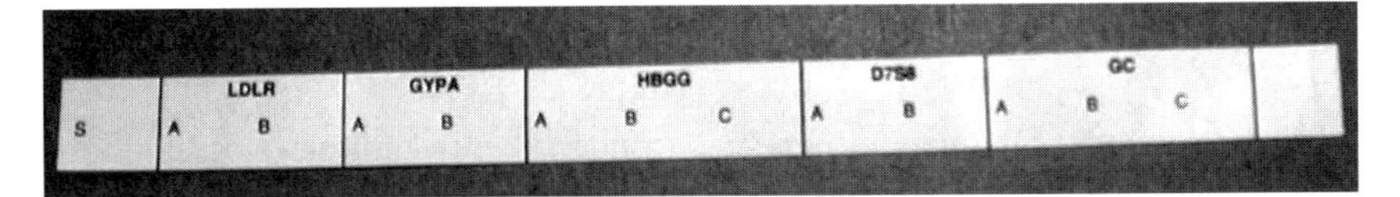

Fig. 3.28 See color insert. Typing strips containing probes for AmpliType PM loci.

GYPA, HBGG, and D7S8) band between these two loci (DQA1 and Gc). As described above, the typing strips contain oligonucleotide probes in the form of dots or circles placed at specific locations on the strip. If the amplified DNA fragments contain sequences that are complementary to the probes, they will hybridize and a chemical reaction is initiated, changing the dot's appearance from colorless to blue (see Fig. 3.29). Both typing strips, HLA DQA1 and AmpliType PM, have internal control dots to ensure that specific amplification takes place. After color development, these control dots should appear blue but less intense than any of the other dots on a strip. One of the recommended positive controls that can be incorporated in an analysis is DNA isolated from the K562 cell line, which produces known genotypes for both HLA-DQA1 (1.1,4.1) and AmpliType PM (LDLR = BB, GYPA = AB, HBGG = AA, D97S8 = AB, and Gc = BB). If a positive control is observed to contain extra alleles, then contamination is a possibility, or if the positive control produces no result, the analyst should determine why the pattern failed to appear. Of course, additional positive and negative controls should be per-

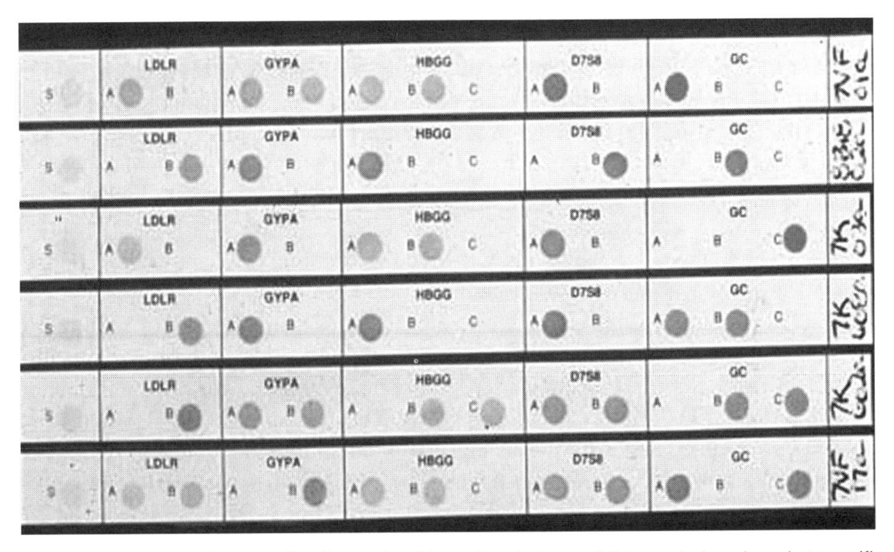

Fig. 3.29 Typing strips contain oligonucleotide probes in form of dots or circles placed at specific locations on strip. If amplified DNA fragments contain sequences that are complimentary to probes, they will hybridize and a chemicaql reaction is initiated changing dot's appearance from colorless to blue.

formed to determine if contamination occurred or if hybridization had produced products that were, in reality, artifacts such as primer-dimers.

Primer-dimers occur when primers bind to each other resulting in a small double-stranded molecule that will be amplified along with the expected amplified products. The primer-dimers, however, will be preferentially amplified because of their small size relative to the alleles for the targeted loci. The result is less amplification of the desired products and more amplification of the primer-dimer products. Another artifact results when the primer binds nonspecifically to the template DNA, causing the amplification of nonspecific products. One can avoid these artifacts either by using "hot start PCR," which means that the amplification is started at elevated temperature (higher than the optimal temperature for primer annealing) or by using a polymerase such as AmpliTaq Gold polymerase, which does not become activated until the temperature is sufficiently high. In both cases, amplification will not start at low temperature (room temperature), and thus there are fewer artifacts created.

One other result that the analyst must be careful to interpret is the development of a weak signal at the dot containing the probe for the 1.1 allele when the individual who provided the specimen is known to have a genotype not containing this allele. The likely explanation for this observation is that there may be some nonspecific amplification of a nearby region known as HLA DXα.

Each of the five AmpliType PM loci consists of either two or three alleles. The polymorphism, and therefore the discriminatory power for each is low but, collectively, can result in very useful information if a match (inclusion) is found. The match probability for both HLA DQA1 and the 5 AmpliType PM loci is approximately 1 in 4000. By comparison to methods in use today, this statistic is not overly impressive. Conclusions of a match can be expressed in the following manner: "The suspect, John Doe, cannot be excluded as the donor of the evidence. The probability of finding a person in the relevant population is 1 in 2500." (A frequency would be provided for each major ethnic/racial group). Statistics are calculated using the Hardy–Weinberg formula and the product rule. However, either of these two kits could produce exclusionary results based on differences in genotype patterns of evidence and suspect. An exclusion is based on reliable genetic evidence and is not debatable.

The great benefits of using a PCR-based procedure in a multiplex manner include: (1) being able to perform multiple tests on the same specimen without the need to consume additional sample (evidence); (2) time and labor required to obtain results are reduced; (3) ability to successfully test samples in limited quantity and of somewhat poor quality (degraded); (4) reducing the chances of contamination by doing a multiplex rather than setting up six independent reactions; (5) the strips can be photographed and the patterns preserved indefinitely; (6) it has been demonstrated that samples recovered from pap smears, semen smear, and postcoital slides produce good typing results (Roy and Reynolds, 1995); (7) HLADQA1 can be co-amplified with amelogenin (see Section 3.3.4.3), to provide information about the gender of the source of sample DNA (Casarino et al., 1995).

Comparison of test results from seven different laboratories demonstrate consistency and reproducibility (Fildes and Reynolds, 1995). Validation studies that have been conducted with the AmpliType PM kit demonstrate its usefulness and reliability (Budowle et al., 1995a).

More recent advances in technology and molecular biology have provided the ability to multiplex 16 or more loci (see Section 3.3.4.3). The *in vitro* process requires 2 primers for each locus, and the amplified products are characterized by size (mass or length). Each can be distinguished because of differences in their *electrophoretic* migration rate. The sophisticated detection systems available today can also differentiate amplified products from different loci that migrate at the same rate because they have similar sizes. Primers that target loci containing alleles of similar size are labeled with different fluorochromes, and the gene-scanning instruments, which are equipped with laser-based detection systems, can differentiate these colors (Fregeau and Fourney, 1993).

3.3.4.2 Analysis of AmpFLPs Realizing the benefits of using PCR to amplify sample DNA that is in limited quantity and the tremendous discrimination power of RFLP analysis, a method was developed that combined the best parts of both procedures. This method is known as amplified fragment-length polymorphism (AmpFLP) analysis. The DNA is extracted (Comey et al., 1994) and purified from the known and questioned specimens as described above for PCR of Amplitype HLA-DQA1 and AmpliType PM. However, the target DNA is a VNTR locus rather than a sequence variant and no restriction enzyme is utilized. Two of the most well-studied VNTR (minisatellite) loci are D1S80 and D17S5. Perkin-Elmer has produced Amplitype kits for each. D1S80 is the locus of choice for forensic analysis (Kasi et al., 1990; Baechtel et al., 1995). It is found on the short arm of chromosome 1 and has a core repeat of 16 bp in length. It is one of the most polymorphic of all forensically useful loci consisting of more than 27 common alleles ranging from 14 repeats (369 bp) to 41 repeats (801 bp). The flanking regions at both sides of the VNTR are a total of 145 bp long. Most of the length polymorphism for this locus occurs in the VNTR itself, but there is also some polymorphism in the flanking regions. The template DNA is amplified using Taq DNA polymerase, the 4 dNTP building blocks, Mg^{2+}, buffer, primers, and bovine serum albumin. Alleles are separated by either agarose or polyacrylamide slab gel electrophoresis, and the amplified products are stained using either silver staining or ethidium bromide (EB), which is an intercalating dye that fluoresces when it combines with double-stranded DNA and is exposed to ultraviolet light. In this way, the banding pattern can be visualized and the gels can then be photographed. EB-stained gels can be photographed using a transilluminator (ultraviolet light source) beneath the gel to cause fluorescence of the bound dye. Polyacrylamide provides better resolution than agarose and is the preferred gel medium for this procedure. Unlike the RFLP method, no probes are used in the AmpFLP technique. An allelic ladder is run together with the questioned specimens, the known specimens, and the positive and negative control specimens. The allelic ladders contain all the common alleles at that locus and span the length of the gel. There is no need to calculate band sizes by comparing positions of questioned bands with bands of the molecular weight markers and per-

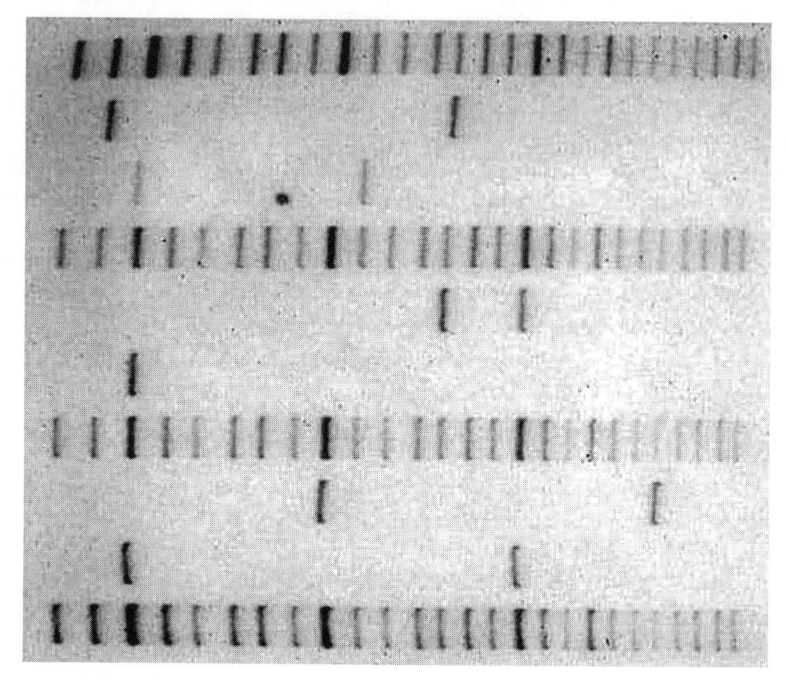

Fig. 3.30 *Analysis of D1S80 illustrating allelic ladders. Top of figure is to the right.*

forming interpolations as there is for RFLP. The experimental bands are visually compared to the bands of the nearby allelic ladder and named accordingly (Budowle et al., 1995a; Cosso and Reynolds, 1995; Sajantila et al., 1992) (see Fig. 3.30). The D1S80 and amelogenin loci can be simultaneously amplified, and individuals can be genotyped using capillary electrophoresis to separate and detect alleles (Isenberg et al., 1996).

3.3.4.2.1 Statistics and Population Genetics Now that the samples have been genotyped at D1S80, together with the genotypes obtained by testing of Amplitype HLA-DQA1 and AmpliType PM, we have a genetic profile over seven loci. The overall frequency of this profile in the relevant population can be determined. Recommendation 4.1 of the 1996 National Research Council report calls for determining the approximate frequencies of the overall genetic profile in the African American, Hispanic, and Caucasian populations. Some laboratories offer profile frequencies for "East Coast" and "West Coast" Hispanics since these two populations are derived from different ancestral groups, and allele frequencies at certain loci can differ significantly between the two groups. It is also reasonable to provide profile frequencies for the Asian American population. Laboratories should maintain allele frequency databases for each of the above populations. DNA databases are discussed more fully in Section 4.11. To illustrate how databases are established and used, let us construct a hypothetical database by sampling 200 African Americans and deter-

TABLE 3.5 Allele Frequencies Used in Statistical Calculations

Gene Number	Number of Observations	Frequency
	Chromosome A, Locus 1	
7	30	0.0750
8	25	0.0625
9	45	0.1125
10	80	0.2000
11	60	0.1500
12	80	0.2000
13	60	0.1500
14	20	0.0500
	$n = 400$	Total 1.000
	Chromosome B, Locus 2	
10	15	0.0375
11	20	0.0500
12	35	0.0875
13	60	0.1500
14	50	0.1250
15	70	0.1750
16	45	0.1125
17	25	0.0625
18	20	0.0500
19	25	0.0625
20	35	0.0875
	$n = 400$	Total 1.0000

mining frequencies of all detected alleles at each locus—one on chromosome A and the other on chromosome B. Two hundred individuals would provide 400 alleles for frequency determinations at each locus. If the total number of alleles at a single locus equals 8 and the range is from 7 to 14, then we can construct a table of allele frequencies that can be used in our statistical calculations (see Table 3.5).

Before a database is used in any statistical calculations, it is first determined that the population is in Hardy–Weinberg equilibrium (Turowska and Sanak, 1995). If it is found to be in equilibrium, then genotype frequencies can be determined. It is also important to determine that the different loci tested are in linkage equilibrium. If they are, then you can use the product rule to calculate the frequency of the overall genetic profile.

The frequency of an individual having the genotype 9,11 at the first locus on chromosome A, using the Hardy–Weinberg formula, would be $2pq = 2 \times 0.1125 \times 0.1500 = 0.0338$ or 338 out of 10,000 African Americans. We can now look at the second locus on chromosome B where the genotype is 18,19. At this locus, the frequencies of the observed. 18 and 19 alleles, are determined to be 0.0500 and 0.0625, respectively. The genotype frequency for this locus is calculated to be $2 \times (0.0500)(0.0625) = 0.0063$. The overall genetic profile frequency (over both loci) is calculated using the product rule, for example, $0.0338 \times 0.0063 = 2.1 \times 10^{-4}$. Thus 2.1 in 10,000 is a very small number, meaning that the genetic profile for these two loci is relatively uncommon and you would have to sample 5000 individuals before finding one with the same profile.

For homozygous loci we do not use the p^2 value of the Hardy-Weinberg rule for the same reason as explained above for RFLP analysis. Unlike RFLP analysis with its binning procedure, PCR analysis of D1S80 leaves little question about distinguishing a homozygote from a heterozygote. Nevertheless, to be conservative, we replace the (less conservative) p^2 value with $2p$ in calculating the frequency of the homozygous genotype.

Calculations are prepared for all loci tested and for each of the racial/ethnic groups indicated above. The judge and jury can then consider the range of statistical calculations in their deliberations. Both allele and genotype frequencies have been reported for various ethnicities in many populations (Pfitzinger et al., 1995)

3.3.4.2.2 Allelic Ladders D1S80 analysis always includes samples composed of allelic ladders. Perkin-Elmer produces ladders for its AmpFLP kits that consist of a mixture of DNA fragments representing the most common alleles at each locus. For D1S80, the alleles supplied in the ladder are 14 and 16 through 41. Of course, there are reports of alleles larger than the 41 allele. In fact a 72 allele has been reported. Allelic ladders are used to identify specific alleles in the experimental specimens (see Fig. 3.30). Allelic ladders are generally supplied with the purchased kit. The best allelic ladder sets include not only the most common alleles within a locus but also some of the rare alleles, especially those that are relatively small and those that are relatively large.

3.3.4.3 PCR—Short Tandem Repeat Analysis The DNA analysis technique that has become the method of choice in forensic laboratories is based on the use of the polymerase chain reaction (PCR) (see Section 3.3.1) to amplify small regions of genomic DNA, known as short tandem repeat (STR) loci (see Section 3.3.4.3.1). PCR amplification produces hundreds of millions of copies of alleles at a number of loci, each of which can be subsequently tested to reveal genetic information about the source of the specimen. In the early 1990s, many crime laboratories invested their resources in developing the capability to perform RFLP and HLA-DQA1, AmpliType PM (Polymarker), and D1S80 to analyze DNA. Today, most crime laboratories in the United States use commercial multiplex STR kits to obtain genetic profiles for the purpose of human identification.

Only 1–2 ng are needed for successful amplification of isolated DNA, but good typing results can sometimes be obtained with levels of isolated DNA below this optimal quantity. In this technique, DNA is purified from a biological specimen, quantified, and then amplified. Isolation and purification methods for PCR reactions have been described previously (Section 3.3.2). The exact method used depends upon the nature of the specimen, for example, tissue, blood, bone, teeth, semen, hair, and so forth. The Chelex method has been found to be one of the easiest and most efficient ways of isolating DNA. Quantification can be done in a number of ways (Waye et al., 1989; Walsh et al., 1992b) (see Sections 3.2.2, 3.3.3, and 3.5.2). Another method has recently been developed that uses the ABI Prism 7700 (Quantifiler, Applied Biosystems, Foster City, CA; see Section 3.3.3). This instrument is used to perform real-time quantitative PCR. It allows a determination of how many copies of a particular allele are present in a biological sample. If the concentration of DNA is too high, the specimen must be diluted. If it is too low then it must be concentrated. Microcon 100 (Millipore) filtration is commonly used for this purpose. When the sample contains the appropriate concentration of DNA, it is then amplified by PCR.

The PCR process begins when the template DNA becomes denatured and the forward primer (XXXXXXX) attaches to the 3′ end of one strand of the original molecule and the reverse primer (YYYYYYY) attaches to the 3′ end of the opposite strand. Primers determine the areas of template DNA that are to be replicated. Only the forward primer is labeled with a fluorescent dye. The most common of these fluorochromes are FAM (blue), JOE (green), TAMRA (yellow), NED (yellow) and ROX (red) (see Table 3.6). Some commercial kits have primers that are labeled with fluorescein (blue) or tetramethyl rhodamine, TMR (yellow). Each dye emits a different color (wavelength) when excited by light of the appropriate wavelength. The ABI Prism sequencers use an argon laser with excitation maxima at 488 and 514 nm. In order to distinguish the emission from each of these dyes, the detector uses filters that minimize overlap in emitted light from each dye. Thus, in a four-dye multiplex (three dyes plus the red calibrator dye), there are four separate filters used to differentiate the emitted colors. In addition, the instrument uses what is referred to as a *matrix file* to distinguish each of the colors. The matrix software calculates the contributions of each dye to the measured emission of a specific dye and subtracts away any unwanted contribution from the other dye(s). This unwanted light is a result of spectral overlap for the fluorochromes that are used. In this way, the machine detects the color (wavelengths of light) emitted from each fluorochrome individually rather than as a mixture of wavelengths from fluorochromes present at or near the position being examined. The matrix file should be reset often to ensure proper analysis of labeled PCR product. If uneven baselines are observed or if there is excessive "pull-up" (see Section 3.3.4.3.2.3), then the matrix file should be recalculated.

TABLE 3.6 Fluorochromes and Their Characteristics

Short Name	Excitation Max (nm)	Emission Max (nm)
FAM	493	522—blue
JOE	528	554—green
TAMRA	560	583—yellow
NED	553	575—yellow
ROX	587	607—red

As explained above, multiplex kits are designed to permit the alleles at a number of loci to be amplified simultaneously. The AmpliType PM (Polymarker) kit is such a kit.

A number of multiplex systems have been designed to allow multiple STR loci to be amplified simultaneously rather than having to amplify only one STR locus at a time (see Fig. 3.31). Loci of different size ranges are chosen so that one marker will not overlap another. Furthermore these multiplex kits allow for the detection instrument (ABI Prism 310, 3100, or 377 or the FMBIO II) to type each of these loci in one run. Were it not for multiplexing kits, the analyst would have to run a separate amplification and typing for each genetic marker (locus) of interest, thereby consuming more and more sample each time a DNA test is done.

Two systems commonly used by crime labs nowadays are known as AmpF*l*STR Profiler Plus and AmpF*l*STR COfiler, both of which are manufactured by Perkin-Elmer Applied Biosystems (PE). The results of a DNA test using the AmpF*l*STR Profiler Plus kit are shown in Figure 3.32. Crime labs can also use two multiplex systems manufactured by Promega Corp., namely PowerPlex 1.1 and PowerPlex 2.1 (Promega, 1999a, 1999b). Two recently developed multiplex kits called

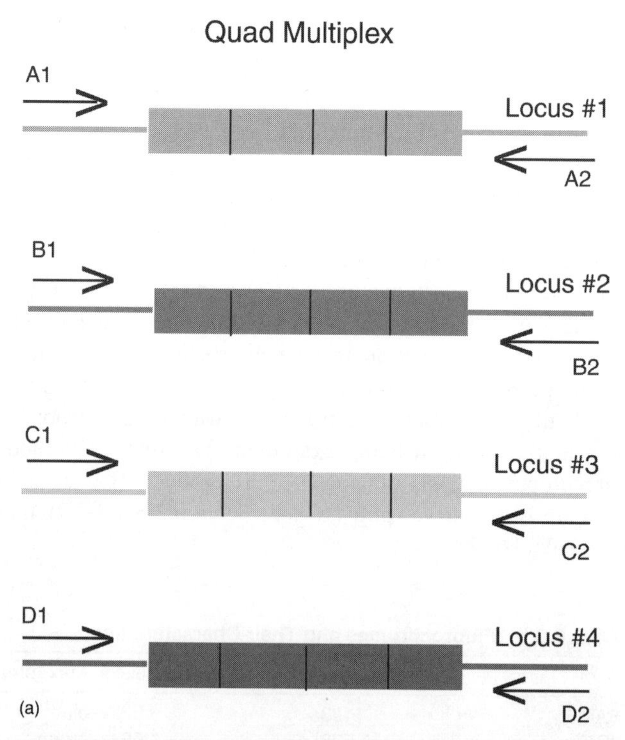

Fig. 3.31 *(a) Multiplex system is designed to allow multiple STR loci to be amplified simultaneously rather than amplifying only one STR loci at a time. (b) [page 93] Allelic ladders are mixture of DNA fragments representing most common alleles at each locus. Ladders included in the Profiler Plus kit are shown. (c) [page 94] Allelic ladders at each locus in the Cofiler kit. Adapted from a power point presentation by Dr. Peter Valone with his permission.*

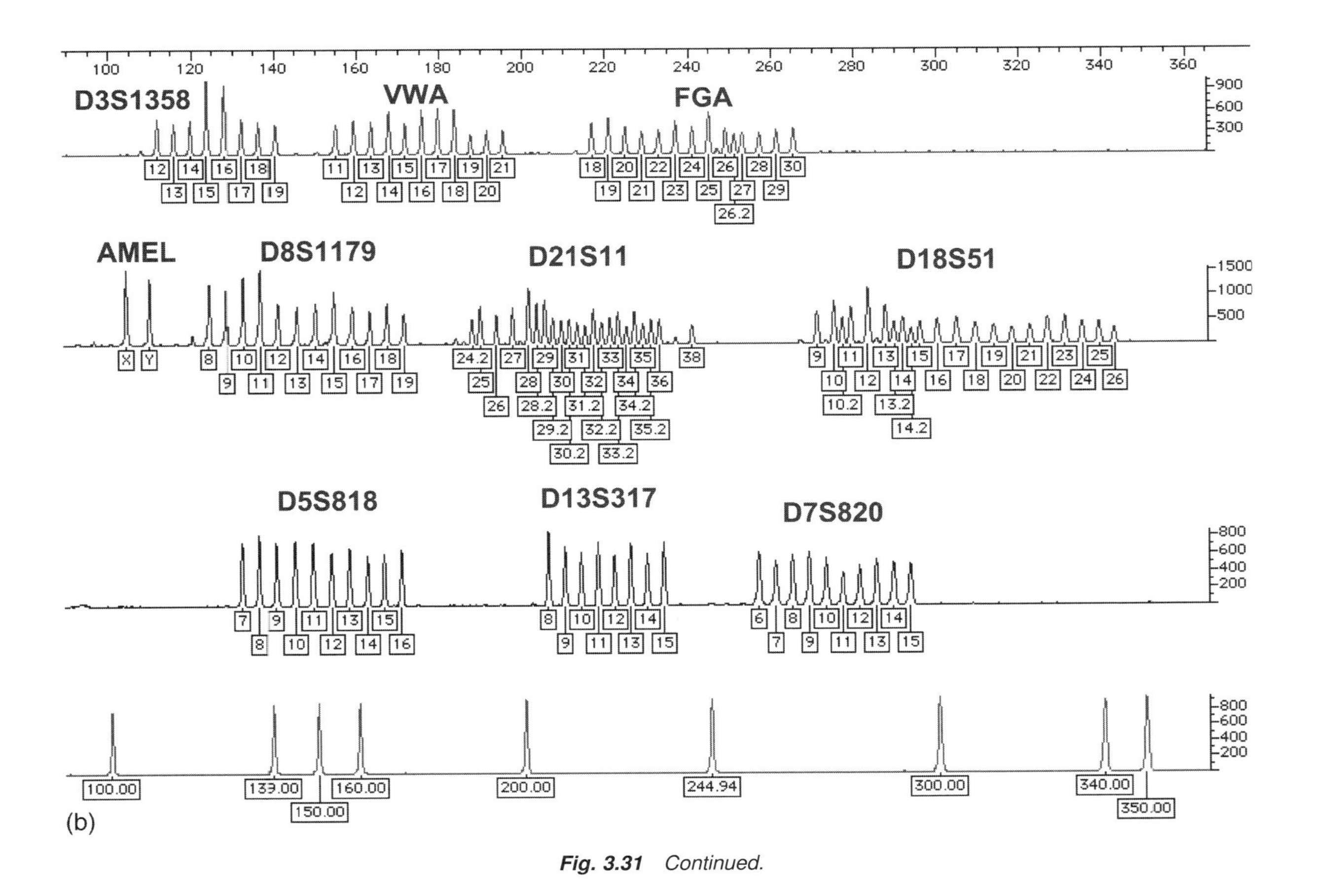

Fig. 3.31 Continued.

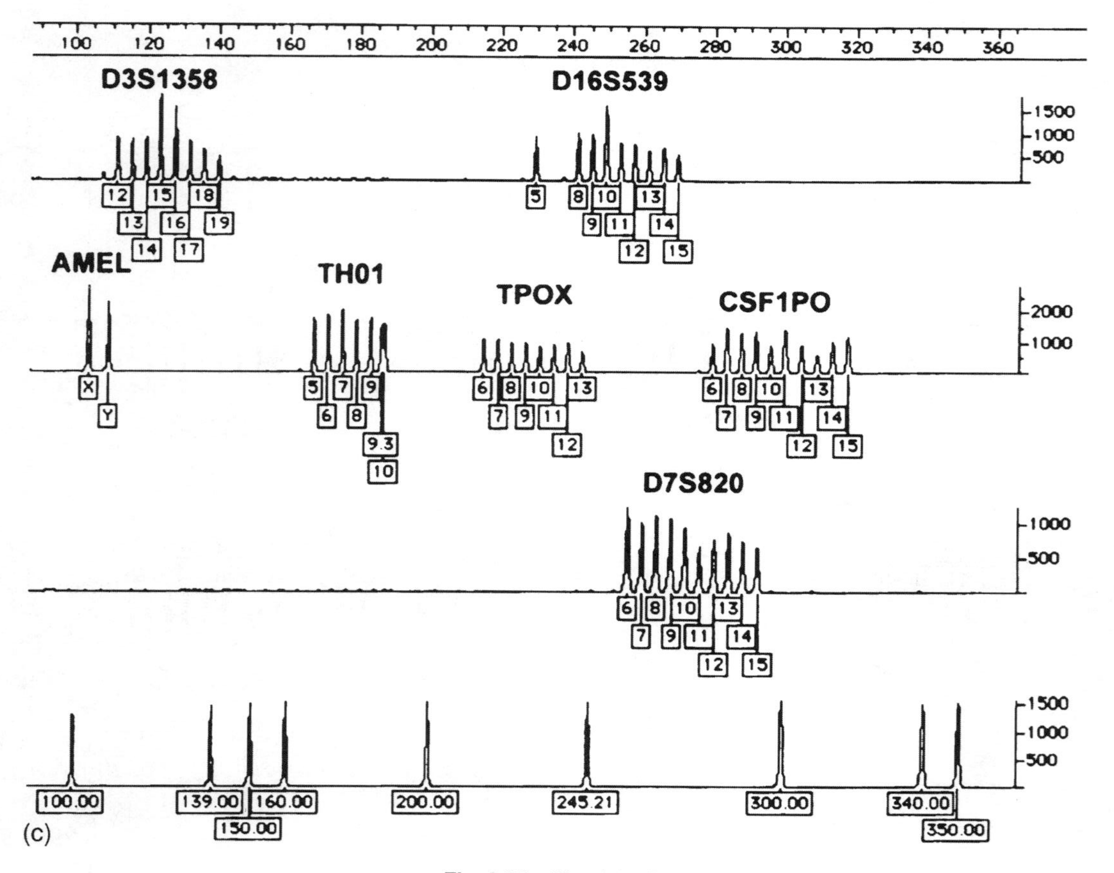

Fig. 3.31 *(Continued).*

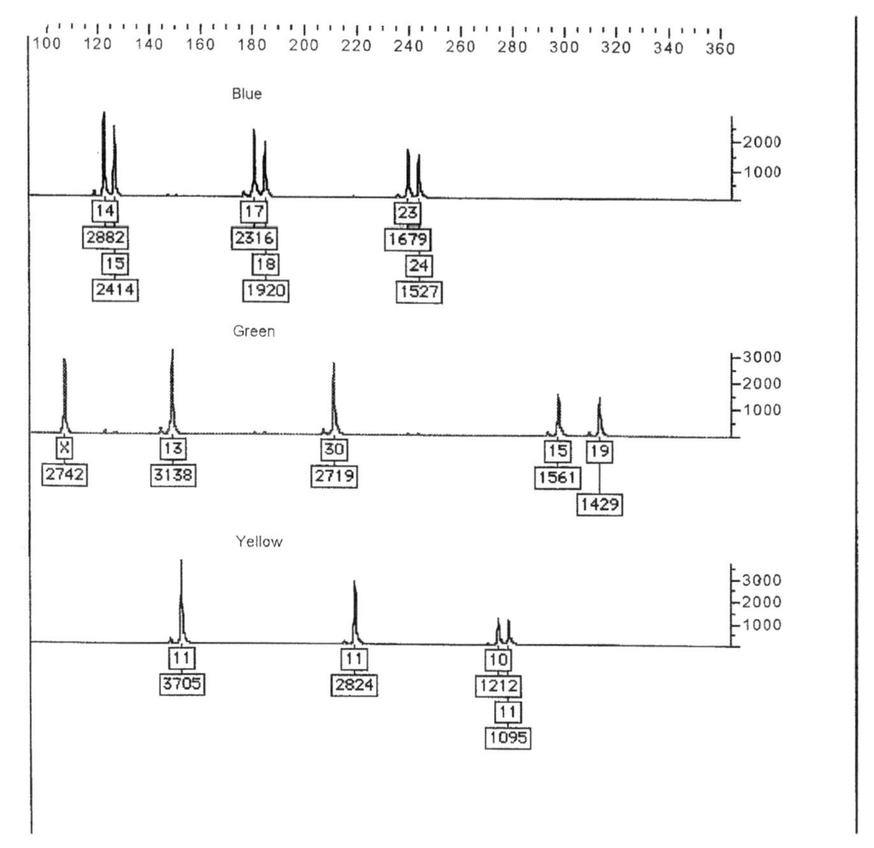

Fig. 3.32 *See color insert. AmpF/STR Profiler Plus is designed to simultaneously type D3S1358, VWA, FGA, Amel, D8S1179, D21S11, D18S51, D5S818, D13S317, and D7S820.*

PowerPlex 16 (Promega Corp.) and Identifiler (PE) are also currently in use (see Figs. 3.33*a* and 3.33*b*). The AmpF*l*STR Profiler Plus kit permits the simultaneous amplification and typing of 9 STR loci while the AmpF*l*STR COfiler kit permits the simultaneous amplification and typing of 6 STR loci. Because there is overlap of 2 loci (D3S1358 and D7S820) in these kits, the combined use of the two kits results in 13 loci. PowerPlex 1.1 permits the simultaneous amplification and typing of 8 STR loci, and PowerPlex 2.1 permits the simultaneous amplification and typing of 9 STR loci. Because there is overlap of 3 loci (vWA, THO1, and TPOX), the combination of loci tested again comes to 14, including the same13 loci obtained with the PE kits plus another locus called Penta E.

The PowerPlex 16 kit produces results at 16 loci simultaneously. Some people refer to this type of kit as a megaplex because of the large number of loci for which it tests. The 16 loci include the same 13 loci described above plus Penta D, Penta E, and amelogenin. The Identifiler kit also types 16 loci, including the same 13 loci plus D2S1338, D19S433, and amelogenin. Although we cannot determine

Current Forensic STR Multiplexes

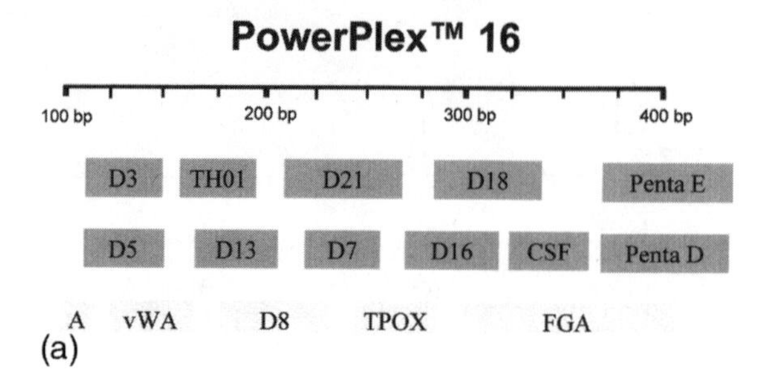

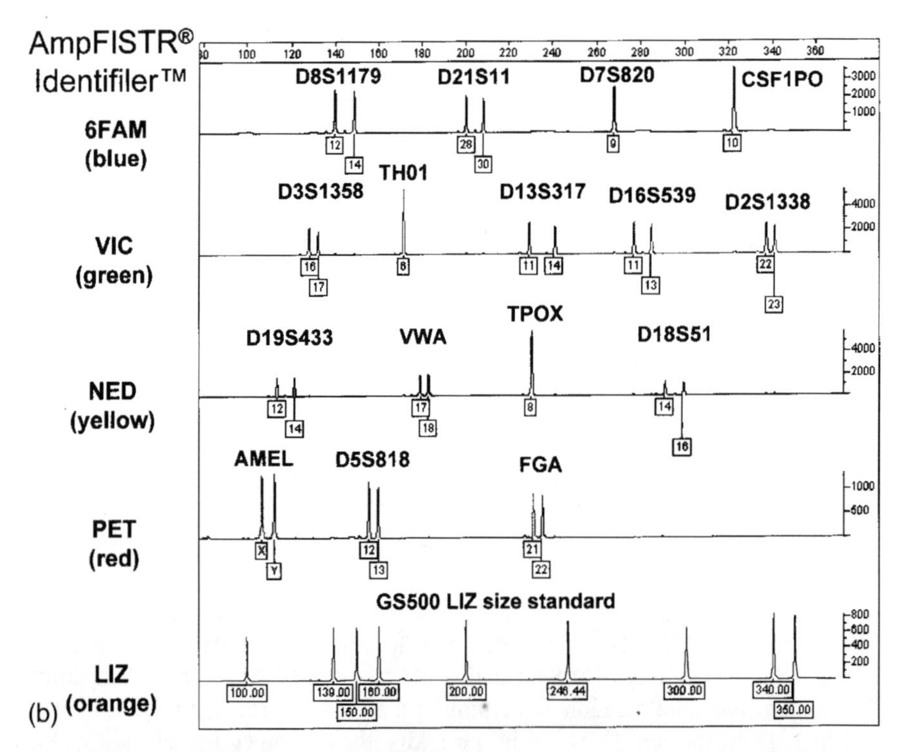

Fig. 3.33 Two multiplex kits have become available: (a) See color insert. Powerplex 16 developed by Promega Corp. and (b) AmpF1STR Identifier by PE Applied Biosystems.

race or ethnicity of the source of the evidentiary sample under study, we can determine gender (Sullivan et al., 1993; Mannucci et al., 1994). Each of the above mentioned kits (except for the PowerPlex 1.1 and 2.1 kits) includes a determination of amelogenin, which provides information regarding gender. Amelogenin is a gene that determines tooth bud formation in the fetus (Salido et al., 1992) and is located on both the X and Y chromosomes (see Fig. 3.34*a* and 3.34*b*). The gene was initially mapped during a study of X-linked *Amelogenesis imperfecta* (Lagerstrom et al.,

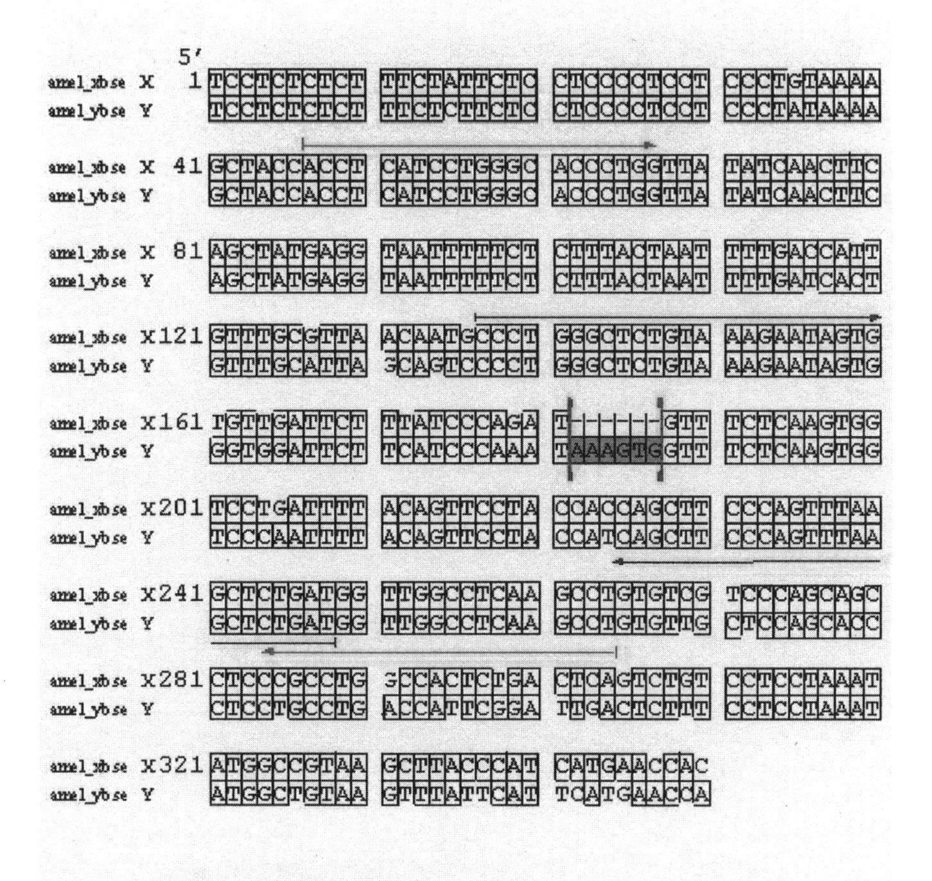

Fig. 3.34 See color insert. Difference in amelogenin sequence on X and Y chromosomes. Reprinted with permission of Dr. John Butler. See STRBase website: http://www.cstl.nist.gov/biotech/strbase/jpg_amel.htm.

1990; Nakahori et al., 1991a, 1991b). It is not an STR locus. However, because the allele on the Y chromosome is slightly longer than the allele on the X, the two alleles can easily be distinguished from each other by using either capillary or gel electrophoresis equipment (see below and Fig. 3.35). It should also be noted that the

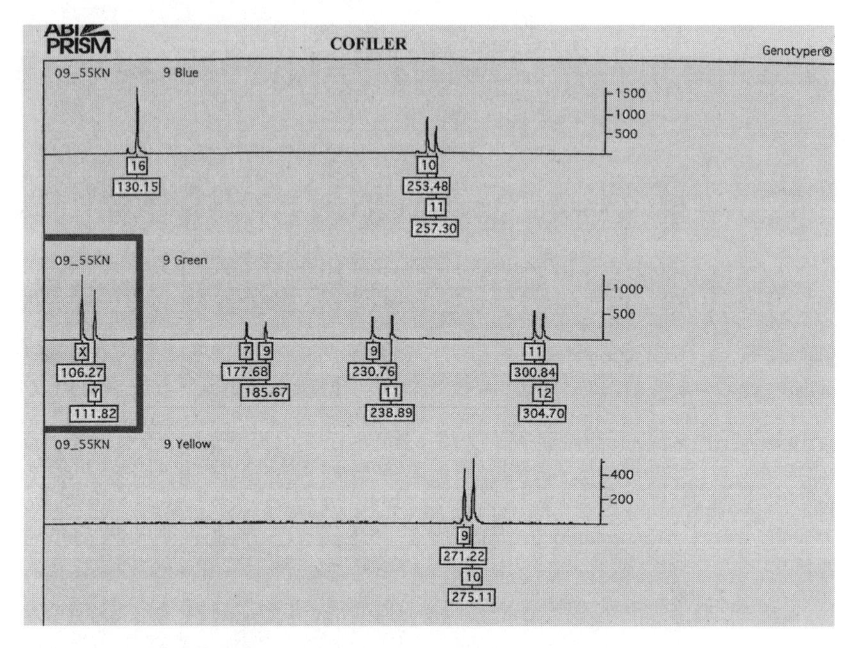

Fig. 3.35 Determination of gender using Cofiler Kit.

primers used to amplify the amelogenin locus are not human specific. It has been reported that a PCR product can be obtained from the DNA of a number of common animals (Buel et al., 1995).

With the primer set used in the Profiler Plus, Cofiler, and Identifiler kits, the amplified allele on the X chromosome is 106 bp; the allele on the Y chromosome is 112 bp. With the primer set used in the PowerPlex kits, the size difference is 212 and 218 bp, respectively. In either case, these size differences can easily be detected at the same time that the alleles at the STR loci are detected and identified.

Along with the test samples, and the positive and negative control samples, the analyst will also run allelic ladders. Figure 3.36*a* illustrates the allelic ladders designed for the Profiler Plus kit while Figure 3.36*b* illustrates the allelic ladders for the Cofiler kit. Promega Corp. and PE Applied Biosystems supply their kits with different allelic ladder systems. For example, Promega's Powerplex 2.1 multiplex kit provides an allelic ladder that includes 13 alleles at VWA (numbers 10–22), whereas the Profiler Plus kit by PE Applied Biosystems contains only 11 alleles (numbers 11–21), and the SGM Plus kit also manufactured by PE Applied Biosystems contains 14 alleles (11–24). The rarer alleles, numbers 10 and 22–24 are not present in the Profiler Plus kit. As a result, if one of these alleles happens to appear in an evidentiary (unknown) sample, the analyst may be less confident in its identification since there is no corresponding allele to directly compare it to.

With AmpF*l*STR Profiler Plus the following loci can be typed: D3S1358, VWA, FGA, amelogenin, D8S1179, D21S11, D18S51, D5S818, D13S317, and

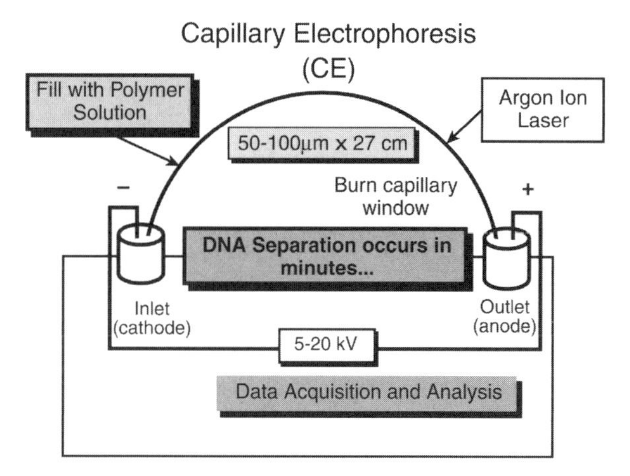

Fig. 3.36 Schematic of capillary electrophoresis setup. Adapted from a power point presentation by Dr. Peter Valone with his permission.

D7S820 (see Fig. 3.32). With AmpF/STR COfiler, the following loci can be typed: D3S1358, D16S539, amelogenin, THO1, TPOX, CSF1PO, and D7S820. When both multiplex kits are used, there are two markers that overlap, D3S1358 and D7S820, so these loci are tested twice. This provides data on the 13 CODIS loci (see Section 4.11) and also provides an "internal" control to examine if there is concordance at these loci for both kits. If these overlapping loci produce the same genotype results, this constitutes a validation of the test kits and the fragment size detection system.

The commercial kits contain reagents that are used as negative controls and a DNA sample to be used as a positive control (known genetic profile). It is recommended that the analyst include a sample of his or her own DNA to verify that the analyst has not contaminated any of the samples. The extraction and amplification negative controls are performed to rule out laboratory-induced contamination. All work involving amplified DNA should be performed under negative pressure laminar flow hoods to minimize the chance of contamination of the evidence. There are now "megaplex" systems being designed that will allow for the simultaneous amplification and typing of 20 distinct loci, and we will surely see advancements in multiplexing capacity in the near future.

3.3.4.3.1 STR Nomenclature After each multiplex amplification, the alleles at each locus must be measured for length. Genes are made up of building blocks called nucleoside triphosphates. These building blocks are linked end-to-end, thereby forming a single strand of the double-stranded DNA molecule. The two strands are held together as a result of hydrogen bonding between the bases of the building block units. We can thus talk about base pairs reflecting each building block base and its complementary partner (A-T and G-C). In most forensic DNA multiplex kits, sequential genes within a short tandem repeat (STR) locus generally differ by 4 nucleotide base pairs. (see Fig. 3.37). Generally, short tandem repeat (microsatellite) DNA consists of core units of 2–7 bases (usually 4) that are repeated one

Locus HUM**THO1**

Located on 11p15-15.5 (small arm of chromosome #11)
(Intron #1 of human tyrosine hydroxylase gene)
Repeat is {AATG}

— AATG AATG AATG AATG AATG AATG AATG —

ALLELE #7

— AATG AATG AATG AATG —

ALLELE #4

Fig. 3.37 *Sequential genes within a short tandem repeat locus (HUMTHO1) generally differ by four nucleotide base pairs. Alleles are named based upon total number of core repeats.*

after the next and tandemly connected to each other (Edwards et al., 1991, 1992). In fact, STR genes are named according to the number of core repeats at a locus. So that if one gene is structured as . . . NNNNNNN/**AATT/AATT/AATT/AATT/AATT/AATT**/NNNNN . . . (where N can be any base), and the bold type illustrates the 4 base repeating pattern we designate the allele number 6 because there are six core units in that repeat sequence. If the allele is structured as . . . NNNNNNN/**AATT/AATT/AATT/AATT/AATT/AATT/AATT/AATT/AATT**/NNNNN . . . , we designate it as allele number 9. We must remember that these alleles, 6 and 9, exist at a specific locus on a specific chromosome. Therefore it is important to note that genes 6 and 9 at *another* locus are completely different from the 6 and 9 genes mentioned above.

Although most genes at a single locus differ by 4 bp units, there are exceptions. For example, at the locus THO1, there are two distinct genes, 9.3 and 10. The difference between the 9.3 and 10 is only a single base. The 9.3 nomenclature means that there are 9 core repeats plus 3 bases of the 4 bases in the core. Similarly at the D18S51 locus, there is a gene designated 13.2, and at the D21S11 locus there are genes designated 24.2, 25.2, 28.2, and 31.2. The designation 13.2 means that there are 13 core repeats plus 2 bases (or 14 core repeats minus 2 bases).

When the amplification is completed, in addition to amplified products there remains in the reaction tube single-stranded primers as well as excess nucleotides. There may also be nonspecific amplification products known as primer-dimers that are formed when primers bind with other primers in the tube because of complementary sequences. DNA smaller than 100 bp can easily be removed from genuine amplified product. These extraneous substances can be removed using a "clean-up" system. One example of this approach is the Wizard MagneSil PCR Clean-Up System manufactured by Promega Corp. Double-stranded PCR products will bind tightly to the MagneSil particles in the presence of guanidine hydrochloride. The

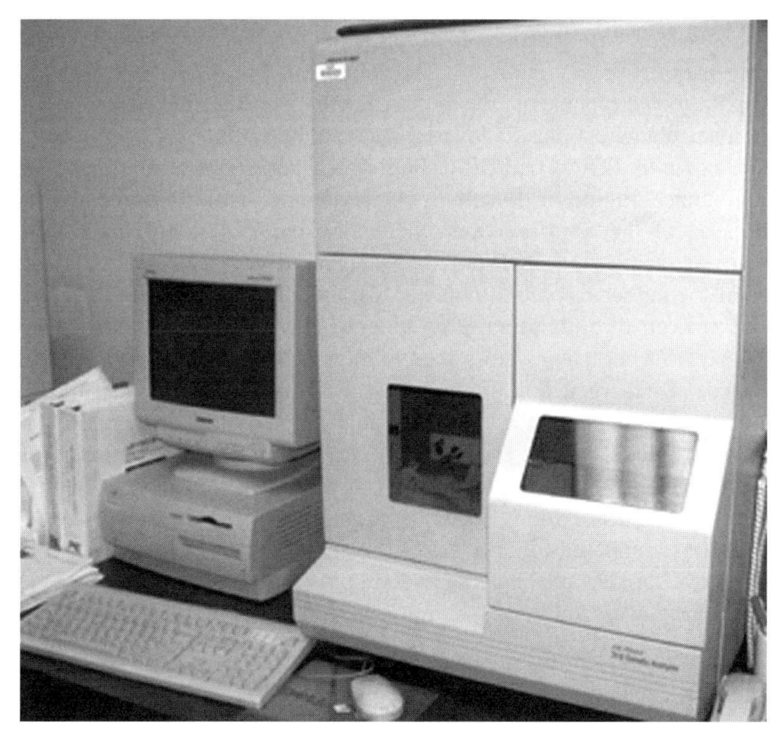

Fig. 3.38 *ABI Prism 310 genetic analyzer capillary electrophoresis instrument.*

particles are then washed to remove the impurities and the bound DNA can then be retrieved by elution with water.

Following PCR amplification of samples, the products are subjected to electrophoresis. Different instruments are available to perform either slab gel electrophoresis (vertical, horizontal, submarine) or capillary electrophoresis (CE) (Fig. 3.38 illustrates the ABI Prism 310 Genetic Analyzer CE instrument).

If performing gel electrophoresis, amplified samples are loaded into wells that are formed in a conductive gel medium (agarose or acrylamide) covering a long rectangular-shaped glass plate. A voltage difference is established across the gel (10 V/cm) with the cathode near the origin (well position) and the anode at the opposite end. This causes the negatively charged DNA fragments to move toward the anode. However, the gel, which contains pores, functions like a sieve in allowing small fragments of DNA to penetrate while large fragments are retarded in their migration. Each amplified DNA molecule contains a fluorochrome-labeled primer. As the fragments cross a window, the fluorochromes are excited by ultraviolet light and a charge-coupled device (CCD) camera detects the emitted light. The instrument detects wavelengths of light corresponding to blue, green, yellow, and red. The data is captured and further analyzed by commercial software that is capable of determining fragment lengths and designating allele numbers. One of the benefits of using gel electrophoresis for the separation of DNA fragments is that many samples can be run simultaneously.

In capillary electrophoresis, the amplified products are injected into a long thin (50 μm inner diameter) plastic-covered glass capillary that has been filled with a polymer known as POP-4 or POP-6. POP-4 is a performance optimized polymer with a 4% concentration of dimethyl-polyacrylamide, while POP-6 contains a 6% concentration of the same polymer. Generally, the POP-4 polymer is used for STR analysis, while the POP-6 polymer is used for DNA sequencing work because of the higher resolution obtained with this media. Both formulations contain urea to keep the migrating DNA fragments in a single-stranded, denatured state. These DNA molecules are forced to migrate within the conductive polymer in the anodal (+charged pole) direction. The DNA fragments must traverse the length of the capillary, 47 cm, from the injection point to the opposite end of the capillary (anode) (Issaq et al., 1997). An electrophoretic buffer is used to supply the ions necessary so that current can be conducted throughout the length of the capillary. The voltage used in capillary electrophoresis is very high (300 V/cm) as compared to gel electrophoresis (10 V/cm). This causes the fragments to move rapidly and separate as they pass through the polymer. Typically, a sample can be analyzed in 30–45 min. The fragments are detected as they pass by a window in the capillary tube that is located 11 cm from the anodal end of the tube (see Fig. 3.36). Therefore, the fragments have to migrate 36 cm until they are detected through the capillary window. A shorter capillary means faster analysis of the STR loci. For sequencing work a longer capillary, 61 cm, provides higher resolution so that the fragments have to migrate 50 cm until they are detected. The drawback to using capillary electrophoresis rather than gel electrophoresis is that with the former only one sample can be studied at a time. Thus, fewer samples are tested in the same time period, and it takes longer to test each sample, but there is better separation of the DNA fragments (Guttman and Cooke, 1991; Siles et al., 1996; Lazaruk et al., 1998). Applied Biosystems now offers a capillary Electrophoresis Instrument designed with 16 capillaries to overcome this problem, the PRISM 3100. An argon laser excites the fluorochrome that had been used to label one of the two primers, at 485 and 514.5 nm. The camera captures the fluoresced light emitted by the excited fluorochrome. The camera collects raw data that is subsequently converted to usable spectral data for each dye when the matrix is applied during GeneScan analysis. Each fragment is displayed on a printout (electropherogram) as a peak. The specific fluorochrome determines the emitted wavelength and color (blue, green, yellow, and red). The peak height (y axis) reflects the intensity of the emitted fluorescence [measured in relative fluorescence units (RFU)] and therefore the amount of amplified DNA that had been injected. The x axis of the electropherogram is an indication of the length of time it took for the fragment to cross the detection point and therefore reflects the size of the fragment. This is also called the scan number. Alleles at particular loci can easily be distinguished from one another because the measuring equipment can distinguish not only the different fluorochrome labels but also DNA fragment sizes differing by as little as one base. The equipment is calibrated using a mixture of DNA fragments in an internal sizing standard.

Peak sizes are determined with GeneScan software, which compares each fragment to the internal standards (Mayrand et al., 1992). There are two different choices for the internal standard. GS500 is a ROX (fluorochrome)-labeled standard manufactured by Applied Biosystems and ILS 600 (Internal Lane Standard) is a CXR-labeled standard produced by Promega Corp. In GS500 the calibrating fragments range in size from 35 to 500 bp, whereas ILS600 calibrating fragments range from 60 to 600 bp. Both ROX and CXR are essentially the same dye and emit light that appears red in color. The sizing algorithm used to determine peak size is called local southern. It determines the size of an unknown STR peak by a calculation using the size of the two internal size standard peaks below and above the questioned peak. GenoTyper software takes the data from the GeneScan software and compares this data to the allelic ladders and identifies specific alleles in the questioned samples. The GenoTyper software recognizes the first allele within each ladder and creates a "bin" around the peak that is limited by ± 0.5 bp. DNA fragments within the unknown samples can then be compared to these peaks and bins and classified accordingly. In the case of THO1 where the 9.3 and 10 alleles are separated by only a single base, the bin window has to be smaller than 0.5 bp to differentiate these two alleles. The window is set at ± 0.48 bp. When questioned alleles fall outside the ladder, or if the allele falls within the ladder but is not represented by any ladder allele, then that allele will be designated by the GenoTyper software as an off-ladder (OL) allele. Such alleles are named according to guidelines established by the International Society for Forensic Haemogenetics. The overall process is based on measurement of each allele by detecting its fluorescence at a particular position in the electrophoretic gel or capillary. The instrument can detect blue, green, yellow, and red fluorescence. Fluorescence occurs when a fluorochrome is excited by light of a higher energy, which can be detected by a sensitive photodetection system. The fluorochromes used in the production of the commercial kits are seen in Table 3.6. Each is attached to primers that become part of newly synthesized DNA strands during the amplification process. Thus, for Profiler Plus, D3S1358, VWA, and FGA are all labeled blue, amelogenin, D8S1179, D21S11, and D18S51 are all labeled green, and D5S818, D13S317, and D7S820 are all labeled yellow. For COfiler, D3S1358 and D16S539 are labeled blue, amelogenin, THO1, TPOX, and CSF1PO are labeled green, and D7S820 is labeled yellow. This system of fluorochrome labeling permits all of the nine loci of Profiler Plus and all of the 6 loci of COfiler to be detected in one run. When using either of these kits, when there is overlap of allele sizes at two different loci, the alleles can be differentiated by differences in fluorochrome labeling. The instrument can also record the fluorochrome, the time it takes for the fragments to appear in the detection window (scan number), the size of sample fragments, peak heights, and peak area. Known and unknown specimens can be compared at each locus to determine if there is a match or an exclusion. The result can also be inconclusive or negative.

The FMBIO II fluorescent scanner is an instrument manufactured by Hitachi Genetic Systems/MiraiBio that can be used to detect fluorochrome-labeled amplified products that have been electrophoresed on an agarose or acrylamide gel. This instrument is designed with four detection channels, and different filters can be placed in these channels enabling the instrument to detect the blue, green, yellow, and red fluorochromes used to label primers in multiplex kits. The filters used are 605, 585, 505, and 650 nm. The StaR software is used to identify alleles in questioned samples.

Another way to analyze the amplified product after electrophoresis is to use silver staining to enhance contrast between the bands of amplified product and background. Bands can be visually checked by comparing their positions on the gel to those bands within an allelic ladder.

3.3.4.3.2 Short Tandem Repeat DNA Short tandem repeats (STRs) are similar to variable number of tandem repeats (VNTRs) in that both have repeating core units that are tandemly connected, and the general principles that govern the two are similar. However, there are also substantial differences in the two types of repetitious DNA. Allele sizes are in general much greater for a VNTR locus than for an STR locus. VNTRs are classified as minisatellites, whereas STRs are classified as microsatellites. VNTRs alleles form a continuum of sizes unlike STRs alleles, which are discrete and easily identifiable.

There are several thousand STRs within the human genome. However, only a small number are suitable for forensic use (Holland et al., 1993; Urquhart et al., 1994; Kimpton et al., 1995; Herd et al., 1999). STRs that are composed of 2 base repeats are generally not satisfactory for human identification due to the high level of stutter peak production (Hauge and Litt, 1993) (see Section 3.3.4.3.2.2). Furthermore, for human identification, STR loci with a high degree of heterogeneity and a high level of polymorphism are preferred. Most forensically useful STR loci have fewer than 30 different alleles. The smaller size of the STR loci (as compared to the larger VNTRs) makes them more suitable for amplification of exceedingly small samples or samples that have been compromised due to environmental factors. PCR amplification of STRs can be used when there is only a small amount of DNA available for testing or when the DNA obtained from a sample is considerably degraded. Most forensic laboratories are performing PCR-STR testing, in part, because less of the evidentiary DNA sample has to be consumed when compared with the amount of DNA necessary to perform RFLP analysis.

The use of STRs allows for a more discriminating type of DNA analysis. STRs consist of a number of discrete alleles at each locus, and therefore, unlike RFLP analysis, there is no need for binning. STR loci generally consist of between 7 and 30 different alleles. Many of these loci have been found to be suitable for use in forensic analysis. The FBI has designated 13 specific STR loci as a "core set" to be used in the determination of whether DNA found at a crime scene matches that obtained from previously tested individuals. Figure 3.39 describes the heterozygosity, number of alleles, and population match probability in a Caucasian American and an African American population. It is no surprise that when analyzing 13 loci, the numbers calculated to describe how common or rare a profile is in the relevant population are astronomically small. With the population of the planet estimated to be in the range of 6 billion people, it is clear that a calculation of a profile that reaches the level of 1 in greater than trillions is essentially absolute identification.

3.3.4.3.3 Interpretation of STR Results To ensure that STR typing results are interpreted properly, it is critical that negative and positive controls are included in every analysis and that the genetic analyzer is properly calibrated and optimized for fluorochrome detection. Internal lane markers (GS 500 for Profiler Plus and Cofiler kits or ILS 600 for the Powerplex kits) must have correct sizes assigned to each component peak since they are used to size unknown peaks. It is also vital to establish guidelines on how to designate alleles in an STR multiplex system (Budowle et al., 2000a; Gill et al., 1997; SWGDAM, 2000). All alleles within the allelic ladders must be correctly sized and identified. All considered peaks must be at the threshold level or above (see Section 3.3.4.3.2.4). All negative controls

		Caucasian American		African American	
Locus	**No alleles**	**Het.**	**Pop. match pr.**	**Het.**	**Pop. match pr.**
CSF1PO	11	0.734	0.112	0.781	0.081
TPOX	7	0.621	0.195	0.763	0.090
TH01	7	0.783	0.081	0.727	0.109
vWA	10	0.811	0.062	0.809	0.063
D16S539	8	0.767	0.089	0.798	0.070
D7S820	11	0.806	0.065	0.782	0.080
D13S317	8	0.771	0.085	0.688	0.136
D5S818	10	0.682	0.158	0.739	0.112
FGA	19	0.860	0.036	0.863	0.033
D3S1358	10	0.795	0.075	0.763	0.094
D8S1179	10	0.780	0.067	0.778	0.082
D18S51	15	0.876	0.028	0.873	0.029
D2S11	20	0.853	0.039	0.861	0.034
Average		**0.7812**		**0.7866**	
Product			**1.738×10^{-15}**		**1.092×10^{-15}**
One in			**5.753×10^{14}**		**9.161×10^{14}**

Data are from the FBI.

Fig. 3.39 Chart showing heterozygosity, number of alleles, and population match probability in a Caucasian American and African American population. Reprinted with permission of the National Criminal Justice Reference Service. The Future of Forensic DNA Testing, Copyright 2000.

TABLE 3.7 Genotypes for 9947A

D3S1358: 14,15	VWA: 17,18	FGA: 23,24	
Amelogenin: X,X	D8S1179: 13,13	D21S11: 30,30	D18S51: 15,19
D5S818: 11,11	D13S317: 11,11	D7S820: 10,11	
D16S539: 11,12	THO1: 8,9.3	TPOX: 8,8	
CSF1PO: 10,12	Penta D: 12,12	Penta E: 12,13	

(reagent blanks and amplification blanks) must exhibit peaks no larger than 50 RFUs. Samples that are observed to produce off-scale (flat top) peaks should be diluted and run again.

Amplification positive controls must provide peaks and overall profiles that are designated correctly. Above all, false inclusions must be prevented (Thompson et al., 2003). Multiplex kits include positive control specimens that must be included in genotyping analyses. Specimen 9947A is a positive control that can be used for Profiler Plus, Cofiler, and all Powerplex multiplex STR kits including Powerplex 16. Table 3.7 illustrates the genotypes at each locus for 9947A.

3.3.4.3.3.1 PLUS A (+A) (ADENYLATION) After amplification, electrophoresis, and gene scanning, sometimes alleles are reported with a base pair size greater by one base than what is expected. Alternately, rather than observing one or two nicely formed peaks, a peak with a shoulder or a split peak appears (see Fig. 3.40). This can lead to the incorrect conclusion that there is a mixture in the original sample. If the other loci and the facts of the case indicate otherwise, the analyst should consider the possibility of an artifact resulting from the phenomenon of adenylation of the 3′-OH end of the amplified product by the Taq polymerase, thereby creating an amplified product longer than expected by a single base (Clark, 1988). The interpretation can be difficult when the observed allele differs from another allele by only one base as in the case of the 9.3 and 10 alleles of the THO1 locus. Obviously, if the 9.3 allele were adenylated, it would appear to be a 10 rather than 9.3. Let us now examine three possible products. In case 1, the expected product is shown with each strand ending in the expected base. However, in case 2, the polymerase has added an adenine to the 3′ side of both strands resulting in the "A+ form." It should be noted that

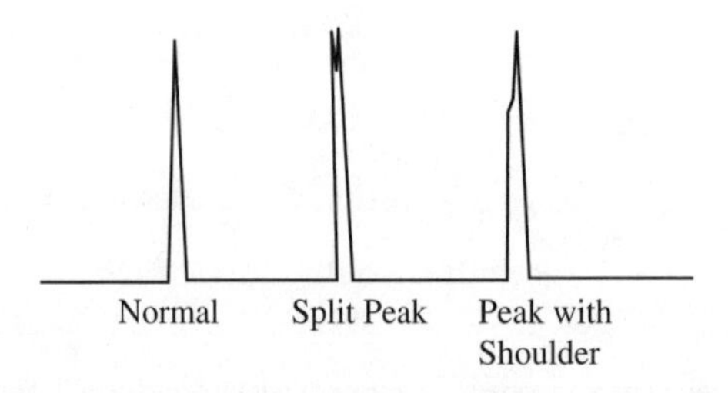

Fig. 3.40 Alleles are observed as either a split peak or a peak with a shoulder.

this added adenine is not dependent on the template itself but rather is an artifact created by the sequence of the primer that has annealed to the 5′ side on the opposite strand. Since only one of the two primers (either the forward or reverse primer) is fluorescently labeled and actually detected by the instrument, the fragment is observed as either a split peak or a peak with the length of the expected number of base pairs plus one.

```
XXXX = Forward primer—labeled with fluorochrome
YYYY = Reverse primer
          5′ *XXXXX → 3′____________
          3′                       5′
          5′_______________________3′
                       3′ ← YYYY5′
```

Case 1 No Adenylation—Fluorochrome-labeled strand is detected.

```
5′*XXXXX________3′    5′__________________3′
3′______________5′    3′__________YYYYY5′
```

Result is a sharp single peak.

Case 2 Complete Adenylation—Fluorochrome-labeled strand is detected.

```
5′*XXXXX__________A3′    5′______________________A3′
3′A_______________5′     3′A____________YYYYY5′
```

Result is a sharp peak with $n + 1$ bases.

Case 3 Partial Adenylation (Combination of adenylation and no adenylation)

```
5′*XXXXX__________A3′ and 5′*XXXXX____________3′
3′A_______________5′      3′__________________5′
                      and
5′________________A3′     5′__________________3′
3′A________YYYYY5′        3′__________YYYYY5′
```

The result is either a split peak or a peak with a shoulder. It is important to note that only labeled strands are detected.

In order to remove the ambiguity caused by having some alleles amplified normally while others are amplified with the added adenylation, the amplification is performed in such a way that either the adenylation is completely blocked or it is fully promoted, thereby causing all amplified products to be adenylated and therefore allele identification is conducted correctly. Perkin-Elmer's recommended amplification protocol includes a final extension step (incubation of product at either 60°C or at 72°C for 30–45 min) after the amplification cycles are completed. This additional step results in complete adenylation of all the amplified products at all loci.

3.3.4.3.3.2 STUTTER PEAKS An artifact created during PCR amplification resulting from "slippage" and incorrect strand base pairing. During the extension phase the polymerase jumps or slips over a repeat unit and continues processing the template. The missed repeat unit takes the shape of a short loop. The result is the production of a sequence containing one (or rarely two) four-base repeat unit(s) less than the actual allele (Walsh et al., 1996). Stutter appears on the electropherogram as a minor peak, usually four bases shorter than the corresponding main allele peak ($n - 4$). The level of observed stutter bands is inversely related to the length of the core repeat unit. One advantage of using Penta STR loci (Promega Corp.), which has a five-base repeat unit, is that there are fewer stutter bands and therefore the allele patterns can be more easily interpreted (Bacher and Schumm, 1998). STRs composed of two-base repeats are generally not satisfactory for human identification due to the high level of stutter production. STR loci with repeat units of four or five bases are preferred. Stutter peaks are more common in longer alleles than shorter ones. Thus, D21S11, D3S1358, FGA, and CSF1PO have a higher percentage of stutter than THO1 or TPOX. Despite the existence of stutter bands (peaks), they generally do not cause confusion in the interpretation of the locus genotype except when mixtures exist. The question that has to be decided is whether the questioned peak is a stutter peak or an authentic minor peak derived from one of the contributors to the mixture. Analysts will compare peak heights of the allele and the questioned peak. Usually a stutter peak is *less than 15% of the larger, adjacent peak height.*

3.3.4.3.3.3 PULL-UP PEAKS This phenomenon is sometimes referred to as "bleed through." It is an artifact seen in the electropherogram for loci analyzed by multiplex kits that use multiple fluorochromes for different primer labeling. For example, it can manifest itself as a high-intensity peak developed in the green channel that also appears in the blue channel at the same size position. Pull-up peaks can be caused by incorporating excessive template DNA into the PCR mixture. The most common type of pull-up is a small green peak that appears in the blue channel. Peak heights can be off-scale (pull-up) or on scale (bleed-through). The result is that the fluorescence emitted by the labeled PCR products will be excessive, and there will be elevated baselines and pull-up of one or more colors corresponding to the peak size of an allele(s) *in other channels.* Pull-up can also be created by the use of an incorrect matrix (see matrix file in Section 3.3.4.3) causing lack of discrimination between the different dye colors. Another problem unrelated to pull-up but that can cause confusion is a peak created by a voltage spike. These spikes are usually much thinner in shape and can usually be easily differentiated from an authentic peak.

3.3.4.3.3.4 PEAKS BELOW THRESHOLD Electropherograms produced during STR analysis often contains small peaks in addition to the allelic peaks present in the subject's genotype. As described above, these peaks are sometimes due to stutter. They may also be due to the phenomenon of pull-up. However, they may also result from mixtures of DNA from two or more individuals where they represent the alleles of the minor component (Perlin and Szabady, 2001). Interpretation of the pattern is facilitated if it can be established that a mixture does or does not exist. One way to handle small peaks is to determine a threshold level for declaring a peak "authentic" or an artifact. Some labs have decided to set that level at 100 RFUs so a designated peak must be at least 100 units in height to be considered conclusive for a match determination. When samples are exposed to environmental insults and/or undergo aging, some loci are seen to be more degraded than others. Usually the larger the amplicon in length, the smaller the peak height. Degraded specimens can pose serious interpretation problems because the larger alleles are also subject to allelic dropout.

Mixtures are sometimes very difficult to interpret. There may be rare times when the specimen analyzed appears to consist of a mixture but it is not. It is the result of a very rare genetic phenomenon known as chimerism. It has been reported that a woman who needed a kidney transplant had undergone human leukocyte antigen (HLA) testing along with her family members. The test results suggested that she was not the biological mother of two of her three children. Nuclear STR DNA testing revealed that she had only one cell line in her peripheral blood but two cell lines in her other tissues. The woman exhibited the phenomenon known as chimerism in which there are two distinct cell lines in one person. Chimerism can happen either through transfusion, transplantation, or it can be inherited. In the above-mentioned case, it appears that two spermatazoa fertilized two ova, and the developing zygotes subsequently fused producing a single individual with mixed cell lines (Yu et al., 2002).

3.3.4.3.3.5 SHOULDERS ON PEAKS Ideally, the electropherogram should display peaks that are symmetrical with peak heights that are above threshold and on-scale. There are a number of possible explanations for the presence of shoulders on peaks. Partial adenylation of amplified product can produce shoulders on peaks as described above. This can be eliminated in a number of ways (see discussion of plus A above). The analysis parameters of the GeneScan software can also be programmed to resolve alleles that differ by only one base such as THO1 9.3 and 10. The bin size is made sufficiently small so as to be able to resolve both of these alleles. GenoTyper software then determines the size of each allele using local southern analysis by comparing it to each of the adjacent alleles in the allelic ladder. The allele may be designated as an "off-ladder" allele by the GenoTyper software if it is not represented in the allelic ladder or if it exists outside a bin created by the GeneScan software.

3.3.4.3.3.6 OFF-LADDER ALLELES Electropherograms will sometimes indicate peaks for certain samples that fall between the alleles present in the allelic ladder. These "unusual" peaks may be artifacts or authentic but rare alleles (Xiao et al., 1998). Artifacts can be ruled out by sequencing analysis. If the peak falls someplace between adjacent ladder alleles Q and S, the peak is sized by the software and then converted into base pairs. If the base pair size is a multiple of 4 units, it is assigned an allelic designation based on the surrounding alleles, Q and S. Thus, it may be assigned the designation R since it lies between Q and S and differs from each by 4 base pairs. If it is off by 1, 2, or 3 bases, from the smaller allele Q, then it is designated as either Q.1 or Q.2 or Q.3. Thus in the locus THO1, the 9.3 allele differs from the 9 allele by 3 bases and from the 10 allele by 1 base. If the off-ladder allele should fall above the largest allelic ladder peak, it is designated greater than (>) that peak, or if it should fall below the smallest allelic ladder peak, it is designated as less than (<) the smallest ladder allele.

3.3.4.3.3.7 MIXED SAMPLES Generally, it can be assumed that the specimen being analyzed is derived from one individual if the electropherogram shows only one or two alleles present at each and every locus examined and if the ratio of the peak heights for heterozygous loci are within expected values. However, samples are sometimes composed of contributions from two or more individuals. The ability to detect a mixture depends, in part, on the specific combination of genotypes at each locus and on the ratio of the DNA components contributed by each person. If we examine loci with limited polymorphism and low levels of heterozygosity, then it is possible that there will be overlapping alleles seen in the genotype for that locus on the electropherogram. In other words, you may not be able to easily determine that there is a mixture. The examination of a significant number of highly polymorphic loci with high heterozygosity makes the determination of a mixture far more likely if one exists. A much more significant problem lies in the detection of mixtures in cases of sexual assault (see Section 2.1.8), especially when there are large differences in the contributions of the donors. Assuming two contributors, if the difference in the relative contributions is large (>10:1), then amplification and subsequent analysis may not reveal the existence of a mixture, even over 13 or more loci. This will also occur when the minor component in the specimen to be amplified is lower than 35 pg. The lesser components are essentially unseen. If the ratio of the amount of DNA is between 1:1 and 10:1, then allele contributions from both can be seen but with markedly different peak heights. Equal contributions by contributors (1:1) will result in peak heights that are more or less the same height. When two individuals share the same allele at a locus, the two alleles overlap and appear as one peak with a peak height that results from a combination of fluorescence from each donor allele.

Genetic analyzers (ABI and FMBIO II) can provide quantitative information regarding peak height RFUs or peak areas (Gill et al., 1998). We can, as a result, examine the differences in peak heights (or peak areas) for all observed peaks at each locus. The following rules generally apply for determining if a mixture exists:

1. Stutter peaks, which are virtually always smaller than the main peak by 4 bp, are usually no more than 15% of the height of the largest peak. Observation of

a small peak (<15%) 4 bp to the right of the observed peak is not a stutter peak and may indicate a mixture.

2. The two peaks of a heterozygous genotype are usually *no more than 30% apart in peak height*. Thus, the smaller of the two peaks is usually greater than 70% of the height of the highest peak. Peak height imbalance is indicative of a mixture.
3. When three or more "above threshold" peaks are present at even one of the loci examined, a mixture should be suspected.
4. It is possible for a mixture derived from two people to result in one, two, three, or four peaks at a single locus. It should also be noted that under unusual circumstances (somatic mutations and chromosomal translocations) three different alleles can be present at a locus (trisomy) in an individual that would result in three peaks for a single individual. These same three peaks would be seen in both exemplar and questioned specimens.

The interpretation of mixed samples is complex and should be done carefully, incorporating all possible combinations, adhering to established laboratory protocols, and being conservative in interpreting the results (Clayton et al., 1998).

3.3.4.3.3.8 SEQUENCE VARIANTS Forensic STR analysis is based on comparing DNA fragment (allele) sizes for known and questioned samples over a sufficient number of predetermined loci. Virtually all loci have alleles that exhibit complex repeat sequences (Puers et al., 1993; Sharma and Litt, 1993; Moller et al., 1994; Brinkmann et al., 1996; Schwartz et al., 1996; Barber and Parkin, 1996; Barber et al., 1996; Iwasa et al., 1997; Momhinweg et al., 1998). For example, the locus VWA has alleles 13, 14, 15, 16 and 18 each with a number of different configurations (Kimpton et al., 1992). There are two configurations for the 13 allele. Both consist of 134 bp, however, one allele has the structure 5′-$(TCTG)_4(TCTA)_8$ TCCA TCTA-3′ and the second allele has the structure 5′-$(TCTA)(TCTG)_4$ $(TCTA)_{10}$-3′. Both of the two 14 alleles are 138 bp in length, however, the structures differ: 5′-$TCTA(TCTG)_4(TCTA)_{11}$-3′ and 5′-TCTA TCTG TCTA $(TCTG)_4$ $(TCTA)_4TCCA(TCTA)_4$-3′. The configuration for 15 has the sequence: 5′-TCTA $(TCTG)_4(TCTA)_{10}$-3′, while the configuration for 15′ has the configuration 5′-$TCTA(TCTG)_3(TCTA)_{11}$-3′. Both have the same length of 142 bp, yet they are entirely different alleles. The VWA 16 allele has the configuration 5′-TCTA $(TCTG)_4(TCTA)_{11}$-3′, whereas 16′ has the configuration 5′-$TCTA(TCTG)_3$ $(TCTA)_{12}$-3′. Both the 16 and 16′ alleles have the same length, 146 bp, but in fact are completely different alleles. The 18 alleles both have the same fragment length but one has the structure 5′-$TCTA(TCTG)_4(TCTA)_{13}$-3′ and the other has the structure 5′-$TCTA(TCTG)_5(TCTA)_{12}$-3′. Typing for length polymorphisms will not resolve differences in allele sequence. Unless sequencing is done, the difference between these alleles goes unnoticed. Each variation described above is a different allele. So if the evidentiary sample is typed as a VWA 15 and the evidence is also typed as a 15, the question remains as to whether we are talking about identical genes (15 and 15) or different genes (15 and 15′). For locus D3S1358 there are two 16 alleles each having a length of 131 bp, yet the structure for one is 5′-$(AGAT)_{12}$ $(AGAC)_2(AGAT)_2$-3′ and the structure for the second is 5′-$(AGAT)_{11}(AGAC)_3$ $(AGAT)_2$-3′. In the case of the alleles having 135 bp lengths and denoted as 17, there are actually two alleles with the structures 5′-$(AGAT)_{13}(AGAC)_2(AGAT)_2$-3′ and 5′-$(AGAT)_{12}(AGAC)_3(AGAT)_2$-3′. For locus D21S11 there are three different alleles, each designated as 29 with 221 bp but with different structures: 5′-$(TCTA)_4(TCTG)_6$......$(TCTA)_{11}$-3′, $(TCTA)_6(TCTG)_5$......$(TCTA)_{10}$-3′, and $(TCTA)_5(TCTG)_6$......$(TCTA)_{10}$-3′. For locus FES/FPS located at 15q.26.1, the same is true (Alper et al., 1995; Barber et al., 1995). There are two alleles designated 10, both having the same size (210 bp); however, they have been determined to have a single base difference in their sequence structure. The same is true for alleles 11 and 11a, both of which are 223 bp in length. In the latter alleles there is a transversion from A to C (in the allele designated A at position 34 in the 5′ flanking region of the sequence). There are also two different 12 alleles having different microvariation in structure but sharing the same length. These types of variation are found in virtually all loci tested. Some variations are very rare (HumFES/FPS allele 12A has a frequency of approximately 0.002, while 12 has a frequency of approximately 0.290) but others are not. In different populations, FES/FPS alleles 11 and 11A have almost identical high frequencies. Out of eight alleles sequenced and analyzed, three (37%) were found to be 10A while five (63%) were found to be 10. Single-base microvariations can be detected either using sequencing techniques or by heteroduplex analysis. As noted in the National Research Council Report of 1996, "The sequence of nucleotides in the genome determines the genetic difference between one person and another." Because we are trying to link a suspect and evidence through genetic testing, it is important that we use the most discriminating method to do that. With the rapid improvements in identification testing that have come about with technological advancements, it is expected that testing will be even more sensitive and even more precise in the near future.

3.3.4.3.3.9 LOW COPY NUMBER DNA Multiplex STR kits are designed to produce optimum results when 1–2 ng of DNA are used to start the amplification reaction, and thermal cycling is conducted with 28–30 cycles. Satisfactory results can be obtained even when the quantity falls between 250 pg and 1 ng. The ability to detect and identify amplified product becomes difficult when starting with only 250 pg of template DNA (Gill et al., 2000; Gill, 2001). In fact, the process may yield no useful results at all. By simply increasing the number of PCR cycles (e.g., from 28 to 34), the quantity of amplified product increases, but the number of artifacts in the resulting profile increases as well. Sometimes, additional alleles can be seen while other alleles that should be seen are not. Contamination of the template DNA with other, lab-borne human DNA becomes a major issue under these conditions. Stutter peaks that can normally be detected and evaluated as such now appear as distinct peaks.

An alternative is to perform a technique known as nested PCR (Picken et al., 1996; Strom and Rechitsky, 1998; Yuan et al., 2003). In this procedure there are two rounds of PCR each using different sets of primers. There is also a modified nested PCR technique in which the same primers are used in both rounds of amplification. The first round is conducted with up to 40 cycles, and a portion of the amplified product is then reamplified using another 20–30 cycles. The problems encountered when testing DNA in low copy number (LCN DNA) include: (1) allelic dropout, (2) allelic drop-in (false alleles resulting from preferential amplification of artifactual products), and (3) contamination. Allelic dropout and allelic drop-in are the result of preferential amplification of one allele over another (Walsh et al., 1992a), not because one allele is longer than the other but, rather, because of stochastic variation due to the low number of template DNA molecules. This means that one molecule is amplified more than another due to chance during the first few cycles of DNA replication. For example, a heterozygote at a specific locus may be typed as a homozygote as a result of one allele being preferentially amplified. A false allele may be produced by preferential amplification of a stutter artifact. Contamination of the template DNA may take place as a result of environmental factors or analyst handling. DNA may be inadvertently transferred to evidentiary specimens. Furthermore, DNA may persist on objects long after they have been handled or exposed to biological substances (Wickenheiser, 2002). Increasing sensitivity of a test is generally considered an improvement or advancement. However, a test that is too sensitive may detect DNA that is irrelevant to the case but had found its way onto the evidence long before the crime had been committed. Thus interpretation of the resulting profile is crucial. Low copy number DNA analysis is not yet considered to be reliable, nor admissible as scientific evidence in the courtroom but can make a big difference in the investigative process. Innocent suspects can be excluded based on this kind of analysis.

3.4 ANALYSIS OF Y-CHROMOSOME STRs

The Y chromosome has been fully sequenced (Skaletsky et al., 2003) and contains about 59 million base pairs and only 161 coding genes. The coding genes are vital to male sex determination, spermatogenesis, and some other male-related functions. They are found on the short arm of this chromosome. The long arm contains virtually no coding genes and consists of noncoding DNA and repetitious sequences. Only about 5% of the Y chromosome can undergo recombination. This happens with sequences located at the telomeres (ends) of the Y chromosome. It does so with homologous regions on the X chromosome. The remainder of the chromosome (95%) is referred to as NRY (nonrecombining Y) (Whitfield et al., 1995).

Recently, Y chromosome STRs have been used to study sexual assault evidence. Most of these noncoding STR loci are located on the long arm (q) while the rest are found on the short arm (p). (see Fig. 3.41). These short repeat units are polymorphic (although less so than the STRs found on the autosomal chromosomes) and inherited in a patrilineal manner, which means that a male will transmit his Y chromosome and all of its genes virtually unchanged to all of his sons who will, in turn, transmit identical copies of the same chromosome to their male offspring. The flow of Y chromosome genes can be used to study the movement of populations over

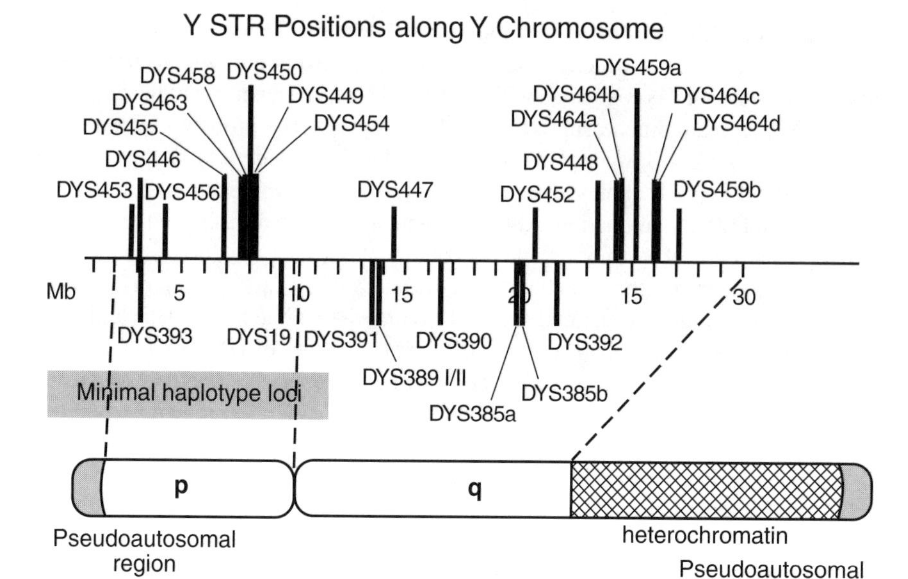

Fig. 3.41 *See color insert. Most of these Y-STR loci are located on the long arm (q) while the remainder are found on the short arm (p). Adapted from a power point presentation by Dr. Peter Valone, with his permission.*

vast time periods and to link ancestral fathers to their male offspring (Casanova et al., 1985; Underhill et al., 2000). They can also be helpful in establishing paternity. More importantly, from a forensic perspective, the STRs found on the Y chromosome are very useful for studying sexual predators. Sexual assault evidence such as vaginal swabs generally will consist of male and female components. In order to isolate the male DNA, a procedure known as a *differential lysis* is performed. In theory, sperm cells and female epithelial cells will be completely separated and then DNA can be isolated from each. However, in practice, the lysis method, at times, fails to separate cells with 100% effectiveness, and as a result female DNA can be present in significant amount in the male DNA fraction. When the ratio of female to male DNA reaches 10:1 or higher, the female (larger) component is preferentially amplified and thus the male (smaller) component is *masked.*

Masking does not occur in a sexual assault case when Y-STR loci are examined. In fact, studies have demonstrated that when the ratio of female DNA to male DNA is as high as 4000 to 1, there is still no apparent masking of the Y chromosome DNA whatsoever. Because female DNA lacks Y chromosome genes, the only contribution of Y chromosome genes must come from the assailant(s). Only the male component will be detected since only this DNA will be amplified. The Y-STR system is especially helpful when more than one male participates in a sexual assault. The mixed pattern in the evidence can help to identify those responsible for the assault.

Analysis of Y-STR loci can also be helpful in cases other than sexual assaults (Gamero et al., 2002). For example, let us assume that an evidentiary bloodstain is tested using PCR-STR multiplex kits, and it appears that there is a mixture of DNA from a female victim and a male suspect with the former contribution being significantly larger than the latter. Further evidence suggests that a second male was involved. Analysis of Y-STR loci reveals that there are indeed Y chromosome profiles from at least two male individuals. The much smaller contribution of the second male to the bloodstain had been masked in the first analysis by the victims DNA but revealed in the second.

The first Y-STR locus was discovered in 1992 (DYS19), but since then more than 44 have been identified, and there are indications that this number will exceed 200 in the near future (Roewer et al., 1992; Roewer and Epplen, 1992; Redd et al., 2002). All of these loci are said to be "linked" since they are all located on the same chromosome. Unlike the autosomal (nonsex chromosome) STR loci, there is no homologous chromosome for the Y chromosome to pair up with during cell division (mitosis), and therefore the alleles that are analyzed at the various Y-STR loci constitute a pattern known as a *haplotype.* Some laboratories are developing multiplex kits in which many such loci can be simultaneously amplified and identified (Meng et al., 2003). Like any other DNA analysis technique, these multiplex test systems must be validated and shown to provide reproducible and reliable results before they can be used in casework (Kayser et al., 1997; Gill et al., 2001).

What has been referred to as a *minimal haplotype* set requires two or three multiplex amplifications and includes DYS19, DYS389 I and II, DYS390, DYS391, DYS392, DYS393, and DYS385 I and II. An *extended haplotype* set consists of all of these plus YCAII a and b.

The Y-STR loci included in the *European minimal haplotype* are DYS19, DYS389 I and II, DYS390, DYS391, DYS392, DYS393, and DYS385 I and II. The Y-STR loci endorsed by the Scientific Working Group on DNA Analysis Methods include DYS19, DYS389 I and II, DYS390, DYS391, DYS392, DYS393, DYS385 I and II, DYS438, and DYS439.

Promega Corp. has recently developed a multiplex known as the PowerPlex Y System. This multiplex will co-amplify 12 Y-STR loci using 3-fluorochrome detection. The amplified loci include DYS19, DYS389 I and II, DYS390, DYS391, DYS392, DYS393, DYS385 I and II, DYS437, DYS438, and DYS439. The benefit of using this multiplex is that the 12 loci included in the kit contain the loci of the European minimal haplotype (see http://www.ystr.org), which has been endorsed by the International Society of Forensic Genetics (Gill et al., 2001), as well as two loci that have been endorsed by the Scientific Working Group on DNA Analysis Methods.

When analyzing autosomal STR loci, you can detect the presence of either 1 allele in a double dose (homozygote) or 2 different alleles, each in a single dose (heterozygote) at any locus. In theory, the analysis of 13 loci could result in the detection of as few as 13 alleles (1 at each locus) or as many as 26 alleles when the individual is heterozygous for all loci tested. The number is usually closer to 26 since STR loci used in forensic casework are chosen, in part, because they demonstrate a high level of heterozygosity. However, the Y chromosome STRs are haploid, which means that there are no heterozygotes (with the exception of Kleinfelter's syndrome, XYY). The finding of more than one allele at any of the Y-STR loci would indicate that more than one male contributed to the sample.

As discussed above, Y-STR alleles can be amplified either as individual systems or in combination as a multiplex or megaplex. ReliaGene Technologies has developed a multiplex in which 11 Y-STR loci are determined together with amelogenin. These are DYS19, DYS385a/b, DYS389I, DYS389II, DYS390, DYS391, DYS392, DYS393, DYS438, and DYS439. This multiplex provides results with only 0.1 ng of male DNA and is unaffected by the addition of as much as 700 ng of female DNA. In a mixed sample, it can detect 0.5 ng of male DNA combined with up to 400 ng female DNA. The simultaneous analysis of 11 loci would provide a determination of 11 alleles.

Twenty different Y-STR loci can be amplified simultaneously using a new multiplex system (20plex). These loci include DYS19, DYS385a/b, DYS388, DYS389 I/II, DYS390, DYS391, DYS392, DYS393, YCAII a/b, DYS426, DYS437, DYS438, DYS439, DYS447, DYS448, DYS460, and Y-GATA-H4.

Recently, a 21-locus Y-STR "Megaplex" system has been developed and validated (Hanson and Ballantyne, 2004). It augments the European Y chromosome "minimal haplotype" by including the following loci: DYS443, DYS444, DYS445, DYS447, DYS448, DYS449, DYS452, DYS453, DYS454, DYS455, DYS456, DYS458, DYS463, DYS464, DYS468, DYS484, DYS522, DYS527, DYS531, DYS557, and DYS588. The megaplex can provide good results with as little as 50 pg of male DNA, but 1 ng has been found to be optimal. For samples consisting of male and female contributors, amplification of female DNA will not take place even when the mixture contains concentrations of female DNA greater than 100 ng. This makes the Y-STR analysis method valuable, especially when autosomal STR analysis fails to provide genetic information about the assailant(s).

Because the mutation rate of Y-STRs is relatively high (average 2.80×10^{-3} per locus, per generation), it has been recommended that analysts should test at least nine loci to establish an inclusion (Kayser and Sajantila, 2001). Exclusion should be based on differences observed for DNA specimens at a minimum of three loci.

If the evidentiary sample Y-STR profile matches the profile of the suspect, a statistic should be determined to illustrate how common or how rare that profile is in the relevant population and what the probability would be of finding such a profile in unrelated individuals within different ethnic groups. Because all Y-STR loci are linked, the analyst cannot use the product rule to determine the frequency of the overall profile. If there is concordance between the evidentiary and suspect specimens, the statistic is calculated using the "counting" method. A database of Y chromosome STR haplotypes in different populations (European American,

African Americans, and Hispanics) is consulted to determine the frequency of the profile (Kayser et al., 2002). If there are 2000 individuals in the database and the results of the present analysis indicates a novel database, then the *frequency* is 1 out of 2001 (assuming that no haplotype is repeated more than once). Despite this statistical issue, as more Y-STRs become included in multiplex kits, the discrimination potential will become better and better. Analysts can use a statistical method based on confidence levels to describe the significance of the findings. See Sections 3.5.6 and 4.6.

3.4.1 Y-Chromosome Single-Nucleotide Polymorphism Analysis

Another approach to human identification by analysis of the Y chromosome is single-nucleotide polymorphism (SNP) analysis. SNPs result from the substitution of single base pairs on the Y chromosome over time. They occur as a result of DNA damage resulting in deamination, depurination, or other alterations of nucleotide bases. The DNA repair mechanism fails to correct the change resulting in the development of an SNP. The SNP mutation rate at various Y chromosome sites is very low (3×10^{-8} per site, per generation), which makes SNP analyses very useful for the study of population migration over long periods of time. Marligen Biosciences has developed the Signet Y-SNP Identification System, which uses multiplexes for both the amplification and detection steps and allows for the analysis of 43 SNPs (see References for website). Essentially, the Y chromosome sequences are amplified using PCR with fluorescently labeled primers, resulting in fluorescently labeled amplified products. These tagged products are allowed to hybridize to microspheres (color-coded beads) to which allele-specific oligonucleotide probes are covalently bound. The Luminex 100 flow reader instrument is used to analyze the bead mixes by recording the spectral characteristics of each bead as well as the amount of labeled product that has become bound to each. The kit also verifies gender using the amelogenin locus on the Y chromosome. The advantages and limitations of using the Y chromosome for human identification are outlined in Table 3.8.

TABLE 3.8 Advantages and Limitations of Y-Chromosome Markers at a Glance

Advantages	Limitations
1. Y-chromosome markers are useful in tracing family relationships among males. This is because all male progeny receive an intact Y chromosome. 2. Y-chromosome markers can be used to measure the relatedness of individuals of common geographic origin. This is because of the patrilineal transmission of the Y chromosome and the lack of recombination.	1. The discriminatory power of the Y-chromosome markers is limited by the size of the database used. This is because there is no recombination among the loci.

Source: National Commission on the Future of DNA Evidence, *The Future of Forensic DNA Testing Predictions of the Research and Development Working Group* (2000).

3.5 ANALYSIS OF MITOCHONDRIAL DNA

3.5.1 The Mitochondrial Genome

Organelles known as mitochondria exist in virtually every cell of the body functioning to convert chemical energy to adenosine triphosphate (ATP), the energy currency that drives most cellular reactions. This process occurs during cellular respiration as electrons are transferred along the cytochrome system and eventually transferred to oxygen. Because there are many similarities of mitochondria to bacteria, some theories suggest that a bacterium entered a primordial cell and established a mutually beneficial relationship (Saccone, 1994). As a result, this relationship was maintained and improved over a long period of time, resulting in the higher cells that we know today that typically contain hundreds of these mitochondria. It is estimated that somewhere between 200 and 400 mitochondria are present in most of our cells, but in some cells there may be thousands. Some nonhuman cells have been found to contain only a single giant mitochondrion. The mitochondrion is formed by two distinct membranes, an outer surface membrane and a highly folded inner membrane. The organelle contains two compartments, the intermembrane space and the matrix within the inner membrane (see Fig. 3.42). Mitochondria contain their own unique form of DNA, mtDNA, and they can replicate autonomously within the cell. There are, on average, 2.2 genomes within the matrix (inner compartment) of each mitochondrion. The genome is a closed circle of double-stranded, antiparallel, double-helical DNA (Fig. 3.43). The two strands of mitochondrial DNA are called "H" for heavy and "L" for light because the heavy strand is rich in the purines, adenine, and guanine whereas the light strand is rich in pyrimidines, cytosine, and thymine. Adenine and guanine have higher molecular weights than cytosine and thymine, which explains the nomenclature of light and heavy. Unlike nuclear DNA, which is passed along to offspring from both mother and father, mitochondrial DNA is maternally inherited (Hutchinson et al., 1974).

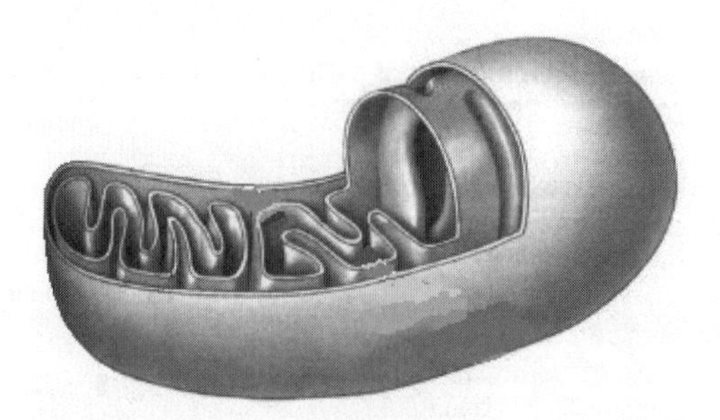

Fig. 3.42 See color insert. Mitochondrion contains two compartments, the intermembrane space and matrix within inner membrane. Courtesy of J. H. Koeslag from http://www.sun.ac.za/medphys/life.htm.

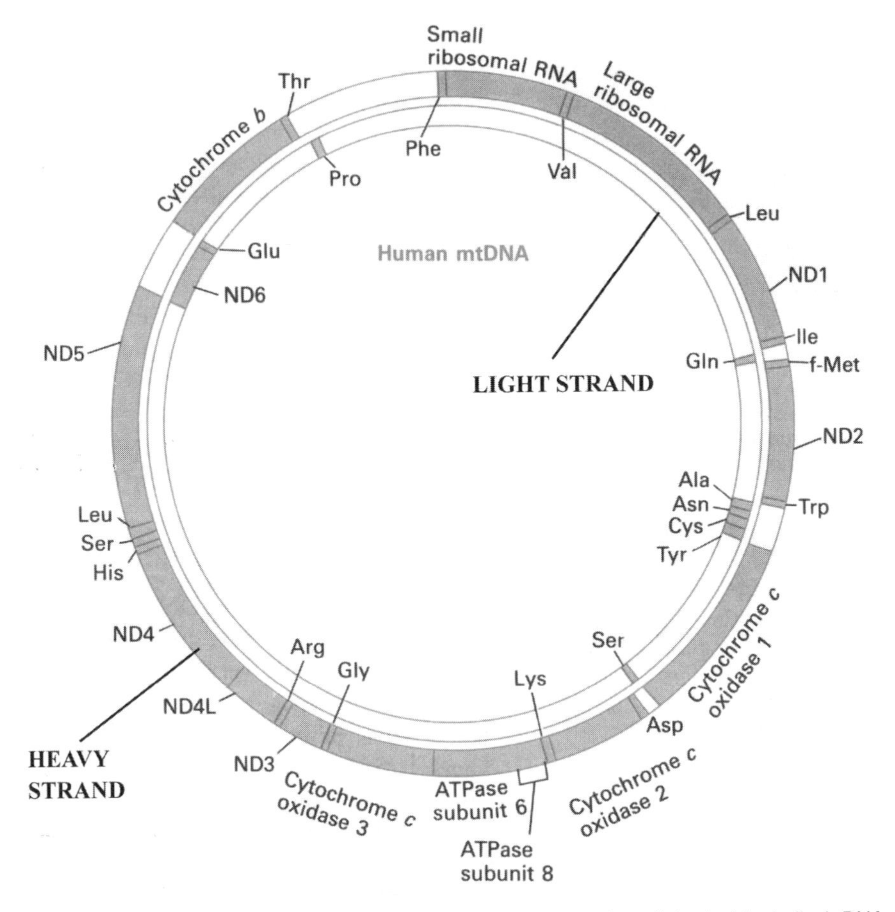

Fig. 3.43 *Genome is closed circle of double-stranded, antiparallel, double helical DNA.* Molecular Cell Biology, 3/e *by Harvey Lodish, et al. Copyright 1986, 1990, 1995 by W. H. Freeman and Company. Used with permission.*

The reason for this is that unlike nuclear DNA, mitochondrial DNA in the spermatozoa is almost never permitted to enter the ovum during the process of fertilization. The result is that mitochondrial DNA is passed along from mother to offspring (male and female) and is, in turn, transmitted from daughter to male and female offspring. This mode of inheritance is maternal. Thus, mitochondrial DNA passes unchanged, with the exception of spontaneous mutations, from mother to daughter from generation to generation. As a result of this matrilineal mode of inheritance, it is possible to determine genetic lineage from grandmother to grandchild even in the absence of the mother. This was most helpful in identifying biological relationships of children of the *Desaparacidos* in Argentina where internal conflict resulted in the death of thousands of people. Many parents simply disappeared, and their children were taken and adopted by military families. For the same reason, mtDNA analysis does not allow one to differentiate between maternal half-siblings nor cousins related through sisters. Unlike nuclear DNA that is transmitted to offspring by both parents resulting in pairs of chromosomes (homologous chromosomes), mitochondrial DNA is said to be *hemizygous* or "monoclonal." Because recombination cannot take place, the genome is transmitted virtually unchanged (Awadalla et al., 1999).

The mitochondrial genome consists of 16,569 bp. The genome contains the information for the synthesis of 13 proteins, 22 transfer RNAs, and 2 ribosomal RNAs. These forms of RNA are used to construct proteins within the mitochondrion. Most of the coding genes are found on the heavy strand. This portion of the mitochondrial genome is known as the coding region. The remainder of the genome contains a relatively small region known as the *control region* that serves to regulate the transcription of genes within the coding region as well as replication of the genome itself. Forensic interest has grown in mitochondrial DNA for two reasons. First, the control region, also known as the D-loop or displacement loop, within each mitochondrial genome is highly polymorphic, making it useful for human identification (Aquadro and Greenberg, 1983; Stoneking et al., 1991). The high level of polymorphism is due, in part, to errors of mitochondrial DNA replication and mutations that are not repaired. In evolutionary terms, mtDNA mutates 5–10 times faster than nuclear DNA. Many population studies have been performed (Melton and Stoneking, 1996; Melton et al., 1997a, 1997b, 1998; Budowle et al., 1999). In unrelated individuals there are approximately 1–2% (9–15) differences in base sequence within this region depending on ethnicity. This region contains approximately 1100 bp and is divided into 2 distinct regions known as hypervariable 1 (HV1) and hypervariable 2 (HV2). HV1 is found between bases 16,024 and 16,365. HV2 is found between bases 73 and 340. (see Fig. 3.44). Thus there are 610 nucleotide bases that exhibit polymorphism. The hypervariable regions of the mitochondrial D-loop can be selectively amplified using a number of different HV1- and HV2-specific primer sets in a multiplex system. After amplification of these regions, the products are sequenced and subsequently compared to a known reference sequence.

The second reason that forensic scientists are interested in mitochondrial DNA is that although it constitutes less than 1% of the total DNA within the cell, its genes exist in very high copy number. This includes, of course, the sequences present in the two hypervariable regions. If we assume that there are approximately 200 mitochondria within each cell (a very low estimate), and on average, each contains 2.2 genomes, then for every nuclear gene that appears in one or two copies, there can be more than 400 times as much DNA for any mitochondrial gene. Because this DNA is present in high copy number, it is very helpful when analyzing tissues that have little nuclear DNA such as hair, a very common item of evidence, especially when little or no root tissue is present to test. Although there is little usable nuclear DNA within the hair shaft, there is usually ample mitochondrial DNA within a centimeter of hair to successfully develop a profile (Wilson et al., 1995b). The examination of mitochondrial DNA in items such as teeth and bone, which are often found to contain degraded nuclear DNA, can still produce good results because of the high copy number of mitochondrial sequences within these cellular organelles (Budowle et al., 2000b; Miller, 2002). There has been a report (Balogh et al., 2003)

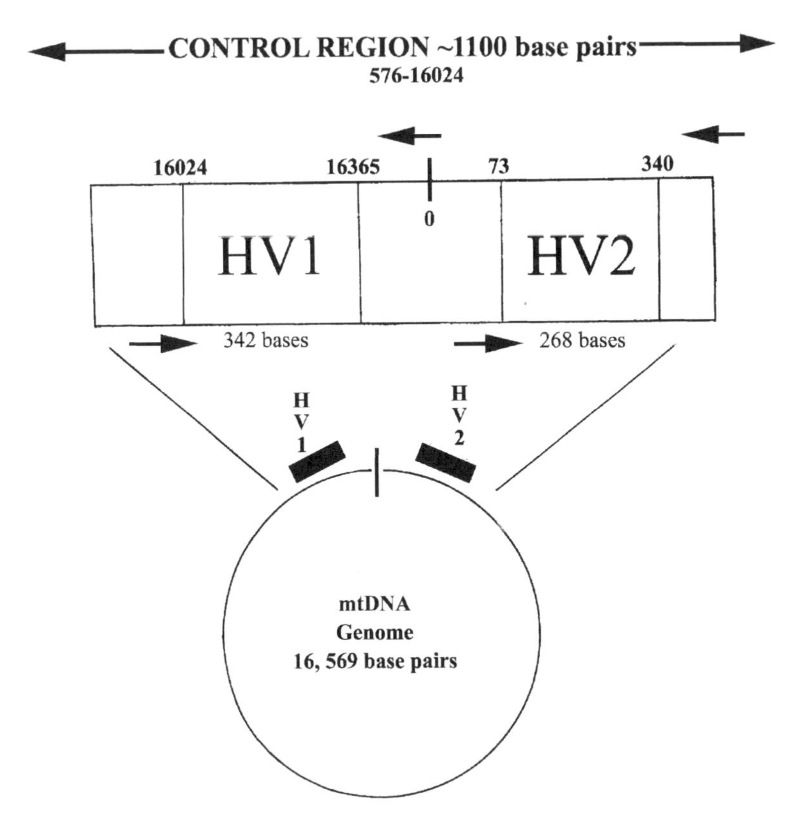

Fig. 3.44 *See color insert. Control region contains approximately 1100 bp and is divided into 2 distinct regions known as hypervariable 1 (HV1) and hypervariable 2 (HV2). HV1 is found between bases 16024 and 16365. HV2 is found between bases 73 and 340. Reprinted from* Forensic DNA Typing *by John M. Butler 1/ed, p. 124. Copyright 2001, with permission from Elsevier.*

that mitochondrial DNA sequencing was used to successfully type a latent fingerprint on paper. Mitochondrial DNA analysis has been performed in a number of high-profile cases (the Boston Strangler, the Green River murders, the Laci Peterson homicide), and in cases of historic importance as well (Russian Czar Nicholas II identification, the identification of the remains in the tomb of the Vietnam Unknown Soldier). In the Boston Strangler case mitochondrial DNA analysis was used to study the fingernails of Mary Sullivan who was murdered in January, 1964. Her body was exhumed in 2000 and her fingernails removed for analysis. MtDNA from the nails was easily differentiated from DNA of contaminating cells (Cline et al., 2003).

Holland and Parsons (1999) performed validation studies for the analysis of casework by mtDNA sequencing. The first use of mitochondrial DNA analysis in a U.S. courtroom took place in August 1996 in Tennessee. Paul Ware, a 27-year old was accused of raping and murdering a 4-year old girl. He had been found drunk and asleep next to the body of the child. None of the victim's blood was found on Ware. Nor was his semen found on the victim. At autopsy, a red hair was found in the throat of the child. There were also several similar hairs found on the bed sheet at the crime scene. MtDNA tests revealed that the hairs matched each other and also matched the defendant. He was convicted as a result of this evidence.

Mitochondrial DNA is studied by isolating the closed circular DNA, and then amplifying it by PCR using primers that are specific for hypervariable regions 1 and 2. Unlike nuclear DNA in which autosomal (nonsex) chromosomes undergo crossing over and recombination, the mitochondrial genome does not recombine and therefore is transmitted intact from mother to offspring as explained above. Mitochondrial DNA, like nuclear DNA, can be isolated and purified in a number of different ways including organic as well as inorganic methods. Nuclear DNA and mtDNA are both amplified. It is what happens after the amplification that makes mtDNA testing very different from nuclear DNA testing. The forensic analysis of STR loci on autosomal chromosomes is performed by amplification and detection of the products by equipment and software that measure fragment size and specific fluorescence emission. On the other hand, analysis of mitochondrial DNA is performed by amplification followed by sequencing of the amplified products using a genetic analyzer (DNA sequencer) such as the ABI Prism 310 or 377. The human mitochondrial genome was completely sequenced in 1981. Its sequence was known as the Anderson reference sequence (Anderson et al., 1981). A 1999 study corrected several minor errors, and the new standard is known as the Cambridge reference sequence (Andrews et al., 1999). Prior to sequencing the hypervariable regions, it is necessary to quantify the amount of amplified product so that the appropriate amount can be used for cycle sequencing. This amount is in the range of 20–35 ng mtDNA.

Known and unknown specimens are amplified using appropriate primers so that, subsequently, the HV1 and HV2 regions can be sequenced and compared to the reference sequence. Specific differences in nucleotide sequence between the reference and the known are noted and recorded. The same procedure is done with the reference and the unknown. The analyst then determines if there is concordance between the known and the unknown specimens by examining the recorded sequence differences. In other words, are the differences the same or not.

Ref:	a b c d e f g h i j k l m n o p
Known:	a b c $\underline{x}$ e f g h $\underline{y}$ j k l $\underline{z}$ n o p
Unknown:	a b c $\underline{x}$ e f g h $\underline{y}$ j k l $\underline{z}$ n o p

There have been a few studies on insect mtDNA, most of which have been attempts to identify the insect species, however, it has been demonstrated that it is possible to identify the human source of blood or tissue found in the gut of blood-sucking or carrion-feeding arthropods (maggots and insects) (Benecke, 1998; Benecke and Seifert, 1999; Wells et al., 2001). In this way an association can be established between a suspect and victim or crime scene.

3.5.2 Quantification

The amount of mitochondrial DNA (HV1and HV2) that has been amplified is determined either by gel electrophoresis or by capillary electrophoresis analysis. The amplification products are cleaned up by filtration using a Microcon YM-100 unit or using a Qiagen spin column (QIAquick). The Microcon unit has a nominal molecular weight limit of 100,000 daltons (Da). This procedure removes any excess primers, primer-dimers, or other small molecules such as dNTPs remaining after PCR amplification. Quantification of purified *amplicons* now becomes important because there must be a sufficient amount of DNA to undergo the next step, cycle sequencing, efficiently. The amplified product can be quantified by gel or capillary electrophoresis. If gel electrophoresis is used, the samples are electrophoresed in a 1% agarose gel containing ethidium bromide, which intercalates in the grooves of double-stranded DNA. Following electrophoresis, the gel is examined over a transilluminator. The position of the fluorescent band is compared to a positive control to determine if the product has the correct size. The amount of product can only be estimated based on the intensity of the fluorescent bands. If using capillary electrophoresis, the amplified products are diluted with water that contains a reference standard in the amount of 384 pg/μL. The mixture also contains ethidium bromide. The mixture is injected, and, as the fragments pass the window in the capillary, the argon laser excites the dye, causing it to emit light that is detected by a camera and a printout is obtained in the form of peaks rather than bands. The area of the mitochondrial DNA peak is divided by the area of the DNA quantity standard peak. The following equation is now applied:

$$\frac{\text{Area of mDNA peak/retention time}}{\text{Area of quantity standard/retention time}} \times \frac{9.6\,\text{ng}}{\mu\text{L}} = \frac{\text{ng mDNA}}{\mu\text{L}}$$

The benefit of using capillary electrophoresis to quantify the product is that you can determine not only its quantity and quality but also if there are any contaminants present such as primer-dimers, excess dNTPs, and so forth (Budowle et al, 2000b).

3.5.3 Sequencing

The amplified product is now ready for sequencing analysis, either manually or using computer-based sequencers. Two methods were developed to sequence nucleic acids in 1977. The first was developed by Maxam and Gilbert and the second by Sanger and Coulson. Although the Maxam–Gilbert method (1997) was easy to perform and favored by many, the Sanger–Coulson method (Sanger et al., 1997) was easily adapted for use with modern sequencing instrumentation. The sequencing reaction requires a denatured template DNA, sequencing primers, the four nucleotide building blocks (dATP, dGTP, dCTP, and dTTP) and four analogs of these nucleotides referred to as dideoxyterminators (ddATP, ddGTP, ddCTP, and ddTTP) each of which is labeled with a different fluorochrome, and finally a DNA polymerase. Because the *dideoxyterminators* lack a 3′OH group, once they become incorporated into a newly synthesized sequence the reaction stops at that point since no additional nucleotides can become linked to it. The reaction proceeds as follows.

The amplification products are denatured to form single-stranded template molecules. Sequencing is performed using only one strand at a time so the amplified product must first be denatured. Upon the addition of the sequencing primer and the DNA polymerase, the primer is extended by the sequential addition of nucleotide subunits complementary to the template strand until a dideoxyterminator is added. Any strand containing such a terminator ceases to elongate. The normal nucleotide subunits compete with the dideoxy analogs for incorporation into the elongating strand, for example, the dATP competes with the ddATP, and the dGTP competes with the ddGTP, and so forth. The process is known as cycle sequencing because during each extension a double-stranded molecule forms. It must be denatured so that another primer can anneal and another round of extension take place. Thus each cycle consists of denaturation, primer annealing, and extension. This continues repeatedly until a dideoxynucleotide is incorporated and the reaction for that newly created molecule ends. The sequencer utilizes a CCD camera that can detect the different fluorochromes present in each newly extended fragment and can easily read the four different fluorochromes simultaneously. The electrophoresis system employed, either gel or capillary, separates the end products of the reaction as they move across the detector. In short, the laser excites the fluorochrome and the detector records the particular fluorescence emitted from labeled fragment as it moves past the window and therefore indirectly determines the actual sequence of the template DNA one base at a time. Cycle sequencing can be used to process a fairly large fragment in hours rather than days or weeks. Mitochondrial DNA analysis is very labor intensive. Automation of the sequencing component would be advantageous for laboratories with a significant casework load (Hopgood et al., 1992).

The same equipment used for gene scanning of STRs (nuclear DNA) can also be used for sequencing analysis. Once the sequences of the two HV regions are determined for both evidence and exemplar, each is compared to the reference mitochondrial sequence (Anderson et al., 1981), and differences at specific positions of the genome are recorded. It is not unusual to find 5–10 nucleotide base differences between one (unrelated) individual and another.

3.5.4 Interpretation of Sequence Data

To declare that two specimens may have a common origin, the analyst must determine *how many* variations each sequence has relative to the Cambridge reference sequence and *which* variations are detected. Thus, if one finds that the questioned and known (exemplar) specimens have exactly the same sequence, there are six variations relative to the Cambridge reference sequence as follows:

Position-BASE
16,257T
16,311C
250C
73G
89C
200G

The sequences are concordant, and therefore you cannot exclude the questioned sample as having a common origin with the exemplar. If there is only one difference in sequence between the questioned and known specimens, then you still cannot exclude the questioned sample. On the other hand, if the analysis shows that there are two or more differences found, then you can exclude the questioned sample. Most forensic labs performing mitochondrial DNA analysis have adopted this principle (SWGDAM Guidelines, 2003).

There are instances where the sequence is ambiguous at a particular site. The sequencer denotes this ambiguity with the designation N as opposed to A, G, C, or T. In the case of heteroplasmy (see below), two different bases can be detected at the same nucleotide site, thus indicating that two populations of mitochondrial DNA are present. R is used to designate that an A and a G were observed; Y is used to designate that a T and a C were observed. As a rule, if one of the two specimens contains an N at a specific site while the other has a specific base called, then one cannot exclude the questioned and exemplar as having a common origin. If one finds, for example, an indication of G/C at a particular site for both exemplar and questioned specimen, this is another example of concordance and the analyst cannot exclude the questioned sample. The sequencing analysis of mtDNA in forensic casework has been validated (Wilson et al., 1995a; Isenberg, 2002).

3.5.5 Heteroplasmy

Generally speaking, when a spermatozoon fertilizes an ovum, the zygote, or product of the fusion, contains only maternal mitochondria. The sperm cell has a single spiraled-shaped mitochondrion within its neck segment located between head and tail. Thus as the zygote divides (cleavage) and develops into an embryo and then a fetus, all of its mitochondria contain copies of the mitochondrial DNA genome that were present in the ovum. Thus all cells of an individual contain the same mitochondrial DNA. However, about 1 in 10,000 times, the sperm head remains in contact with the ovum long enough for the male mitochondrion to enter the ovum along with the nucleus. Now the ovum contains two distinct forms of mitochondrial DNA. As a result of this phenomenon, cells of an individual can contain two distinct mitochondrial genomes (Bendall et al., 1997; Wilson et al., 1997; Morris and Lightowlers, 2000). It is also possible to have somatic mutations of mtDNA that result in two distinct populations of mtDNA in certain cells, tissues, or organs (Marchington et al., 1997). This is referred to as sequence *heteroplasmy*. It is possible to compare mitochondrial DNA sequences for two scalp hair shafts from the same individual and find that the sequence differs. Indeed recent studies show that heteroplasmy is even more common than first estimated. It is therefore very important to interpret sequence differences properly and correctly. It is all too easy to conclude that there is an exclusion when in reality, the two items have common origin. It is therefore advantageous for multiple exemplar specimens to be examined to determine if heteroplasmy is responsible for the apparent difference in sequence between known and unknown hairs.

For example, upon sequencing the L strand of a questioned and exemplar specimen, one finds that at a particular site, two different fluorescence signals for the questioned specimen are detected. The site is designated 300 G/A whereas the exemplar is designated 300A. The first sample is classified N since it is ambiguous but the second is designated A. In this case, there is no exclusion since the explanation can be attributed to heteroplasmy. If the second specimen is seen as 300 G/A, then there is certainly no exclusion.

There is a second type of heteroplasmy known as length heteroplasmy (Hauswirth and Clayton, 1985; Bendall and Sykes, 1995; Wilson et al., 2002). It results from slippage of the DNA polymerase during replication of a repetitious sequence. The result is that the replicated genome can contain either deleted bases or additional bases that have been inserted within the repeated sequence of bases. Within the control loop, there is a stretch of "polyC" (polyG on the opposite strand) that is subject to this phenomenon. Thus, if an error of replication occurs that does not get repaired, then the individual can have two distinct populations of mitochondrial DNA with one genome slightly larger than the other. Thus if there were three extra cytosine bases included following site 312, the type would be designated "312.3C." It should be noted that heteroplasmy within a single individual generally occurs at no more than one or two sites, whereas differences in sequences between individuals generally occurs at six to eight sites. It has been reported that heteroplasmy in the HV2 segment differs with tissue type examined. Furthermore, as the individual ages, there is an exponential accumulation of somatic mtDNA rearrangements and base substitutions (Calloway et al., 2000; Coskun et al., 2003). Because of this observation, some have theorized that mtDNA functions as the aging clock for humans and that a specific mutation of the mitochondrial genome is responsible for longevity.

3.5.6 Statistics

Mitochondrial DNA sequence profiles are stored in a database known as CODIS^{mt}, which contains data on more than 5000 individuals from all ethnic groups in the United States (Monson et al., 2002). Like any other database, the unknown profile can be searched to determine if it has ever been reported before, and if so, then how many times. It must be remembered that mtDNA sequences in HV1 and HV2 are not unique to a single individual. Using the information developed from the database allows the analyst to describe how rare the profile is in the relevant population. Obviously, if the database is very small, then the statistical frequency will not carry much weight or significance, but the larger the database becomes, the more significant the statistic that can be reported to the jury.

In the case of a match between evidence and exemplar, the statistical conclusion must be provided. Unlike the statistical calculations using nuclear DNA, the analyst is unable to use the product rule since all nucleotides are located (linked) on the same chromosome. The significance of a match is expressed as 1 in the total number of mitochondrial DNA profiles that have been assembled in the database thus far. For example, if there are 2000 distinct profiles in the database and the present testing produces an unreported profile, then the frequency of this profile would be expressed as 1 in 2001 (see Section 4.6). If 5 such profiles have been reported in the database, then the statistic becomes 6 out of 2001.

When an exclusion cannot be established, some laboratories report their mitochondrial sequence statistics using the likelihood ratio while others use other statistical approaches, for example, Bayesian statistics. In the United States, laboratories that perform mitochondrial analyses prefer to present their results using confidence intervals to describe the significance of the match. Using this approach, the analyst can inform the jury that the questioned and exemplar sequences are concordant and therefore may have a common origin. Furthermore, that this particular profile would be seen in no more than 0.1% of the Hispanic population in the United States. This is described further in Section 4.6, which deals with statistics for Y-chromosome STRs and mtDNA.

3.5.7 SNP Analysis of Mitochondrial DNA

Sequencing of mitochondrial DNA is a laborious, time-consuming procedure (see Table 3.9). The development of a screening procedure would decrease the number of specimens requiring complete processing (Stoneking et al., 1991). Although there are a number of labs trying to develop rapid screening procedures, currently no validated assay exists for use in criminalistics. One approach to develop such a system (see Marligen Biosciences in the References) is to use fluorescently labeled

TABLE 3.9 Advantages and Limitations of Mitochondrial DNA Testing (mtDNA) at a Glance

Advantages	Limitations
1. Information can be obtained from extremely small amounts of DNA. This is because there are multiple mitochondria per cell, and this type of analysis can be used in conjunction with PCR methods.	1. Mitochondrial DNA analysis cannot be used to distinguish among members of a sibship or maternal relatives because results from analysis of all descendants through the female line will be identical.
2. mtDNA molecules are small and do not degrade as fast as nuclear DNA.	2. The discrimination power of mtDNA is limited by the size of the database because there is very little, if any, recombination.
3. mtDNA is useful in tracing family lineages due to the fact that mtDNA is inherited from one's mother. Each of the mother's children receives this mtDNA.	3. The occurrence of more than one mitochondrial type in a single cell or in a single person (also known as heteroplasmy) can complicate mtDNA analysis.
4. The discriminating power of mitochondrial DNA analysis is larger than a single nuclear locus. This is because most mitochondrial haplotypes are found only once in a database.	4. The discrimination power of mtDNA analysis is further limited by the size of the database because the entire mitochondrial genome is inherited as a unit. Because of this method of inheritance of the mtDNA genome, it is equivalent to a single nuclear locus.

Source: National Commission on the Future of DNA Evidence, The Future of Forensic DNA Testing Predictions of the Research and Development Working Group (2000).

allele-specific oligomers (ASO) bound to microspheres that can hybridize with fluorochrome-labeled amplified HV1 and HV2 product. The complex can be analyzed in an automated flow reader instrument (Luminex 100) that uses optical (fluorescence) detectors. In this screening technique, PCR amplifications of the two HV regions would be performed, and a significant number (30–40) of polymorphic sites within these regions would be analyzed using the ASO-bound microspheres and a database consisting of a relatively large number of individuals in each major population. A similar approach might have the allele-specific oligonucleotide probes immobilized on nylon membrane strips through a sequence of poly-T, much like the design of the HLA DQA1 test kits (Stoneking et al., 1991). The primers would be covalently linked at the 5′ end to biotin, and therefore the amplified products would be labeled with biotin. A streptavidin–horse radish peroxidase enzyme conjugate would be added, and the complex would enzymatically convert a colorless substrate such as tetramethyl benzidine to a blue-colored precipitate. This type of analysis will result in a number of individuals with unique SNP types as well as those who share mitotypes (Brandstatter et al., 2003). The power of discrimination for this SNP detection system approaches 99% when the appropriate number of specific SNP sites are analyzed. An additional method used to screen mtDNA for SNPs has recently been developed. It utilizes the SnaPshot ddNTP Primer Extension kit (Applied Biosystems, Foster City, CA) in two multiplex reactions. This method is based on the use of dideoxyterminator single-base extension of unlabeled oligonucleotide primers. The sample DNA is first amplified using PCR and then cleaned up to remove excess primers and dNTPs. The cleanup may be performed using the ExoSAP-IT kit (Amersham BioSciences, Piscataway, NJ), which contains two hydrolytic enzymes, exonuclease I and shrimp alkaline phosphatase, which will degrade any excess deoxyribonucleotides and primers that are present. SNP-specific primers and the SnaPshot-IT mix are then added to the amplified products. Cycle sequencing is then performed to enable SNP extension. The 3′ OH terminus of an unlabeled primer is extended by a single fluorochrome-labeled ddNTP. (There are no dNTPs in the reaction so that no further extension can take place.) Each dideoxyterminator is labeled with a different fluorochrome. The primers are designed to hybridize to the template strand immediately adjacent to the SNP. As a result the SNP becomes fluorochrome-labeled and the extension ends. The product is now cleaned up again using the ExoSAP IT enzymes and electrophoresed using a genetic analyzer. The software collects and analyzes the data resulting in SNP typing. SNPs can be detected not only in the control region but also in the coding region (93%) of the mitochondrial genome.

3.6 PROBLEMS WITH PCR

3.6.1 Contamination

One of the most important concepts in forensic science is the Locard exchange principle (Locard, 1934). Edmond Locard, born in France in 1877, is respected by forensic scientists around the world for his insights into crime and physical evidence. His most famous observation (translated from French) is that "It is impossible for a criminal to act, especially considering the intensity of a crime, without leaving traces of this presence." In other words, when two objects come into physical contact, there is an exchange of material between them. The quantity of material that is exchanged may be large or small, and, depending on the amount transferred, it may or may not be detectable. Biological evidence produced at a crime scene may be subject to contamination depending on its physical state, location, and prevailing environmental conditions. Blood, for example, is subject to contamination from the moment it leaves the body. Blood droplets or spatter may be deposited on soil, concrete, glass, hair, vegetable matter, synthetic or natural fibers, and the like. Substrates may be flat and hard or soft and porous. Blood may pool or be absorbed or adsorbed to the substrate. It will coagulate and become dry as the fluid portion of the blood evaporates. Depending on the nature of the substrate, this may or may not adversely affect the ability to obtain genetic information for individualization. Such a specimen could easily become contaminated simply by coming into contact with a substrate, for example, blood spatter on an item of clothing worn by an individual who had been profusely sweating (McNally et al., 1989a, 1989b).

Another source of contamination may occur from contact with bacteria or fungi, which are ubiquitous in the environment and which will grow in large numbers if environmental conditions are favorable and nutrients are available. These kinds of contamination can occur between the time that the crime was committed and the biological evidence was produced and the time that the scene investigators recognize and take possession of it. The danger of microbial contamination of biological evidence is that these organisms contain enzymes that are capable of destroying nucleic acid polymers and therefore may convert genomic DNA into a multitude of useless oligomers, short nucleic acid fragments. Bacterial and fungal DNA will not become amplified by PCR since the primers recognize only primate DNA.

Evidence may also become contaminated at the crime scene by the investigators themselves. Experienced crime scene investigators are trained to avoid such contamination by using gloves, shoe covers, protective garments, head coverings, and so forth and protecting the evidence so that there is no inadvertent transfer of material to or from the evidence. It is not uncommon to have both crime scene investigators and detectives working together at the scene. Their efforts should be coordinated so as to avoid contaminating the scene or the evidence at the scene. Each item of evidence should be individually packaged. Combining items in the same container can result in the transfer of material from one to the other, producing results that may confuse the investigation.

Contamination can also occur in the laboratory. DNA analysts should perform their procedures using a laminar flow hood and with instruments such as pipetting devices, pipette tips, and microtube openers designed to avoid airborne contamination. Evidence handling should be performed in a location isolated from the site where amplification is performed or where amplified product is utilized for testing. If amplified product contaminates an evidentiary specimen, it will be preferentially amplified owing to its relatively small size (Walsh et al., 1992a). Amplified products are very small and can easily become airborne. If these airborne molecules found their way into another amplification reaction, they would be preferentially amplified over the DNA that the analyst had intended to amplify and type.

The result of this "crossover" contamination of evidentiary DNA with another human DNA prior to amplification may result in a mixture in which the ratio of contaminant to evidence is either very low, very high, or someplace in between. Under these circumstances, the experimental observations may indicate either a "clean" specimen or a mixture and the conclusion may be incorrect. This is one reason for the use of extensive controls when performing experimentation or casework analysis. The inclusion of negative and positive controls together with the evidentiary and exemplar specimens can often resolve these questions. Once a laboratory is contaminated with amplified product, it is time consuming and difficult to clean up.

3.6.2 Degradation

Isolated DNA is a very long and very thin polymeric molecule. Because of this characteristic it is very fragile and will easily fragment. Care must be taken to isolate DNA in its unbroken, native, state. The earliest form of DNA analysis for legal purposes was restriction fragment-length polymorphism (RFLP) analysis (see Section 3.2) used by Dr. Alec Jeffreys to solve two murders in Leicester, England. The technique used to require a minimum of 100–200 ng of high-quality, high-molecular-weight (HMW) genomic DNA to produce a result. During the early 1990s RFLP was made more sensitive and could be performed with only 50 ng of HMW DNA. HMW DNA generally consists of fragments of DNA that are around 10–20 kb long. Without HMW DNA, RFLP results are incomplete since some genetic markers will provide results while others will not. Sometimes degraded DNA will produce artifacts that are hard to interpret and sometimes no useful results are produced at all. Biological contamination can sometimes result in degradation of HMW DNA.

On the other hand PCR can provide results even with moderately degraded DNA. The regions that are amplified are very small (hundreds of bases) compared to the thousands of bases needed to perform RFLP. This explains, in part, why PCR of DNA obtained from archeological specimens produces useful information. DNA degradation poses far more of a problem for RFLP than it does for PCR. Interestingly, soil contamination of a DNA specimen can cause total breakdown of genomic DNA. It is likely that the humic acid component of soil is responsible for this activity. It has also been reported that genomic DNA can be obtained from archived PCR product mixes and successfully multiplexed to analyze additional loci (Patchett et al., 2002).

In cases of mass fatalities, such as the World Trade Center (9/11/2001) attacks or Swissair Flight 111 crash (1998), identification of bodies and body parts is critically important. However, genetic profiling cannot always successfully identify victims. Severe degradation of DNA poses the most serious problem. Partial genetic profiles become the rule rather than the exception. Family members are asked to provide DNA samples either from the victim (hair on a hairbrush or check cells on toothbrushes, and so forth) or from themselves so that kinship can be established. A software application known as Bloodhound, now commercially available, is designed to perform kinship analysis with very large numbers of either complete or partial STR genetic profiles (Ananomouse Corp, San Francisco, CA). The software can also handle missing-persons cases at a national level.

3.6.3 Sunlight

When biological evidence is found at an outdoor crime scene, environmental conditions, aging, and substrate become factors in the ability to amplify DNA from different specimens using PCR techniques. Ultraviolet radiation is a component in sunlight that can directly interfere with the ability of the DNA polymerase enzyme to replicate DNA. It does so by forming thymine dimers (T-T) wherever thymine nitrogenous bases are situated adjacent to one another in a DNA sequence. Such *dimerization* prevents proper replication of the template sequence and therefore can prevent successful genotyping of evidentiary DNA.

3.6.4 Inhibitors

There are a number of substances known to inhibit the PCR reaction. Hematin (a breakdown product of hemoglobin), soil, calcium, leather, and a chemical component present in blue jeans interfere with some aspect of DNA template amplification (Akane et al., 1994). The extraction procedure serves two purposes in that it isolates DNA in preparation for amplification, and it also removes inhibitors that may be present in the specimen itself or in the substrate upon which the specimen has been deposited. There are a number of methods that can be used to remove such inhibitors prior to amplification. For example, the DNA IQ system developed by Promega uses a paramagnetic resin in the form of particles. DNA becomes bound to these particles and any impurities (inhibitors) can be removed by washing. Chelex 100 extraction is a good way to remove hematin present in a bloodstain.

3.6.5 Allelic Dropout—Null Alleles

One of the potential problems associated with PCR amplification of a DNA sequence is the failure to amplify one of the two alleles present in a heterozygote. Thus the individual will appear to be homozygous at that locus. Allelic dropout can occur if the site on the DNA template where one of the two primers normally binds becomes mutated. A single base pair change at the primer annealing site will result in failure to amplify that allele.

This problem became evident when an attempt was made to amplify the DNA of an individual who is heterozygous at a locus (e.g., VWA), using two different sets of primers, each set binding at different flanking sequences adjacent to the core repeats. With one set of primers, both alleles were observed, but with the second set of primers only one of the two alleles was detected. The missing allele is called a "null" allele since it is present but goes undetected.

Before allelic frequencies can be used to determine genotype frequencies in various populations, it must be determined that the population is in Hardy–Weinberg equilibrium. One way to determine this is to see if there is an excess of homozygotes observed at a specific locus relative to what is expected. The presence of null alleles can create the appearance that the population is out of equilibrium. For these reasons, for human identification testing, it is best to use loci that are not only highly polymorphic but also loci where their flanking region sequences are stable and not prone to mutation. Now that many laboratories are performing PCR STR analysis and entering their data into the CODIS databases, it is important that the primers used in each of these labs are standardized to avoid entering genetic profiles that may be incorrect because of the presence of null alleles.

Allelic dropout can also be a result of using too little template (<150 pg) in the PCR reaction mix resulting in selective amplification of one of the two alleles present in a heterozygote. This is due to stochastic effects that occur when only a small number of template molecules are vying for primer annealing. Again the result is an overestimate of homozygotes at that locus.

Allelic dropout can also occur when there is sample degradation at the locus being examined; there is a large size difference in the two alleles of a heterozygote. The polymerase enzyme preferentially amplifies the shorter allele, resulting in the failure to detect the large allele.

3.6.6 Human Error

Analysis of DNA has had a profound impact on the investigation of violent crime. It is used to develop a profile or signature of the individual from which the DNA originated. The information is digitized and entered into local and state database systems and ultimately into a national database. As a result of the establishment of state and national databases, an individual who may or may not be a suspect can sometimes be associated with biological evidence found at a crime scene. DNA evidence and expert testimony generally have a significant influence on jurors since it is widely accepted that DNA testing is highly reliable and, usually, the expert's conclusions are accepted as fact. Because of the extraordinary impact of such evidence on the outcome of a trial, it is fair to ask about the probability of a coincidental match. Perhaps an even more important question is the likelihood of a false positive (Kwok and Higuchi, 1989) when comparing a questioned (unknown) specimen with an exemplar (known specimen).

A false positive occurs when two specimens originate from different individuals but are reported to have matching genetic profiles. This can occur in a number of ways. For example, a written report may erroneously state that the two samples have identical profiles when they do not. Other possible problems that can lead to a false match include contamination during evidence collection or handling, sample labeling errors, pipetting errors, and even misinterpretation of test results. There is an extensive literature on the subject of databases, the use of the product rule, population statistics, and random match probabilities. However little has been published regarding the frequency or probability of human error, nor how much of a significant factor it really is in the determination of inclusion or exclusion. Every effort must be made to minimize the occurrence of human error.

False positives have been documented in proficiency testing records as well as for casework. In one instance, an analyst working on rape kit evidence accidentally switched an exemplar specimen from the victim and suspect resulting in a conclusion and a report that wrongfully stated that the suspect could have been a potential donor of the seminal stain evidence. In fact, he should have been excluded as the source of the evidence. A similar incident occurred in a 1995 case in which an analyst working for a private company switched two specimens and again included an individual as a potential semen donor when he should have been excluded. False-positive results can also occur when using a test with relatively low resolution. For example, many laboratories used a commercially available kit known as HLA DQ alpha (or A1). The earlier kits were able to detect the presence of alleles 1.1, 1.2, 1.3, 2, 3, and 4. However, they were unable to distinguish the subgroups of allele 4, which consists of three distinct alleles, 4.1, 4.2, and 4.3. If DNA evidence and a suspect's exemplar DNA are amplified at this locus and each was genotyped as "2, 4," then a match would be declared. Based on allelic frequencies, a statistic illustrating the frequency of this genotype in the population would be calculated and presented either in a report or in testimony to show the significance of the match. Had the more advanced HLA kit been used, the evidence might be typed as a 2, 4.1 and the suspect as a 2, 4.2/4.3. This would have been a clear exclusion. In the more advanced kit, alleles 4.2 and 4.3 remained indistinguishable; so the argument could still be made that had the evidence been 2, 4.2 and the suspect been 2, 4.3 a match would have been declared and a statistic calculated despite the fact that the suspect could not possibly have contributed the evidentiary specimen. Furthermore, studies conducted by the World Health Organization have determined that more than 20 alleles within the HLA DQA1 system exist. The commercial kits were designed to provide information about only 7 or 8 alleles. Clearly, the use of a technology that has low resolution can lead to highly misleading conclusions. In any case, testing only a single locus is insufficient to declare a meaningful inclusion but unquestionably can be used effectively if an exclusion has been found.

Laboratory-induced contamination can also impact negatively on test results. For example, if amplified DNA from a previous PCR amplification finds its way into known and unknown specimens, this DNA will preferentially become amplified and can result in confusing results that could be misinterpreted and lead to false conclusions. Because of the small size of amplified products, they compete favorably with larger DNA fragments for primer binding and overtake the reaction, with the result that the final product consists almost completely of copies of the contaminant.

Even if questioned and exemplar specimens are properly collected, handled, and labeled, a pipetting error in the field or in the laboratory can result in a sample being placed into the wrong tube or well of an electrophoretic gel, resulting in a banding pattern that appears to be perfectly acceptable to the analyst and to any other individual asked to view the resulting electrophoretic gel or electropherogram. The broad statement that "there are no false positives" is simply not the case since every case is unique, and, although the testing methodology has been found to be reliable and trustworthy, there is no guarantee that in a specific case the test results are infallible (Koehler, 1993; Thompson, 1995, 1997; Thompson et al., 2003; 2004, 2004; Brenner and Inman, 2004; Giannelli, 1998). False-positive results were obtained in proficiency tests as well as in actual casework in the early 1990s. The cause of these errors was either misinterpretation of test results or accidental switching of samples prior to analysis.

Proficiency testing is one way to find out how significant the issue of human error is in the analysis of casework evidence (TWGDAM, 1990). The purpose of proficiency testing is for the laboratory director to determine if analysts are performing techniques correctly and interpreting experimental results properly. Failure of a proficiency test would result in additional training and supervision. In a proficiency test, the analyst is provided with mock casework specimens and is asked to perform the analysis. The results could easily be checked to determine if an error had occurred. Blind proficiency testing is a preferred variation on the theme. In this procedure, the analyst is not told that he is being tested with mock evidence. Thus, there is less chance that the mock evidence will be treated any differently from authentic casework evidence. Blind proficiency is one way to address the issue of human error rates. Testing in this manner has been recommended by the National Research Council in its 1992 report and is also discussed in the TWGDAM guidelines and in the federal DNA Identification Act of 1994. Proficiency testing should be performed twice a year according to TWGDAM (SWGDAM) guidelines. Another way to address the issue of human error is to retain sufficient evidence whenever possible so that an independent analysis can be conducted and the results compared.

3.7 UNDERLYING FACTS AND ASSUMPTIONS IN FORENSIC DNA TESTING

There are a number of facts and assumptions made when using DNA analysis methods in a scientific-legal setting. Some of these include:

- Testing of *length-specific polymorphisms* generally is not complemented by testing of amplified fragment sequence. Thus, when two fragments at the same locus (amplified using the same primers) have the same length, they are considered identical. No confirmation is done to verify that the sequence of bases in the two fragments are, in fact, identical (see Section 3.3.4.3.2.8).
- Testing of sequence-specific polymorphism uses probes that can detect all common alleles at each locus (Ziegle et al., 1992).
- DNA obtained from nonhuman sources (other than primate) will not be amplified (Crouse and Schumm, 1995).
- Mutation (germ line and somatic) of the DNA at the loci tested does not take place at a significantly rapid rate. Statistical analysis and use of the product rule require stable gene frequencies.

- All tissues of an individual will contain the same DNA and therefore will provide the same analytical result. Mutation in tissues within certain malignant tumors and heteroplasmy of certain tissues seen when performing mitochondrial DNA analysis are exceptions to the rule.
- The DNA specimen donors have not undergone recent transplantation or blood transfusion.
- Human error is not a significant issue (see Section 4.17.3)
- Databases are appropriately constructed and sufficiently large so that even some rare alleles are included.
- Corrections in statistical calculations are made to compensate for the existence of subpopulations.
- The population is in Hardy–Weinberg equilibrium and there is linkage equilibrium for all loci tested.
- Statistics will be provided for all the major ethnic/racial groups in the relevant population in the relevant geographical area. Differences in gene frequencies between subgroups within a larger (racial) population are not greater than differences in these same gene frequencies observed between the different major racial groups.
- Suspects do not have identical twins or family members that may be involved (siblings and close family members have much in common genetically). When they become suspects, additional testing may be required.
- Testing results are reproducible and correct.
- Gene scanning results can be edited providing that there is reason to believe that there is an error on the electropherogram, that is, the existence of stutter patterns or "pull-up" peaks (see Section 3.3.4.3.2.3).
- When a match or inclusion is determined, the conclusion should always be expressed in terms of probabilities rather than certainties.
- In criminal matters, a nonmatch at even one locus should be interpreted as an exclusion over the entire genetic profile. Paternity determinations differ and require two or more differences for an interpretation of nonparentage).
- The number of forensically useful loci tested should be sufficient so as to minimize the possibility of a false inclusion.
- If two tests of differing resolution are available, then the higher resolution test should be performed with the second used only as a confirmatory test.
- Heteroplasmy observed in mitochondrial DNA does not make mtDNA testing unreliable but should be discussed, when relevant, in the report and/or when providing testimony (see Section 3.5.5). The proper interpretation of the observed results is critical.
- Mixtures can be detected and properly interpreted (Sections 3.3.4.3.2.7 and 4.13).
- Artifacts can be identified, interpreted, and edited out of electropherogram reports.

REFERENCES

Abrahamson, S. S. Chair of National Commission on the Future of DNA Evidence, The Future of Forensic DNA Testing, Predictions of the Research and Development Working Group. National Institute of Justice NCS 183697 November 200. Available online: http://www.ojp.usdoj.gov/nij/pubs-sum/1833697.htm.

Akane, A., K. Matsubara, H. Nakamura, S. Takahashi, and K. Kimura. Identification of the Heme Compound Copurified with Deoxyribonucleic Acid (DNA) from Bloodstains, a Major Inhibitor of Polymerase Chain Reaction (PCR) Amplification. *J. Forensic Sci.*, **39**(2), 362–372 (1994).

Akane, A., A. Seki, H. Shiono, H. Nakamura, M. Hasegawa, M., Kagawa, K. Matsubara, Y. Nakahori, S. Nagafuchi, and Y. Nakagome. Sex Determination of Forensic Samples by Dual PCR Amplification of an X-Y Homologous Gene. *Forensic Sci. Int.*, **52**(1), 43–148 (1992).

Akane, A., H. Shiono, K. Matsubara Y. Nakahori, A. Seki, S. Nagafuchi, M. Yamada, and Y. Nakagome. Sex Identification of Forensic Specimens by Polymerase Chain Reaction (PCR): Two Alternative Methods. *Forensic Sci. Int.*, **49**, 81–88 (1991).

Allen, M., T. Saldeen, U. Pettersson, and U. Gyllensten. Genetic Typing of HLA Class II Genes in Swedish Populations: Application to Forensic Analysis. *J. Forensic Sci.*, **38**(3), 554–570 (1993).

Alper, B., E. Meyer, M. Schurenkamp, and B. Brinkmann. HumFES/FPS and HumF13B: Turkish and German Population Data. *Int. J. Legal Med.*, **108**(2), 93–95 (1995).

Anderson, S., A. T. Bankier, B. G. Barrell, M. H. de Bruijn, A. R. Coulson, J. Drouin, I. C. Eperon, D. P. Nierlich, B. A. Roe, F. Sanger, P. H. Schreier, A. J. Smit, R. Staden, and I. G. Young. Sequence and Organization of the Mitochondrial Genome. *Nature*, **290**, 457–465, 1981).

Andrews, R. M., I. Kubacka, P. F. Chinnery, R. N. Lightowlers, D. M. Turnbull, and N. Howell. Reanalysis and Revision of the Cambridge Reference Sequence for Human Mitochondrial DNA. *Nature Genet.*, **23**(2), 147 (1999).

Anker, R., T. Steinbrueck, and H. Donis-Keller. Tetranucleotide Repeat Polymorphism at the Human Thyroid Peroxidase (hTPO) Locus. *Hum. Mol. Genet.*, **1**(2), 137 (1992).

Aquadro, C. F. and B. D. Greenberg. Human Mitochondrial DNA Variation and Evolution: Analysis of Nucleotide Sequences from Seven Individuals. *Genetics* **103**, 287–312 (1983).

Awadalla, P., A. Eyre-Walker, and J. M. Smith. Linkage Disequilibrium and Recombination in Hominid Mitochondrial DNA. *Science*, **286**. 2524–2525 (1999).

Bacher, J. and J. W. Schumm. Development of Highly Polymorphic Pentanucleotide Tandem Repeat Loci with Low Stutter. *Profiles in DNA*, Promega Corp., 3–5 (1998).

Baechtel, F. S., K. W. Presley, and J. B. Smerick. D1S80 Typing of DNA from Simulated Forensic Specimens. *J. Forensic Sci.*, **40**(4), 536–545 (1995).

Balogh, M. K., J. Burger, K. Bender, P. M. Schneider, and K.W. Alt. STR Genotyping and MtDNA Sequencing of Latent Fingerprint on Paper. *Forensic Sci. Int.*, **137**(2–3), 188–195 (2003).

Barber, M. D. and B. H. Parkin. Sequence Analysis and Allelic Designation of the Two Short Tandem Repeat Loci D18S51 and D8S1179. *Int. J. Legal Med.*, **109**, 62–65 (1996).

Barber, M. D., B. J. McKeown, and B. H. Parkin. Structural Variation in the Alleles of a Short Tandem Repeat System at the Human Alpha Fibrinogen Locus. *Int. J. Legal Med.*, **108**, 180–185 (1996).

Barber, M. D., R. C. Piercy, J. F. Andersen, and B. H. Parkin. Structural Variation of Novel Alleles at the Hum vWA and Hum FES/FPS Short Tandem Repeat Loci. *Int. J. Legal Med.*, **108**, 31–35 (1995).

Bendall, K. E., V. A. Macaulay, and B. C. Sykes. Variable Levels of a Heteroplasmic Point Mutation in Individual Hair Roots. *Am J. Hum. Genet.*, **61**, 1303–1308 (1997).

Bendall, K. E. and B. C. Sykes. Length Heteroplasmy in the First Hypervariable Segment of the Human MtDNA Control Region. *Am. J. Hum. Genet.*, **57**, 248–256 (1995).

Benecke, M. Random Amplified Polymorphic DNA (RAPD) Typing of Necrophageous Insects (Diptera, Coleoptera) in Criminal Forensic Studies: Validation and Use in Praxi. *Forensic Sci. Int.*, **98**, 157–168 (1998).

Benecke, M. and B. Seifert. Forensische Entomologie am Beispiel eines Tötungsdeliktes. Eine kombinierte Spuren- und Liegezeitanalyse. (Forensic entomology in a high profile murder case: a combined stain, and post mortem interval analysis.). *Archiv Kriminologie*, **204**, 52–60 (1999).

Brandstatter, A., T. J. Parsons, and W. Parson. Rapid Screening of MtDNA Coding Region SNPs for the Identification of West European Caucasian Haplogroups. *Int. J. Legal Med.*, **117**(5), 291–298 (2003).

Brenner and K. Inman. Commentary on the Thompson article. *J. Forensic Sci.*, **49**(1), 192–193 (2004).

Brinkmann, B., A. Sajantila, H. W. Goedde, H. Matsumoto, K. Nishi, K., and P. Wiegand. Population Genetic Comparisons Among Eight Populations Using Allele Frequency and Sequence Data from Three Microsatellite Loci. *Eur. J. Hum. Genet.*, **4**, 175–182 (1996).

Brookfield, J. F. Statistical Issues in DNA Evidence. *Electrophoresis*, **16**(9), 1665–1669 (1995).

Budowle, B. and A. M. Giusti. Fixed Bin Frequency Distributions for the VNTR Locus D5S110 in General United States Reference Databases. *J. Forensic Sci.*, **40**(2), 236–238 (1995).

Budowle, B., G. Carmody, R. Chakraborty, and K. L. Monson. Source Attribution of a Forensic DNA Profile. *Forensic Sci. Commun.*, **2**(3) (July 2000a). Available online: http://www.fbi.gov/hq/lab/fsc/bookissn/july2000/index.htm

Budowle, B., J. Smith, T. Moretti, and J. Dizinno. DNA Typing Protocols: Molecular Biology and Forensic Analysis. In *Biotechniques Books*. Eaton, Natick, MA. 2000b.

Budowle, B., M. R. Wilson, J. A. diZinno, C. Stauffer, M. A. Fasano, M. M. Holland, and K. L. Monson. Mitochondrial DNA Regions HVI and HVII Population Data. *Forensic Sci. Int.*, **103**, 23–35 (1999).

Budowle, B., F. S. Baechtel, J. B. Smerick, K. W. Presley, A. M. Giusti, G. Parsons, M. C. Alevy, and R. Chakraborty. D1S80 Population Data in African Americans, Caucasians, Southeastern Hispanics, Southwestern Hispanics, and Orientals. *J. Forensic Sci.*, **40**(1), 38–44 (1995a).

Budowle, B., F. S. Baechtel, C. T. Comey, A. M. Giusti, and L. Klevan. Simple Protocols for Typing Forensic Biological Evidence: Chemiluminescent Detection for Human DNA Quantification and RFLP Analyses and Manual Typing of PCR Amplified Polymorphisms. *Electrophoresis* **16**(9), 1559–1567 (1995b).

Budowle, B., J. A. Lindsey, J. A. DeCou, B. W. Koons, A. M. Giusti, and C. T. Comey. Validation and Population Studies of the Loci LDLR, GYPA, HBGG, D7S8, and GC (PM Loci), and HLA DQ Alpha Using a Multiplex Amplification and Typing Procedure. *J. Forensic Sci.*, **40**(1), 45–54 (1995c).

Budowle, B., K. L. Monson, A. M. Giusti, and B. L. Brown. The Assessment of Frequency Estimates of *Hae*III-Generated VNTR Profiles in Various Ethnic Databases. *J. Forensic Sci.*, **39**(2), 319–352 (1994a).

Budowle, B., K. L. Monson, A. M. Giusti, and B. L. Brown. Evaluation of Hinf I-Generated VNTR Profile Frequencies Determined Using Various Ethnic Databases. *J. Forensic Sci.*, **39**(4), 988–1008 (1994b).

Budowle, B., K. L. Monson, and A. M. Giusti. A Reassessment of Frequency Estimated of PvuII-Generated VNTR Profiles in a Finnish, an Italian, and a General U.S., Caucasian Database: No Evidence for Ethnic Subgroups Affecting Forensic Estimates. *Am. J. Hum. Genet.*, **55**(3), 533–539 (1994c).

Budowle, B., A. M. Giusti, J. S. Waye, F. S. Baechtel, R. M. Fourney, D. E. Adams, L. A. Presley, H. A. Deadman, and K. L. Monson. Fixed-Bin Analysis for Statistical Evaluation of Continuous Distributions of Allelic Data from VNTR Loci, for Use in Forensic Comparisons. *Am. J. Hum. Genet.*, **48**(5), 841–855 (1991).

Buel, E., G. Wang, and M. Schwartz. PCR Amplification of Animal DNA with Human X-Y Amelogenin Primers Used in Gender Determination. *J. Forensic Sci.*, **40**, 641–644 (1995).

Calloway, C. D., R. L. Reynolds, G. L. Herrin, and W. W. Anderson. The Frequency of Heteroplasmy in the HVII Region of MtDNA Differs across Tissue Types and Increases with Age. *Am. J. Hum. Genet.*, **66**, 1384–1397 (2000).

Casanova, M., P. Leroy, C. Boucekkine, J. Weissenback, C. Bishop, M. Fellous, M. Purello, G. Fiori, and M. Siniscalco. A Human Y-Linked DNA Polymorphism and its Potential for Estimating Genetic and Evolutionary Distance. *Science*, **230**(4732), 1403–1406 (1985).

Casarino, L., F. De Stefano, A. Mannucci, and M. Canale. HLA-DQA1 and Amelogenin Coamplification: A Handy Tool for Identification. *J. Forensic Sci.*, **40**, 456–458 (1995).

Cetus Corporation, AmpliType User Guide for the HLA DQα Forensic DNA Amplification and Typing Kit, 1990, Section—Interpretation, Cetus Corporation, Emeryville, CA.

Clark, J. Novel Non-Templated Nucleotide Addition Reactions Catalyzed by Prokaryotic and Eukaryotic DNA Polymerases. *Nucleic Acid Res.*, **16**, 9677–9686 (1988).

Clayton, T. M., J. P. Whitaker, R. Sparkes, and P. Gill. Analysis and Interpretation of Mixed Forensic Stains Using DNA STR Profiling. *Forensic Sci. Int.*, **91**, 55–70 (1998)

Cline, R. E., N. M. Laurent, and D. R. Foran. The Fingernails of Mary Sullivan: Developing Reliable Methods for Selectively Isolating Endogenous and Exogenous DNA from Evidence. *J. Forensic Sci.*, **48**(2), 328–333 (2003).

Comey, C. T., B. W. Koons, K. W. Presley, J. B. Smerick, C. A. Sobieralski, D. M. Stanley, and F. S. Baechtel. DNA Extraction Strategies for Amplified Fragment Length Polymorphism Analysis. *J. Forensic Sci.*, **39**, 1254–1269 (1994).

Comey, C. T., B. Budowle, D. E. Adams, A. L. Baumstark, J. A. Lindsey, and L. A. Presley. PCR Amplification and Typing of the HLA DQ Alpha Gene in Forensic Samples. *J. Forensic Sci.*, **38**(2), 239–249 (1993).

Coskun, P. E., E. Ruiz-Pesini, and D. C. Wallace. Control Region MtDNA Variants: Longevity, Climatic Adaptation, and a Forensic Conundrum. *Prof. Nat'l. Acad. Sci.*, **100**(5), 2174–2176 (2003).

Cosso, S. and R. Reynolds. Validation of the AmpliFLP D1S80 PCR Amplification Kit for Forensic Casework Analysis According to TWGDAM Guidelines. *J. Forensic Sci.*, **40**(3), 424–434 (1995).

Crouse, C. A. and J. Schumm. Investigation of Species Specificity Using Nine PCR-Based Human STR Systems. *J. Forensic Sci.*, **40**, 952–956 (1995).

Duewer, D. L., L. A. Currie, D. J. Reeder, S. D. Leigh, H. K. Liu, and J. L. Mudd. Interlaboratory Comparison of Autoradiographic DNA Profiling Measurements, Measurement Uncertainty and Its Propagation. *Anal. Chem.*, **67**(7), 1220–1231 (1995).

DNA Commission Recommendations. 1992 Report Concerning Recommendations of the DNA Commission of the International Society for Forensic Haemogenetics Relating to the Use of PCR-Based Polymorphisms. *Int. J. Legal Med.*, **105**(1), 63–64 (1992).

Edwards, A., H. A. Hammond, L. Jin, C. T. Caskey, and R. Chakraborty. Genetic Variation at Five Trimeric and Tetrameric Repeat Loci in Four Human Population Groups. *Genomics* **12**, 241–253 (1992).

Edwards, A., A. Civitello, H. Hammond, and C. T. Caskey. DNA Typing and Genetic Mapping with Trimeric and Tetrameric Tandem Repeats. *Am. J. Hum. Genet.*, **49**, 746–756 (1991).

Erlich, H. A. (ed.). *PCR Technology: Principles and Applications for DNA Amplification.* Stockton, New York, 1989.

Erlich, H. A., D. Gelfand, and J. J. Sninsky. Recent Advances in the Polymerase Chain Reaction. *Science* **252**, 1643–1651 (1991).

Fildes, N. and R. Reynolds. Consistency and Reproducibility of AmpliType PM Results Between Seven Laboratories: Field Trial Results. *J. Forensic Sci.*, **40**(2), 279–286 (1995).

Fregeau, C. J. and R. M. Fourney. DNA Typing with Fluorescently Tagged Short Tandem Repeats: A Sensitive and Accurate Approach to Human Identification. *Biotechniques*, **15**, 100–119 (1993).

Gamero, J., J. Romero, J. Gonzalez, M. Caralho, M. Anjos, F. C. Real, D. Vieira, and M. Vide. Y-Chromosome STR Haplotypes in Central-West African immigrant in Spain Population Sample. *J. Forensic Sci.*, **47**(2), 421–423 (2002).

Giannelli. P. C. Defense Tactics for DNA Litigation. *Profiles in DNA*, **2**(3), (1998).

Gill, P. Application of Low Copy Number DNA Profiling. *Croatian Med. J.*, **42**(3), 229–232 (2001).

Gill, P., C. Brenner, B. Brinkmann, B. Budowle, E. Mayr, M. Jobling, P. de Knijff, M. Kayser, M. Krawczak, N. Morling, B. Olaisen, V. Pascali, M. Prinz, L. Roewer, P. M. Schneider, A. Sajantila, and C. Tyler-Smith. International Society of Forensic Genetics Recommendations on Forensic Analysis Using Y-Chromosome STRs. *Int. J. Legal Med.*, **114**, 305–309 (2001).

Gill, P., J. Whitaker, C. Flaxman, N. Brown, and J. Buckleton. An Investigation of the Rigor of Interpretation Rules for STRs Derived from Less Than 100 pg of DNA. *Forensic Sci. Int.* **112**(1), 17–40 (2000).

Gill, P., R. Sparkes, R, Pinchin, T. Clayton, J. Whitaker, and J. S. Buckleton. Interpreting Simple STR Mixtures Using Allele Peak Areas. *Forensic Sci. Int.*, **91**, 41–53 (1998).

Gill, P., R. Sparkes, and C. Kimpton. Development of Guidelines to Designate Alleles Using an STR Multiplex System. *Forensic Sci. Int.*, **89**, 185–197 (1997).

Gill, P., A. J. Jeffreys, and D. J. Werrett. Forensic Applications of DNA Fingerprints, *Nature*, **318**, 577–579 (1985).

Griffin, H. G. and A. M. Griffin, (eds.). *PCR Technology: Current Innovations.* CRC Press, Boca Raton, FL, 1994.

Guttman, A. and N. Cooke. Effect of Temperature on the Separation of DNA Restriction Fragments in Capillary Gel Electrophoresis. *J. Chromatogr.*, **559**, 285–294 (1991).

Haff, L., J. G. Atwood, J. DiCesare, E. Katz, E. Picossa, J. F. Williams, and T. Woudenberg. A High-Performance System for Automation of the Polymerase Chain Reaction. *BioTechniques* **10**, 102–112 (1991).

Hanson, E. K. and J. Ballantyne. A Highly Discriminating 21 Locus Y-STR "Megaplex" System Designed to Augment the Minimal Haplotype Loci for Forensic Casework. *J. Forensic Sci.*, **49**(1), 40–51 (2004).

Hauge, X. Y. and M. Litt. A study of the Origin of "Shadow Bands" Seen When Typing Dinucleotide Repeat Polymorphisms by the PCR. *Hum. Mol. Genet.*, **2**(4), 411–415 (1993).

Hauswirth, W. W. and D. A. Clayton. Length Heterogeneity of a Conserved Displacement Loop Sequence in Human Mitochondrial DNA. *Nucleic Acid Res.* **13**, 8093–8104 (1985).

Herd, K., K. Irving, and A. Day. A Short Primer on STRs: Why Do Prosecutors Need to Learn about STRs? *Silent Witness Newsletter*, **4**(5), (1999).

Hicks, J. W. DNA Profiling: A Tool for Law enforcement. *FBI Law Enforcement Bull.*, **57**(8), 3 (1988).

Holland, M. M., D. L. Fisher, D. A. Lee, C. K. Bryson, and V. W. Weedn. Short Tandem Repeat Loci: Application to Forensic and Human Remains Identification. In *DNA Fingerprinting: State of the Science*, S. D. J. Pena, R. Chakraborty, J. T. Eppleton, and A.J. Jeffereys (eds). Birkhauser, Basel, Switzerland, 1993, pp. 267–274.

Holland, M. M. and T. J. Parsons. Mitochondrial DNA Sequence Analysis—Validation and Use for Forensic Casework. *Forensic Sci. Revi*, **11**, 21–49 (1999).

Hopgood, R., K. M. Sullivan, and P. Gill. Strategies for Automated Sequencing of Human Mitochondrial DNA Directly from PCR Products. *Biotechniques*, **13**(1), 82–92 (1992).

Hutchinson, C. A., J. E. Newbold, S. S. Potter, and M. H. Edgell. Maternal Inheritance of Mammalian Mitochondrial DNA. *Nature* **251**, 536–538 (1974).

Isenberg, A. R. Forensic Mitochondrial DNA Analysis. *FBI Law Enforcement Bull.*, **71**(8), 16–21 (2002).

Isenberg, A. R., B. R. McCord, B. W. Koons, B. Budowle, and R. O. Allen. DNA Typing of a Polymerase Chain Reaction Amplified D1S80/Amelogenin Multiplex Using Capillary Electrophoresis and a Mixed Entangled Polymer Matrix. *Electrophoresis*, **17**, 1505–1511 (1996).

Issaq, H., K. Chan, and G. Muschik. The Effect of Column Length, Applied Voltage, Gel Type and Concentration on the Capillary Electrophoresis Separation of DNA Fragments and Polymerase Chain Reaction Products. *Electrophoresis*, **18**, 1153–1158 (1997).

Iwasa, M., P. Wiegand, S. Rand, M. Schurenkamp, S. Atasoy, and B. Brinkman. Genetic Variation at 5 STR Loci in Subpopulations Living in Turkey. *Int. J. Legal Med.*, **110**(3), 170–172 (1997).

Jeffereys, A. J., V. Wilson, and S. L. Thein. Individual-Specific Fingerprints of Human DNA. *Nature* **316**, 76–79 (1985a).

Jeffreys, A. J., V. Wilson, and S. L. Thein. Hypervariable "Minisatellite" Regions in Human DNA. *Nature*, **314**, 67–73 (1985b).

Kasi, K., Y. Nakamura, and R. White. Amplification of a Variable Number of Tandem Repeats (VNTR) Locus (pMCT118) by the Polymerase Chain Reaction (PCR) and its Application to Forensic Science. *J. Forensic Sci.*, **35**, 1196–1200 (1990).

Kayser, M. and A. Sajantila. Mutations at Y-STR Loci: Implications for Paternity Testing and Forensic Analysis. *Forensic Sci. Int.*, **118**, 116–121 (2001).

Kayser, M., S. Brauer, S. Willuweit, H. Schadlich, M. A. Batzer, J. Zawacki, M. Prinz, L. Roewer, and M. Stoneking. Online Y-Chromosomal Short Tandem Repeat Haplotype Reference Database (YHRD) for U.S. Populations. *J. Forensic Sci.*, **47**, 513–519 (2002).

Kayser, M., A. Caglia, D. Corach, N. Fretwell, C. Gehrig, G. Graziosi, F. Heidorn, S. Hermann, B. Herzog, M. Hidding, K Honda, M. Jobling, M. Krawczak, K. Leim, M. Meuser, E. Meyer, W. Oesterreich, A. Pandya, W. Parson, A. Piccinini, A. Perez-Lezaun, M. Prinz, C. Schmitt, P.M. Schneider, R. Szibor, J. Teifel-Greding, Ge. Weichhold, P. de Knijff, and L. Roewer. Evaluation of Y Chromosome STRs—A Multicenter Study. *Int. J. Legal Med.*, **110**(3), 125–133 (1997).

Kimpton, C., P. Gill, E. D'Aloja, J. F. Andersen, W. Bar, S. Holgersson, S. Jacobsen, V. Johnsson, A. D. Kloosterman, M. V. Lareu. Report on the Second EDNAP Collaborative STR Exercise, European DNA Profiling Group. *Forensic Sci. Int.*, **171**(2), 137–152 (1995).

Kimpton, C. P., J. A. Walton, and P. Gill. A Further Tetranucleotide Repeat Polymorphism in the vWF Gene. *Hum. Mol. Genet.*, **1**(4), 287 (1992).

Koehler, J. J. Error and Exaggeration in the Presentation of DNA Evidence at Trial. *Jurimetics J.*, **34**, 21–39 (1993).

Krawchak, M., J. Reiss, J. Schmidtke, and U. Rosler. Polymerase Chain Reaction: Replication Errors and Reliability of Gene Diagnosis. *Nuc. Acids. Res.* **17**, 2197–2201 (1989).

Kunkel, T. A. DNA Replication Fidelity. *J. Biol. Chem.* **267**, 18251–18254 (1992).

Kwok, S. and R. Higuchi. Avoiding False Positives with PCR. *Nature* **339**, 237–238 (1989).

Laber, T. L., S. A. Giese, J. T. Iverson, and J. A. Liberty. Validation Studies on the Forensic Analysis of Restriction Fragment Length Polymorphism (RFLP) on LE Agarose Gels without Ethidium Bromide: Effects of Contaminants, Sunlight, and the Electrophoresis of Varying Quantities of Deoxyribonucleic Acid (DNA). *J. Forensic Sci.*, **39**(3), 707–713 (1994).

Lagerstrom, M., N. Dahl, L. Iselius, B. Backman, and U. Pettersson. Mapping of the Gene for X-Linked Amelogenesis Imperfecta by Linkage Analysis. *Am. J. Hum. Genet.* **46**, 120–125 (1990).

Lazaruk, K., P. Walsh, F. Oaks, D. Gilbert, B. Rosenblum, S. Menchen, D. Scheibler, H. Wenz, J. C. Holt, and J. Wallin. Genotyping of Forensic Short Tandem Repeat (STR) Systems Based on Sizing Precision in a Capillary Electrophoresis Instrument. *Electrophoresis*, **19**, 86–93 (1998).

Locard, E. La Police et les méthodes scientifiques (1934) page 8.

Mannucci, A., K. M. Sullivan, P. L. Ivanov, and P. Gill. Forensic Application of a Rapid and Quantitative DNA Sex Test by Amplification of the X-Y Homologous Gene Amelogenin. *Int. J. Legal Med.*, **106**, 190–193 (1994).

Marchington, D. R. E., G. M. Hartshorne, D. Barlow, and J. Poulton. Homopolymeric Tract Heteroplasmy in MtDNA from Tissues and Single Oocytes: Support for a Bottleneck. *Am. J. Hum. Genet.* **60**, 408–416 (1997).

Marligen Biosciences. Signet™ Mitochondrial DNA Screening System; available at http://www.marligen.com/products/signetmito.htm.

Marligen Biosciences. Signet™ Y-SNP Identification System; available at http://www.marligen.com/products/signetysnp.htm. This site lists hundreds of publications related to Y chromosome SNPs.

Maxam, A. M. and W. Gilbert. A New Method for Sequencing DNA. *Proc. Nat'l Acad. Sci.*, **74**, 560–564 (1997).

Mayrand, P. E., K. P. Corcoran, J. S. Ziegle, J. M. Robertson, L. B. Hoff, and M. N. Kronick. The Use of Fluorescence Detection and Internal Lane Standards to Size PCR Products Automatically. *Appl. Theor. Electrophoresis*, **3**, 1–11 (1992).

McNally, L., R. C. Shaler, M. Baird, I. Balazs, P. R. DeForest, and L. Kobilinsky. Evaluation of Deoxyribonucleic Acid (DNA) Isolated from Human Bloodstains Exposed to Ultraviolet Light, Heat, Humidity, and Soil Contamination. *J. Forensic Sci.*, **34**(5), 1059–1069 (1989a).

McNally, L., R. C. Shaler, M. Shaler, M. Baird, I. Balazs, L. Kobilinsky, and P. R. DeForest. The Effects of Environment and Substrata on Deoxyribonucleic Acid (DNA): The Use of Casework Samples from New York City. *J. Forensic Sci.* **34**(5), 1070–1077 (1989b).

Melton, T. and M. Stoneking. The Extent of Heterogeneity in Mitochondrial DNA of Ethnic Asian Populations. *J. Forensic Sci.* **41**, 587–598 (1996).

Melton, T., M. Batzer, S. Clifford, J. Martinson, and M. Stoneking. Evidence for the proto-Austronesian homeland in Asia: Mitochondrial and nuclear DNA variation in Taiwanese aboriginal tribes. *Am. J. Hum. Genet.*, **63**, 1807–1823 (1998).

Melton, T., C. Ginther, G. Sensabaugh, H. Soodyall, and M. Stoneking. The Extent of Heterogeneity in Mitochondrial DNA of Sub-Saharan African Populations. *J. Forensic Sci.*, **42**, 580–590 (1997a).

Melton, T., M. Wilson, M. Batzer, and M. Stoneking. Extent of heterogeneity in mitochondrial DNA of European populations. *J. Forensic Sci.*, **42**, 437–446 (1997b).

Meng, C. W., K. Both, and L. A. Burgoyne. Casework: A Y-STR Triplex for Use after Autosomal Multiplexes. *Forensic Sci. Commun.*, **5**(2), (2003). Available online: http://www.fbi.gov/hq/lab/fsc/bookissu/april2003//index.htm

Meselson, M. and R. Yaun. DNA Restriction Enzyme from E. coli. *Nature*, **217**, 1110–1114 (1968).

Miller. K. A. Identifying Those Remembered. *Scientist*, **16**(12), 40–42 (2002).

Mills, K. A., D. Even, and J. C. Murray. Tetranucleotide Repeat Polymorphism at the Human Alpha Fibrinogen Locus (FGA). *Hum. Mol. Genet*, **1**, 779 (1992).

Moller, A., E. Meyer, and B. Brinkmann. Different Types of Structural Variation in STRs: HumFES/FPS, HumVWA and HumD21S11. *Int. J. Legal Med.*, **106**, 319–323 (1994).

Momhinweg, E., C. Luckenbach, R. Fimmers, and H. Ritter. D3S1358: Sequence Analysis and Gene Frequency in a German Population. *Forensic Sci. Int.*, **95**(2), 173–178 (1998).

Monson, K. L., K. W. P. Miller, M. R. Wilson, J. A. DiZinno, and B. Budowle. The MtDNA Population Database: An Integrated Software and Database Resource for Forensic Comparison. *Forensic Sci. Commun.*, **4**(2), (2002); available at http://www.fbi.gov/fbilibrary/forensic science communications/backissues.

Morris, A. A. M., and R. N. Lightowlers. Mitochondrial DNA Recombination. *Lancet* **356**, 941 (2000).

Nakahori, Y., K. Hamano, M. Iwaya, and Y. Nakagome. Sex Identification by Polymerase Chain Reaction Using X-Y Homologous Primers. *Am. J. Med. Genet.*, **39**, 472–473 (1991a).

Nakahori, Y., O. Takenaka, and Y. Nakagome. A Human X-Y Homologous Region Encodes "Amelogenin." *Genomics*, **9**(2), 264–269 (1991b).

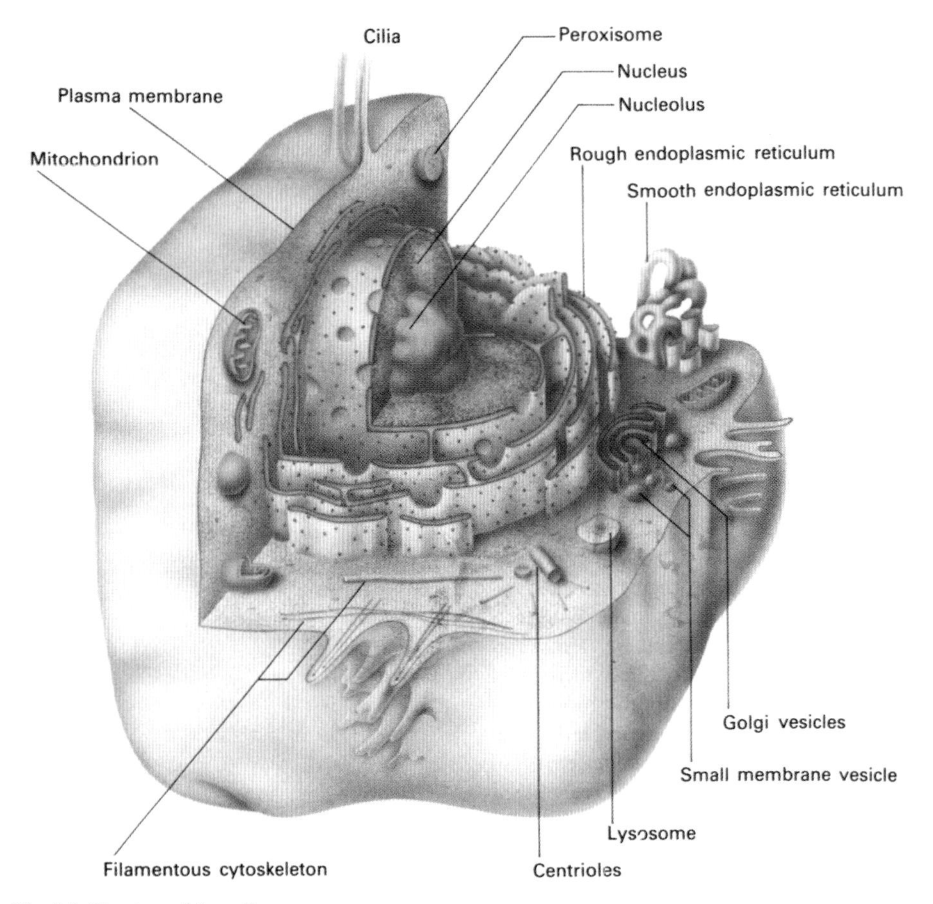

Fig. 1.1 Structure of the cell.

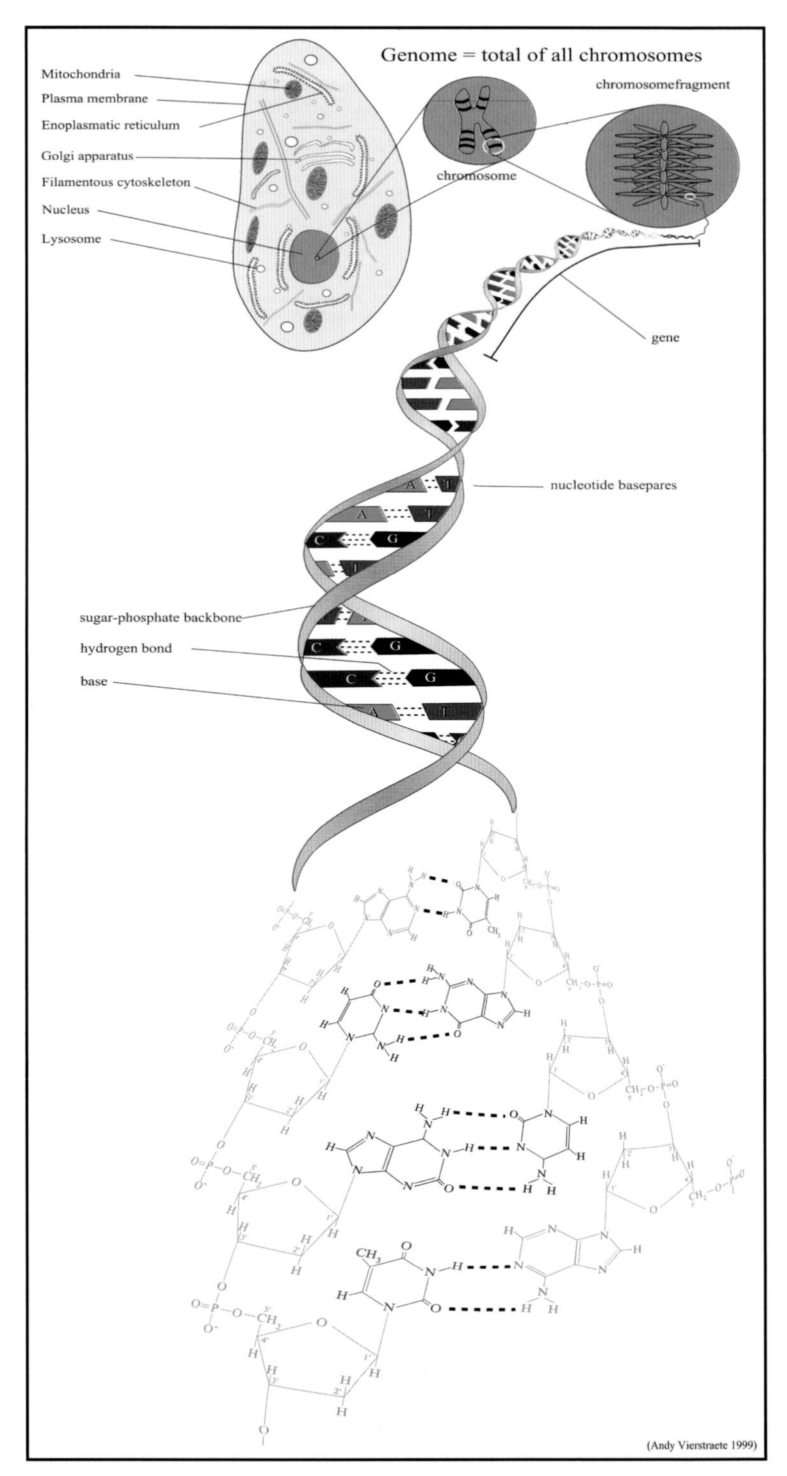

Fig. 1.5 Bases are held together by hydrogen bonds, two of which hold the A-T pair together while three bonds hold the G-C pair together.

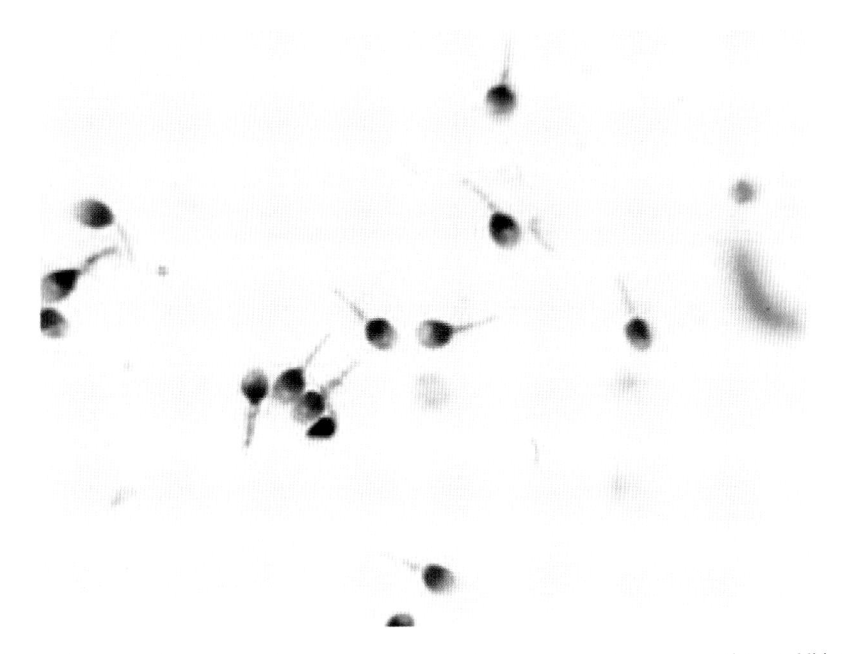

Fig. 2.6 Detection of spermatozoa in evidentiary stain eliminates need to perform presumptive or additional confirmatory testing since only semen contains these cells.

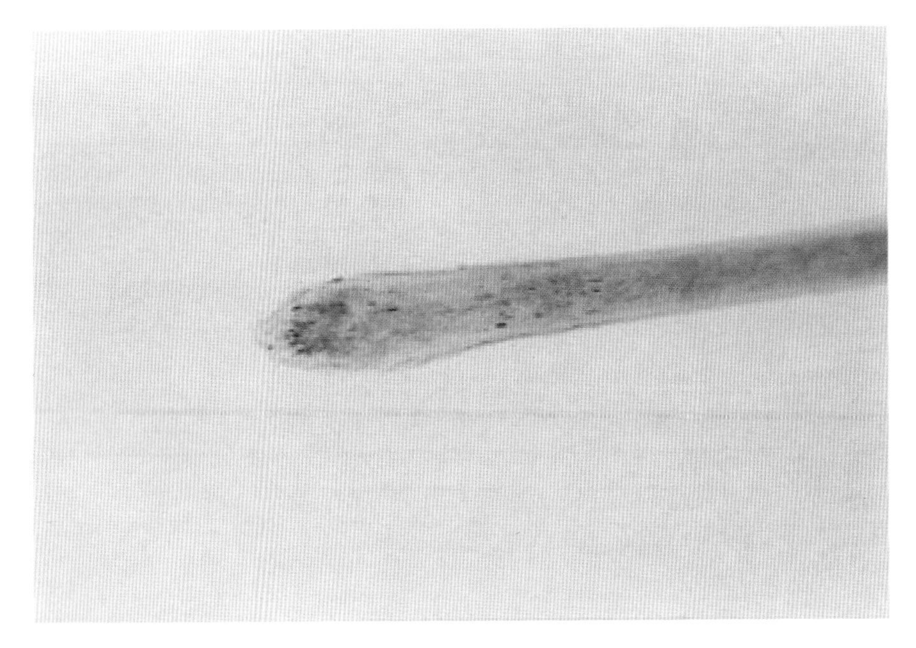

Fig. 2.7 Appearance of shaft and root of telogen hair.

Dot Blot Format

The amplified product is dotted onto the typing strip and the labeled allele specific probe is added to the strip. Where the probe recognizes the corresponding allele, a blue color will develop.

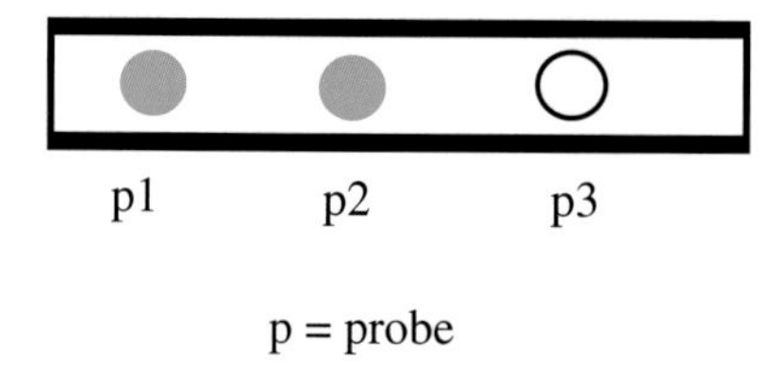

Reverse Dot Blot Format

Oligonucleotide probes that are specific for each allele are covalently bound to the typing strip. The labeled amplified product is added to the strip. If the probe hybridizes to the amplified alleles, a blue color will develop.

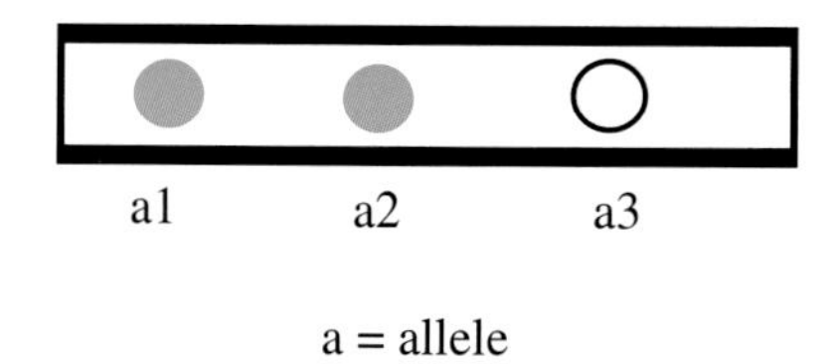

Fig. 3.25 Reverse dot blot format of testing.

HLA DQA1 TYPE

ALLELES 1 2 3 4 C 1.1 1.2, 1.3 & 4 1.3 All but 1.3 4.1 4.2 4.3

2, 3

1.1, 3

Fig. 3.27 (a) Examination of pattern of blue dots reveals phenotype (and genotype) of source of DNA specimen.

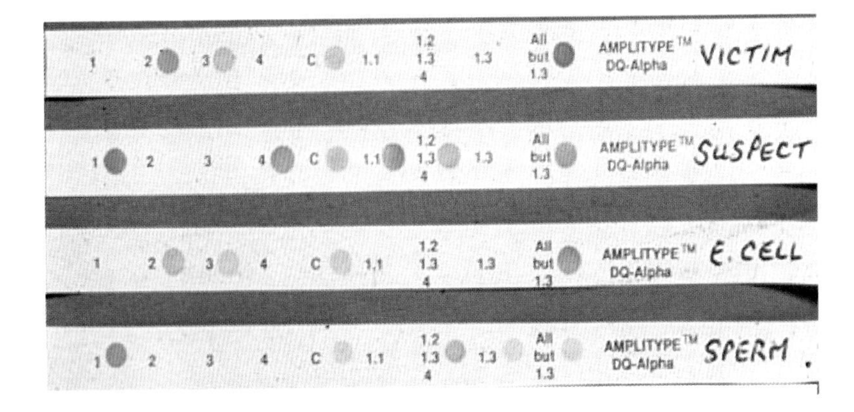

Fig. 3.27 (b) Comparison of patterns can be made to determine if suspect can be excluded.

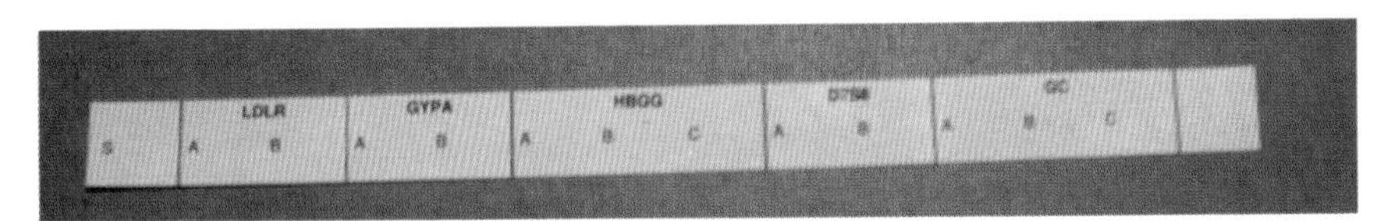

Fig. 3.28 Typing strips containing probes for AmpliType PM loci.

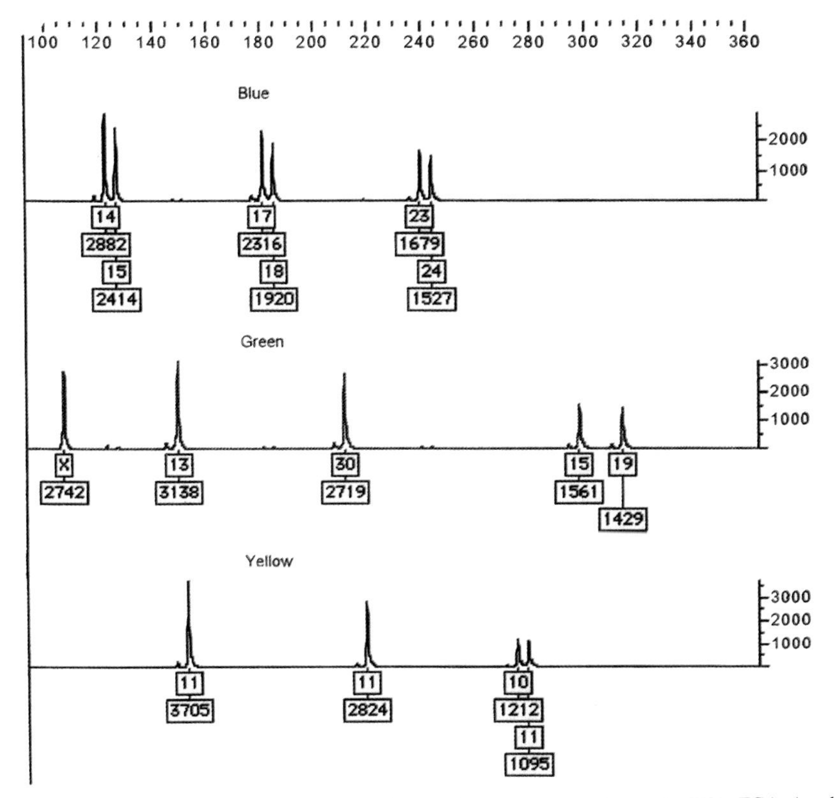

Fig. 3.32 AmpF/STR Profiler Plus is designed to simultaneously type D3S1358, VWA, FGA, Amel, D8S1179, D21S11, D18S51, D5S818, D13S317, and D7S820.

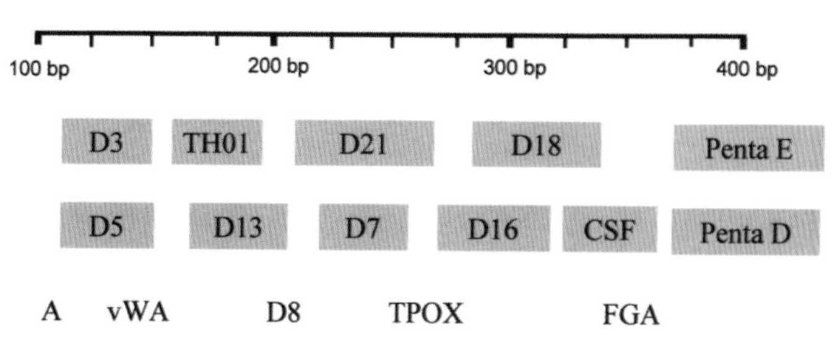

Fig. 3.33 Powerplex 16 developed by Promega Corp.

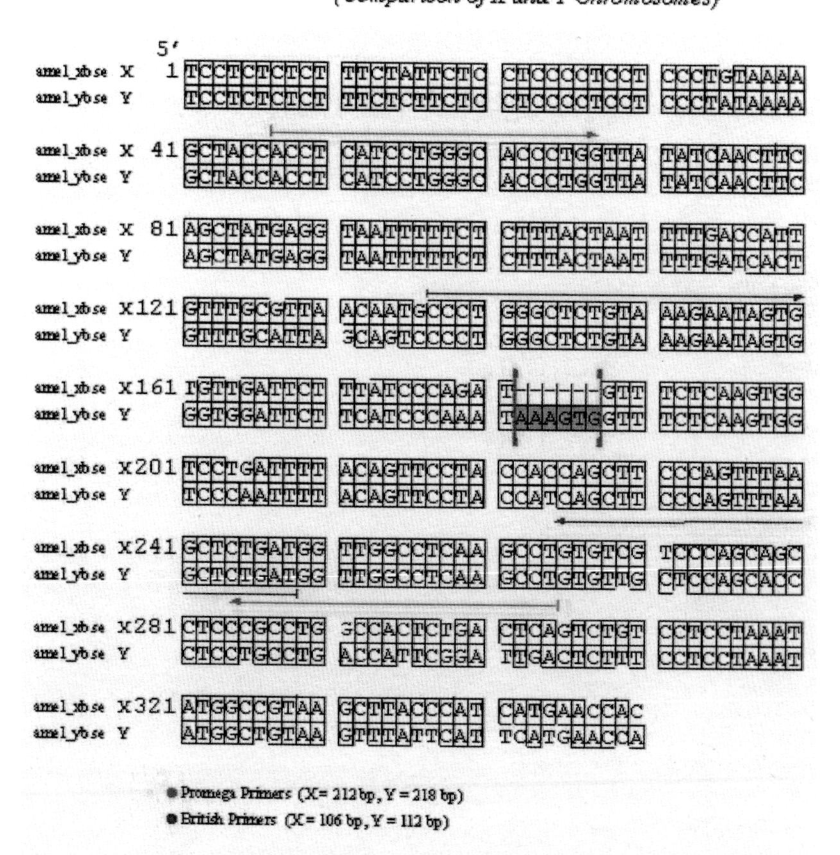

Fig. 3.34 Difference in amelogenin sequence on X and Y chromosomes.

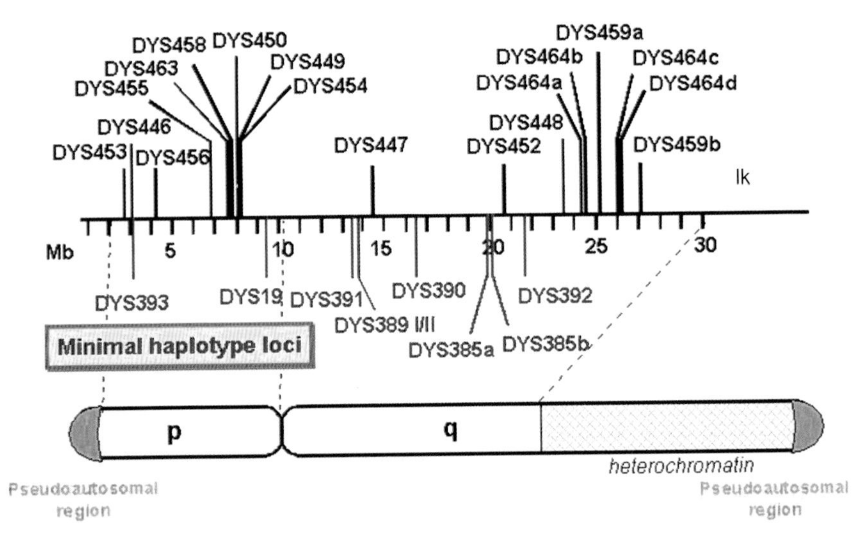

Fig. 3.41 Most of these Y-STR loci are located on the long arm (q) while the remainder are found on the short arm (p).

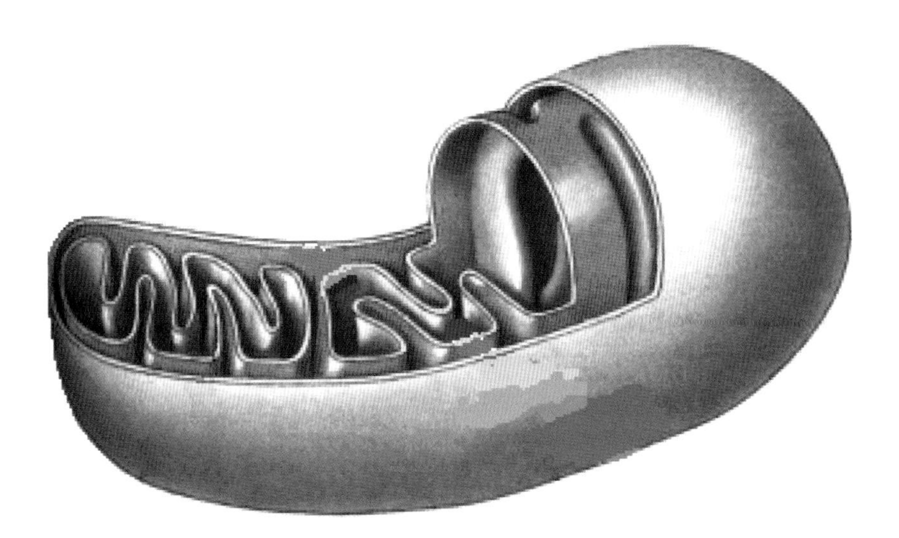

Fig. 3.42 Mitochondrion contains two compartments, the intermembrane space and matrix within inner membrane.

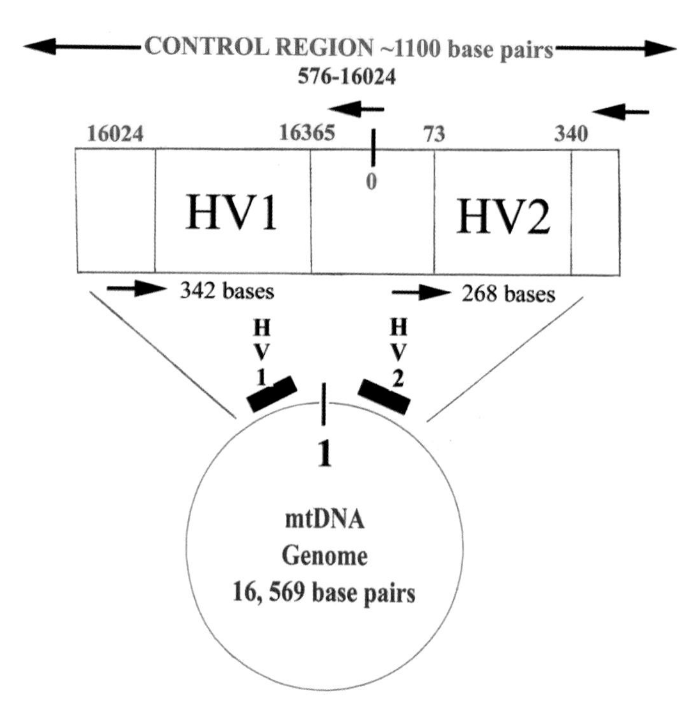

Fig. 3.44 Control region contains approximately 1100 bp and is divided into 2 distinct regions known as hypervariable 1 (HV1) and hypervariable 2 (HV2). HV1 is found between bases 16024 and 16365. HV2 is found between bases 73 and 340.

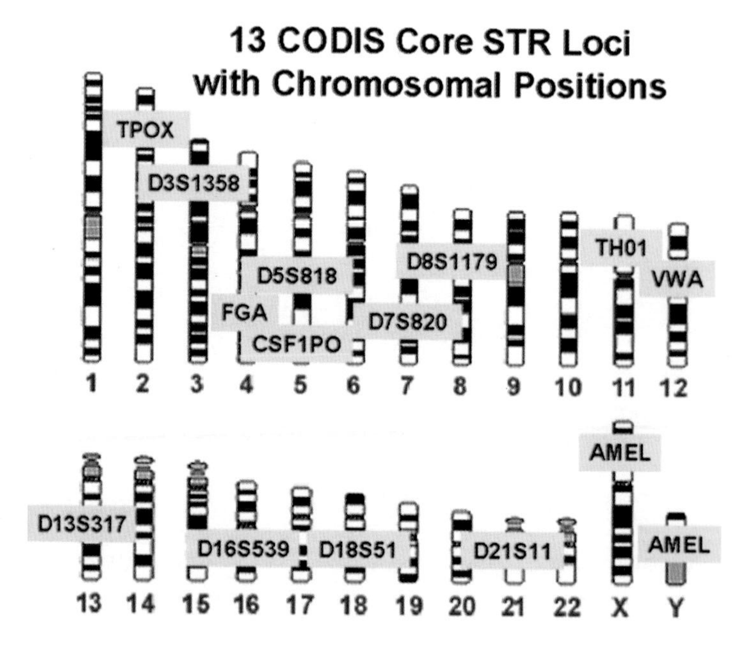

Fig. 4.3 The 13 CODIS loci.

National Commission on The Future of DNA Evidence. The Future of Forensic DNA Testing. Prediction's of The Research and Development working Group. National Institute of Justice NCJ 183697, November 2000. Available online: http://www.ojp.usdoj.gov/nij/pubs-sum/183697.htm

National Research Council, *The Evaluation of Forensic DNA Evidence*, Washington, D.C., National Academy Press, 1996. p. 63

Patchett, K. L., K. J. Cox, and D. M. Burns. Recovery of Genomic DNA from Archived PCR Product Mixes for Subsequent Multiplex Amplification and Typing of Additional Loci: Forensic Significance for Older Unsolved Criminal Cases. *J. Forensic Sci.*, **47**(4), 786–796 (2002).

Perlin, M. W. and B. Szabady. Linear Mixture Analysis: A Mathematical Approach to Resolving Mixed DNA Samples. *J. Forensic Sci.*, **46**(6), 1372–1378 (2001).

Pfitzinger, H., B. Ludes, P. Kintz, A. Tracqui, and P. Mangin. French Caucasian Population Data for HUMTH01 and HUMFES/FPS Short Tandem Repeat (STR) Systems. *J. Forensic Sci.*, **40**(2), 270–274 (1995).

Picken, M. M., R. N. Picken, D. Han, Y. Cheng, and F. Strle. Single-Tube Nested Polymerase Chain Reaction Assay Based on Flagellin Gene Sequences for Detection of Borrelia Burgdorferi Sensu Lato. *Eur. J. Clin. Microbiol. Infect. Dis.*, **15**(6), 489–498 (1996).

Presley, L. A., A. L. Baumstark, and A. Dixon. The Effects of Specific Latent Fingerprint and Questioned Document Examinations on the Amplification and Typing of the HLA DQ Alpha Gene Region in Forensic Casework. *J. Forensic Sci.*, **38**(5), 1028–1036 (1993).

Promega. GenePrint™ PowerPlex™ 1.1 System. Technical Manual. Part # TMD008, Madison, WI, 1999a.

Promega. GenePrint™ PowerPlex™ 1.2 System. Technical Manual. Part # TMD011, Madison, WI, 1999b.

Puers, C., H. Hammond, L. Jin, C. Caskey, and J. Schumm. Identification of Repeat Sequence Heterogeneity at the Polymorphic Short Tandem Repeat Locus HUMTHO1 [AATG]n and Reassignment of Alleles in Population Analysis by Using a Locus-Specific Allelic Ladder. *Am. J. Hum. Genet*, **53**, 953–958 (1993).

Recommendations of the DNA Commission of the International Society for Forensic Haemogenetics Relating to the Use of PCR-Based Polymorphisms. *Forensic Sci. Int.*, **1**, 1–3 (1992).

Redd, A. J., A. B. Agellon, V. A. Kearney, V. A. Contreras, T. Karafet, H. Park, P. de Knijff, J. M. Butler, and M. F. Hammer. Forensic Value of 14 Novel STRs on the Human Y-Chromosome *Forensic Sci. Int.*, **130**(2–3), 97–111 (2002).

Risch, N. J. and B. Devlin. On the Probability of Matching DNA Fingerprints. *Science*, **255**, 717–720 (1992).

Roberts. R. J. Restriction and Modification Enzymes and Their Recognition Sequences. *Nucleic Acids Res.*, **11**, r135–167 (1983).

Roewer, L., J. Arnemann, N. K. Spurr, K.-H. Grzeschik, and J. T. Epplen. Simple Repeat Sequences on the Human Y Chromosome are Equally Polymorphic as Their Autosomal Counterparts. *Hum. Genet.*, **89**, 389–394 (1992)

Roewer, L. and J. T. Epplen. Rapid and Sensitive Typing of Forensic Stains by PCR Amplification of Polymorphic Simple Repeat Sequences in Case Work. *Forensic Sci. Int.* **53**, 163–171 (1992).

Roffey, P.E., C. I. Eckhoff, and J. L. Kuhl. A Rare Mutation in the Amelogenin Gene and its Potential Investigative Ramifications. *J. Forensic Sci.*, **45**(5), 1016–1019 (2000).

Roy, R. and R. Reynolds. AmpliType PM and HLA DQ Alpha Typing From Pap Smear, Semen Smear, and Postcoital Slides. *J. Forensic Sci.*, **40**(2), 266–269 1995.

Saccone, C. The Evolution of Mitochondrial DNA. *Curr. Opin. Genet. Develop.*, **4**, 875–881 (1994).

Saiki, R. K., S. Scharf, F. Faloona, K. B. Mullis, G. T. Horn, H. A. Erlich, and N. Arnheim. Enzymatic Amplification of β-globin Sequences and Restriction Site Analysis for Diagnosis of Sickle Cell Anemia. *Science* **230**, 1350–1354 (1985).

Sajantila, A., P. Pacek, M. Lukka, A. C. Syvanen, P. Nokelainen, P. Sistonen, L. Peltonen, and B. Budowle. A Microsatellite Polymorphism in the von Willebrand Factor Gene: Comparison of Allele Frequencies in Different Population Samples and Evaluation for Forensic Medicine. *Forensic Sci. Int.*, **68**(2), 91–102 (1994).

Sajantila, A., M. Strom, B. Budowle, P. J. Tienari, C. Ehnholm, L. Peltonen [H. A. Erlich (ed)]. The Distribution of the HLA DQ Alpha Alleles and Genotypes in the Finnish Population as Determined by the Use of DNA Amplification and Allele Specific Oligonucleotides. *Int. J. Legal Med.*, **104**(4), 181–184 (1991).

Sajantila, A., B. Budowle, M. Strom, V. Johnsson, M. Lukka, L. Peltonen, and C. Ehnholm. PCR Amplification of Alleles at the DIS80 Locus: Comparison of a Finnish and a North American Caucasian Population Sample, and Forensic Casework Evaluation. *Am. J. Hum. Genet.* **50**(4), 816–825 (1992).

Salido, E. C., P. H. Yen, K. Koprivnikar, L. C. Yu, and L. J. Shapiro. The Human Enamel Protein Gene Amelogenin is Expressed from Both X and the Y Chromosomes. *Am. J. Hum. Genet.*, **50**, 303–316 (1992).

Sanger, F., S. Nicklen, and A. R. Coulson. DNA Sequencing with Chain-Terminating Inhibitors. *Proc. Nat'l. Acad. Sci. USA*, **74**, 5463–5468 (1997).

Schlesinger, L. B. *Sexual Murder: Catathymic and Compulsive Homicides*, CRC Press, Boca Raton, FL, 2004.

Schneider, P. M., R. Fimmers, S. Woodroffe, D. J. Werrett, W. Bar, B. Brinkmann, B. Eriksen, S. Jones, A. D. Kloosterman, B. Mevag, V. L. Paseali, C. Rittner, H. Schmitter, J. A. Thomson, and P. Gill. Report of a European Collaborative Exercise Comparing DNA Typing Results Using a Single Locus VNTR Probe. *Forensic Sci. Int.*, **49**(1), 1–15 (1991).

Schwartz, D. W. M., E. M. Dauber, B. Glock, and W. R. Mayr. AMPFLP-Typing of the D21S11 Microsatellite Polymorphism: Allele Frequencies and Sequencing Data in the Austrian Population. *Adv. Forensic Haemogenet.* **6**, 6222–6625 (1996).

Scientific Working Group on DNA Analysis Methods (SWGDAM). Guidelines for Mitochondrial DNA (MtDNA) Nucleotide Sequence Interpretation. *Forensic Sci. Commun.*, **5**(2) (2003). Available online: http://www.fbi.gov/hq/lab/fsc/bookissuapril2003/index.htm

Scientific Working Group on DNA Analysis Methods (SWGDAM). Short Tandem Repeat (STR) Interpretation Guidelines. *Forensic Sci. Commun.*, **2**(3) (2000). Available online: http://www.fbi.gov/hq/lab/fsc/backissujuly2000/index.htm

Sensabaugh, G. F. and C. von Beroldingen. The Polymerase Chain Reaction: Application to the Analysis of Biological Evidence. In *Forensic DNA Technology*, M. A. Farley and J. J. Harrington (eds.). Lewis, Chelsea, MI, 1991.

Sharma, V. and M. Litt. Tetranucleotide Repeat Polymorphism at the D21S11 Locus. *Hum. Mol. Genet.*, **53**, 953–958 (1993).

Sharma, B. R., M. Thompson, J. R. Bolding, Y. Zhong, L. Jin, and R. A. Chakraborty. Comparative Study of Genetic Variation at Five VNTR Loci Three Ethnic Groups of Houston, Texas. *J. Forensic Sci.*, **40**(6), 933–942 (1995).

Siles, B., G. B. Collier, D. J. Reeder, and W. E. May. The Use of a New Gel Matrix for the Separation of F: A Comparison Study Between Slab Gel Electrophoresis and Capillary Electrophoresis. *App. Theor. Electrophor.*, **6**, 15–22 (1996).

Skaletsky, H., T. Kuroda-Kawaguchi, P. J. Minx, H. S. Cordum, L. Hillier, L. G. Brown, S. Repping, T. Pyntikova, J. Ali, T. Bieri, A. Chinwlla, A. Delehaunty, K. Delehaunty, H. Du, G. Fewell, L. Fulton, R. Fulton, T. Graves, S. F. Hou, P. Latrielle, S. Leonard, E. Mardis, R. Maupin, J. McPherson, T. Miner, W. Nash, C. Nguyen, P. Ozersky, K. Pepin, S. Rock, T. Rohlfing, K. Scott, B. Schultz, C. Strong, A. Tin-Wollam, S. P. Yang, R. H. Waterston, R. K. Wilson, S. Rozen, and D. C. Page. The Male-Specific Region of the Human Y Chromosome Is a Mosaic of Discrete Sequence Classes. *Nature*, **423**(6942), 825–837 (2003).

Smith, J. C., R. Anwar, J. Riley, D. Jenner, A. F. Markham, and A. J. Jeffreys. Highly Polymorphic Minisatellite Sequences: Allele Frequencies and Mutation Rates for Five Locus-Specific Probes in a Caucasian Population. *J. Forensic Sci. Soc.*, **30**(1), 9–32 (1990).

Stoneking, M., D. Hedgecock, R. G. Higuchi, L. Vigilant, and H. A. Erlich. Population Variation of Human MtDNA Control Region Sequences Detected by Enzymatic Amplification and Sequence Specific Oligonucleotide Probes. *Am. J. Hum. Genet.*, **48**, 370–382 (1991).

Strom, C. M. and S. Rechitsky Use of Nested PCR to Identify Charred Human Remains and Min Amounts of Blood. *J. Forensic Sci.*, **43**(3), 696–700 (1998).

Sullivan, K. M., A. Mannucci, C. P. Kimpton, and P. Gill. A Rapid and Quantitative DNA Sex Test: Fluorescence Based PCR Analysis of X-Y Homologous Gene Amelogenin. *Biotechniques*, **15**(40), 636–642 (1993).

Sullivan, K. M., P. Gill, D. Lingard, and J. E. Lygo. Characterization of HLA DQ Alpha for Forensic Purposes, Allele and Genotype Frequencies in British Caucasian, Afro-Caribbean and Asian Populations. *Int. J. Legal Med.*, **105**(1), 17–220 (1992).

Technical Working Group on DNA Analysis Methods (TWGDAM). Guidelines for a Proficiency Testing Program for DNA Restriction Fragment Length Polymorphism Analysis. *Crime Lab Digest*, **17**(2), 50–60 (1990).

Technical Working Group on DNA Analysis Methods (TWGDAM). Guidelines for a Quality Assurance Program for DNA Restriction Fragment Length Polymorphism Analysis. *Crime Lab Digest*, **16**(2), 40–59 (1989).

Thompson. W. C. Subjective Interpretation, Laboratory Error and the Value of Forensic DNA Evidence: Three Case Studies. *Genetica*, **96**(1–2), 153–168 (1995).

Thompson, W. C., F. Taroni, and C. G. G. Aitken. Author's response. *J. Forensic Sci.*, **49**(1), 194–195 (2004).

Thompson, W. C., F. Taroni, and C. G. G. Aitken. How the Probability of a False Positive Affects the Value of DNA Evidence. *J. Forensic Sci.*, **48**(1), 47–54 (2003).

Thompson, W. C. Accepting Lower Standards: The National Research Council's Second Report on Forensic DNA Evidence. *Jurimetrics*, **37**, 405–424 (1997).

Thompson. W. C. Examiner Bias in Forensic RFLP Analysis. *Scientific Testimony*—An online journal, available at http://www.scientific.org.

Tindall K. R. and T. A. Kunkel. Fidelity of DNA Synthesis by the Thermus aquaticus DNA Polymerase. *Biochemistry*, **27**, 6008–6013 (1988).

Turowska, B. and M. Sanak. D1S80 VNTR Locus Genotypes in Population of South Poland; Meta-Analysis Pointer to Genetic Disequilibrium of Human Populations. *Forensic Sci. Int.*, **75**(2–3), 207–216 (1995).

Underhill, P. A., P. Shen, A. A. Lin, L. Jin, G. Passarino, W. H. Yang, E. Kauffman, B. Bonne-Tamir, J. Bertranpetit, P. Francalacci, M. Ibrahim, T. Jenkins, J. R. Kidd, S. Q. Mehdi, M. T. Seielstad, R. S. Wells, A. Piazza, R. W. Davis, M. W. Feldman, L. I. Cavalli-Storza, and P. I. Oefner. Y Chromosome Sequence Variation and the History of Human Populations. *Nat. Genet.*, **26**, 358–361 (2000).

Urquhart, A., C. P. Kimpton, T. J. Downes, and P. Gill. Variation in Short Tandem Repeat Sequences—A Survey of Twelve Microsatellite Loci for Use as Forensic Identification Markers. *Int. J. Legal Med.*, **107**, 13–20 (1994).

Walsh, P. S., N. J. Fildes, and R. Reynolds. Sequence Analysis and Characterization of Stutter Products at the Tetranucleotide Repeat Locus vWA. *Nucleic Acids Res.*, **24**(14), 2807–2812 (1996).

Walsh, P. S., H. A. Erlich, and R. Higuchi. Preferential PCR Amplification of Alleles: Mechanisms and Solutions. *PCR Meth. Applic.* **1**, 241–250 (1992a).

Walsh, P. S., J. Varlaro, and R. Reynolds. A Rapid Chemiluminescent Method for Quantification of Human DNA. *Nucleic Acids Res.* **20**, 5061–5065 (1992b).

Waye, J. S. and R. M. Fourney. Agarose Gel Electrophoresis of Linear Genomic DNA in the Presence of Ethidium Bromide: Band Shifting and Implications for Forensic Identity Testing. *Appl. Theor. Electrophor.*, **1**(4), 193–196 (1990).

Waye, J. S., M. Richard, G. Carmody, and P. J. Newall. Allele Frequency Data for VNTR Locus D17S79: Identification of an Internal HaeIII Polymorphism in the Black Population. *Hum. Mut.*, **3**(3), 248–253 (1994).

Waye, J. S., D. Michaud, J. H. Bowen, and R. M. Fourney. Sensitive and Specific Quantification of Human Genomic Deoxyribonucleic Acid (DNA) in Forensic Science Specimens: Casework Examples. *J. Forensic Sci.*, **36**(4), 1198–1203 (1991).

Waye, J. S., L. A. Presley, B. Budowle, G. G. Shutler, and R. M. Fourney. A Simple and Sensitive Method for Quantifying Human Genomic DNA in Forensic Specimen Extracts. *BioTechniques* **7**(8), 852–855 (1989).

Webb, M. B., N. J. Williams, and M. D. Sutton. Microbial DNA Challenge Studies of Variable Number Tandem Repeat (VNTR) Probes Used for DNA Profiling Analysis. *J. Forensic Sci.*, **5**, 1172–1175 (1993).

Weber, J. and P. May. Abundant Class of Human DNA Polymorphisms Which Can Be Typed Using the Polymerase Chain Reaction. *Am J. Hum. Genet.*, **44**, 388–396 (1989).

Wells, J. D., F. Introna, G. Di Vella, C. P. Capobasso, J. Hayes, and F. A. H. Sperling. Human Maggot and Insect Mitochondrial DNA Analysis from Maggots. *J. Forensic Sci.*, **46**(3), 685–687 (2001).

Whitfield, S., J. E. Sulston, and P. N. Goodfellow. Sequence Variation of the Human Y Chromosome. *Nature*, **378**, 379–380 (1995).

Wickenheiser. R. A. Trace DNA: A Review, Discussion of Theory, and Application of the Transfer of Trace Quantities of DNA Through Skin Contact. *J. Forensic Sci.*, **47**(3), 442–450 (2002).

Wilson, M. R., M. W. Allard, K. L. Monson, K. W. P. Miller, and B. Budowle. Further Discussion of the Consistent Treatment of Length Variants in the Human Mitochondrial DNA Control Region. *Forensic Sci. Commun.*, **4**(4) (2002).

Wilson, M. R., D. Polanskey, J. Replogle, J. A. DiZinno, and B. Budowle. A Family Exhibiting Heteroplasmy in the Human Mitochondrial DNA Control Region Reveals Both Somatic Mosaicism and Pronounced Segregation of Mitotypes. *Hum. Genet.*, **100**, 167–171 (1997).

Wilson, M. R., J. A. DiZinno, D. Polanskey, J. Replogle, and B. Budowle. Validation of Mitochondrial DNA Sequencing for Forensic Casework Analysis. *Int. J. Legal Med.*, **108**(2), 68–74 (1995a).

Wilson, M. R., D. Polanskey, J. Butler, J. A. DiZinno, J. Replogle, and B. Budowle. Extraction, PCR Amplification, and Sequencing of Mitochondrial DNA from Human Hair Shafts. *BioTechniques*, **18**, 662–669 (1995b).

Wilson, R. B., J. L. Ferrara, H. J. Baum, and R. C. Shaler. Guidelines for Internal Validation of the HLA DQ Alpha DNA Typing System. *Forensic Sci. Int.*, **66**(1), 9–22 (1994).

Xiao, F. X., A. Gilissen, J. J. Cassiman, and R. Decorte. Quadruplex Fluorescent STR Typing System (HUMVWA, HUMTHO1, D21S11 and HPRT). with Sequence-Defined Allelic Ladders. Identification of a New Allele at D21S11. *Forensic Sci. Int.*, **94**, 39–46 (1998).

Yuan, H., A. Shahidi, S. Park, D. Guilfoyle, and I. Hirshfield. Detection of Extra-Hepatic C Virus Replication by a Novel Highly Sensitive Single Tube Nested Polymerase Chain Reaction. *Am. J. Clin. Path.*, **119**(1), 95–100 (2003).

Yu, N., M. Kruskall, J. J. Yunis, J. H. M. Knoll, L. Uhl, S. Alosco, M. Ohashi, O. Clavijo, Z. Husain, E. Yunis, J. Yunis, and E. J. Yunis. Disputed Maternity Leading to Identification of Tetragametic Chimerism. *New Engl. J. Med.*, **346**(20), 1545–1552 (2002).

Ziegle, J. S., Y. Su, K. P. Corcoran, L. Nie, E. Mayrand, L. B. Hoff, L. J. McBride, M. N. Kronick, and S. R. Diehl. Application of Automated DNA Sizing Technology for Genotyping Microsatellite Loci. *Genomics*, **14**, 1026–1031 (1992).

GENERAL REFERENCES

Budowle, B., J. Smith, T. Moretti, and J. A. DiZinno. DNA Typing Protocols: Molecular Biology and Forensic Analysis. In *Biotechniques Books*. Eaton, Natick, MA, 2000.

Butler, J. M. *Forensic DNA Typing: Biology & Technology behind STR Markers*. Academic, New York, 2001.

DNA Recommendations—1994 report concerning further recommendations of the DNA Commission of the ISFH regarding PCR-based polymorphisms in STR (short tandem repeat) systems. *Int. J. Legal Medicine*, **107**(3), 159–160 (1994).

National Research Council. *The Evaluation of Forensic DNA Evidence*. National Academy Press, Washington, D.C., 1996.

Forensic Science Review. The subject of Y-chromosome analysis is treated thoroughly in **15**(2), July, 2003, which is devoted to the Workshop on Y-chromosome Analysis and Its Application in Forensic Science, which was held at the 3rd European Academy of Forensic Science Triennial meeting in September 2003 in Istanbul, Turkey.

SPECIFIC REFERENCES

STR Analysis

Buel, E., M. B. Schwartz, and M. LaFountain. Capillary Electrophoresis STR Analysis: Comparison to Gel-Based Systems. *J. Forensic Sci.*, **43**, 164–170 (1998).

Lindsey, S., R. Hertwig, and G. Gigerenzer. Communicating Statistical DNA Evidence. *Jurimetrics J.*, **43**, 147–163 (2003).

Moller, A., E. Meyer, and B. Brinkmann. Different Types of Structural Variation in STRs: Hum FES/FPS, Hum VWA and Hum D21S11. *Int. J. Legal Med.*, **106**, 319–323 (1994).

Polymeropoulos, M. H., D. S. Rath, H. Xiao, and C. R. Merrill. Tetranucleotide Repeat Polymorphism at the Human Tyrosine Hydroxylase Gene (TH). *Nucleic Acid Res.*, **19**, 3753 (1991).

Prinz, M., A. Ishii, H. J. Coleman, H. Baum, and R. C. Shaler. Validation and Casework Application of a Y Chromosome Specific STR Multiplex. *Forensic Sci. Int.*, **120**, 177–188 (2001).

4

Genetics, Statistics, and Databases

4.1 HUMAN GENETICS, POPULATION GENETICS, AND STATISTICS

4.1.1 Power of Forensic DNA Analysis: How Significant Is the Match?

The significance of a DNA match is expressed in the form of a statistic. Statistics are calculated after determining that the genetic profile of the evidentiary specimen matches that of the suspect. These statistics reflect how rare the overall profile is within the relevant population under consideration. What are some factors that can limit the relevant population in a given situation? In the case of a male rapist, there are a significant number of individuals who because they are either too young or too old are excluded from consideration. All women regardless of age are obviously also excluded. Interest must be focused on those individuals who could have committed the assault. The geographical distribution of potential offenders is another factor in deciding on the relevant population. In practice, when statistics are developed in a particular case, they are provided for each of the major ethnic groups in the *relevant* population. The jury is made aware not only of the finding of a match (inclusion) but also the frequency statistics associated with the genetic profile. Calculated statistics inform the jury about the significance of the match by providing an understanding of the rarity of the genetic profile in the relevant population. The major population groups in the United States include Caucasians, Blacks (including Caribbean), West Coast Hispanics, East Coast Hispanics, and

DNA: Forensic and Legal Applications, by Lawrence Kobilinsky, Thomas F. Liotti, and Jamel Oeser-Sweat
ISBN 0-471-41478-6

Asians. Some laboratories will combine West Coast and East Coast Hispanics into one group. There has been a great deal of controversy over the years about how to calculate the random match probability after testing reveals that the evidence and suspect have matching genetic profiles (Koehler et al., 1995; Slimowitz and Cohen, 1993; Mueller, 1993). That controversy came to a head in a series of articles written by Eric Lander and Bruce Budowle. In 1994, Budowle and Monson reported their findings that there were greater differences in DNA profile frequencies estimated from racial groups than from ethnic groups (Budowle and Monson, 1994). The dispute ended in 1994 with both in agreement about how statistical calculations should be made (Lander and Budowle, 1994; Lempert, 1997).

4.1.2 Genetics and Statistics

Whether the technique used is RFLP, sequence-specific oligonucleotide probes, PCR-VNTRs, or PCR-STRs, statistics are calculated using the product rule (Klitz et al., 2000). The exceptions to this statement are when mitochondrial DNA analysis or Y-chromosome analysis are performed (see Section 4.6). Based on the rules of probability, when there are x number of events that occur independently of each other, the frequency of their simultaneous occurrence can be determined by simply multiplying the probability of each event. This is known as the product rule. For example, a coin flip has a 50:50 chance of landing head or tail. The chance of flipping 2 coins simultaneously and landing 2 heads or 2 tails is 25% or $\frac{1}{2} \times \frac{1}{2} = \frac{1}{4}$. Similarly, the probability of flipping 3 coins and landing 3 heads or 3 tails is calculated as $\frac{1}{2} \times \frac{1}{2} \times \frac{1}{2} = 1/8$ or 12.5% and so on. If we analyze 3 loci simultaneously and have the following frequencies for each genotype, $1/100$, $1/200$, and $1/300$, then the probability of having all 3 genotypes in the same individual is $1/(6 \times 10^6)$ or 0.0000167%. It is obvious that as more loci are tested, more frequencies (fractions) will be multiplied by one another, resulting in a total probability that becomes smaller and smaller. Analysis of the CODIS loci results in the determination of 13 genotypes in total. If there is no exclusion, then the probability for the entire genetic profile can be calculated using the product rule. Typically, the number that results is smaller than 1 in trillions of individuals.

Genes at STR loci can be thought of as co-dominant autosomal markers of inheritance. This is very much like the polymorphic protein markers used for so many years by forensic serologists. Databases at each locus are constructed by sampling a sufficient number of individuals and determining the frequency of each allele that has been detected. Thus, let us take the example of a hypothetical locus, XYZ, and let us sample 100 individuals to construct our database. One hundred individuals will have a total of 200 genes at any STR locus. The locus consists of 5 different alleles. Fifteen different genotypes can be observed. Homozygotes will have two copies of the same alleles while heterozygotes will have copies of two different alleles. All alleles are tallied and Table 4.1 reflects the results.

The Hardy–Weinberg law can be used to predict the frequency of genotypes in a population under a set of ideal conditions (no immigration, random mating with respect to the locus in question, no mutation, etc., see below). Essentially, p, q, r,

TABLE 4.1 Locus XYZ: Caucasians (or Other Group)

	Allele	Number of Alleles	Frequency	Designation
	1	20	0.10	p
	2	40	0.20	q
	3	80	0.40	r
	4	30	0.15	s
	5	30	0.15	t
TOTAL	5	200	1.00	

s, and t are used to designate the frequencies for alleles 1, 2, 3, 4, and 5, respectively, in the population. The Hardy–Weinberg law states that $(p+q+r+s+t)=1.0$ and that $(p+q+r+s+t)^2=1.0$. The latter calculation is a binomial expansion that indicates that the frequency of all of the genotypes add up to 100%. With 5 alleles there are $5+4+3+2+1=15$ genotypes in the population. To determine if the population under study is in Hardy–Weinberg equilibrium, a chi square analysis is performed to determine if there is a difference between the observed genotype frequencies and the expected. The chi square statistic is

$$\chi^2 = \sum \frac{(O-E)^2}{E}$$

where $O=$ number observed and $E=$ number expected. The larger the χ^2 value, the less likely the population is in equilibrium. A table of χ^2 values (which contains significance levels at different degrees of freedom) will reveal if the differences exceed the 5% significance level. In the example provided above, the degrees of freedom is $5-1=4$, where 5 represents the number of alleles. Although some small difference can be expected as a result of the small sampling size, a large difference would indicate that the population is out of equilibrium, perhaps a result of selection pressure affecting (a) specific genotype(s). If the difference is not great, then the population is said to be in equilibrium, and the analyst can use the allele frequencies in calculations. Let us assume that the population is in equilibrium. Let us also assume that biological evidence is found at a crime scene, a suspect is in hand, and an analysis of locus XYZ reveals that both share the genotype 1,2. This heterozygote genotype would be expected to be found with a frequency of $2pq$, which equals 2 (0.10)(0.2) $=0.04$ or 4 in 100 or 1 in 25. If the genotype had been 1,1 or 2,2, then the frequency of these genotypes would have been $(0.1)^2=0.01$ (1 in 100) or $(0.2)^2=0.04$ (4 in 100), respectively.

The Hardy–Weinberg rule is used to calculate whether the observed genotype frequencies are those expected based on allele frequencies or whether the observed genotype frequencies are being influenced by environmental pressures such as selection, migration, nonrandom mating, and so on. Only when the observed genotype frequencies are found to be in agreement with the expected frequencies can the data for the particular genotype be used in calculating the population frequency of a genetic profile (Brookfield, 1995a, 1995b). The final statistic can be used to demonstrate the rarity of the overall genetic profile. The basis for the genotype frequency calculations is derived from Mendel's laws of heredity, which describe the rules whereby alleles are inherited from one's parents.

4.1.3 Mendel's Laws of Genetics

4.1.3.1 Independent Segregation and Independent Assortment Male and female gametes (spermatozoa and ova), respectively, contain 23 chromosomes each. This is the haploid number of chromosomes for the human species. The number arises as a result of a process known as meiosis. In meiosis the diploid number is reduced from 46 to 23. Upon fertilization of the ovum by a spermatozoan, that number is restored to the diploid number. The 46 chromosomes are composed of 22 pairs of autosomes and 2 sex chromosomes, XX in the female and XY in the male. Each member of a pair of chromosomes is referred to as a sister chromosome. The paired sets of "homologous" chromosomes share the same loci. Since an ovum may contain

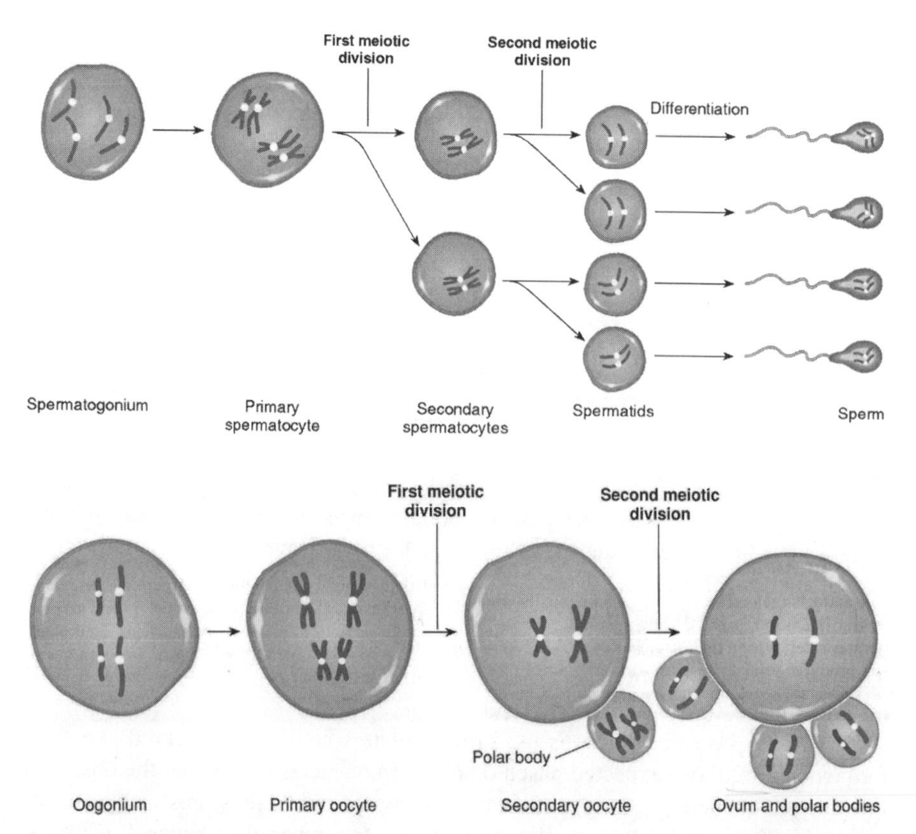

Fig. 4.1 Meiosis. (a). Spermatogenesis and (b). Oogenesis. Reprinted with permission of The McGraw-Hill Companies, Inc., Principles of Genetics 3/e *by R. Tamarin. (Copyright 1991 The McGraw-Hill Companies, Inc.)*

only a single X chromosome but a sperm cell can contain either an X or a Y chromosome, it is the sperm cell that determines the sex of the offspring.

4.1.4 Meiosis

Genes, which are located on sister (homologous) chromosomes, are distributed in such a manner during meiosis that they are transferred to gametes (sperm and egg) independently of each other (See Fig. 4.1*a* and 4.1*b*). Put another way, a gamete receives only one chromosome from a pair of chromosomes. This is referred to as *independent segregation.* If the genes of two loci are located on the same chromosomes, they are said to be *linked*, and, if they are located very close to each other, then they will generally be inherited as a unit. Genes of different loci that are linked but distant from each other on the chromosome tend to become separated during meiosis, gamete formation. If they are far enough apart, they will become separated as if they had been on different chromosomes. Genes located on *different* chromosomes will be assorted to sperm and egg independently of each other. This is known as *independent assortment* (see Fig. 4.2). Under normal conditions, an individual will never inherit both chromosomes of a pair from one parent. Genes can become recombined during the process of *crossing over*, which occurs during "synapsis" of meiotic metaphase I. (Meiosis consists of two stages.) This crossing

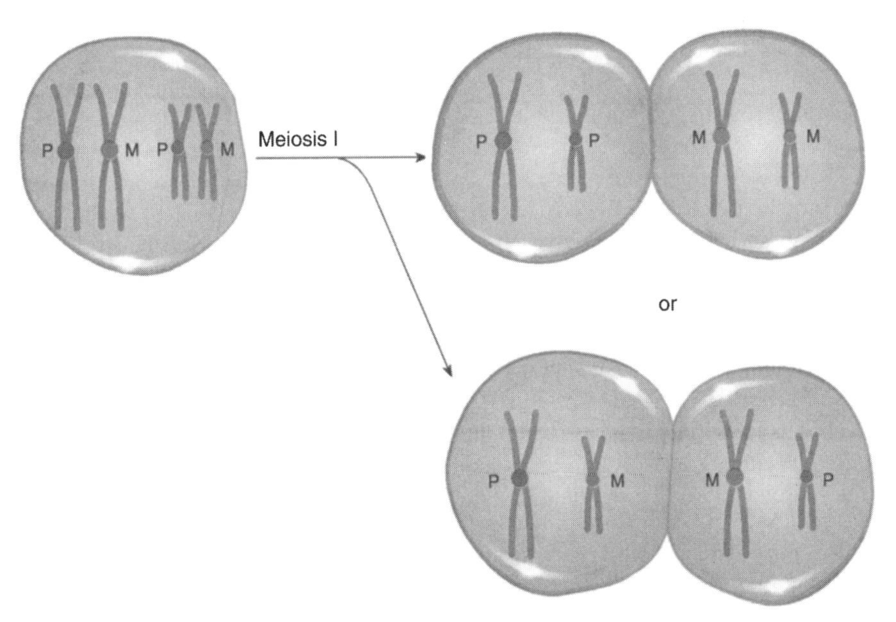

Fig. 4.2 *Independent assortment. During two divisions of meiosis, maternal and paternal chromosomes are distributed independently of each other; thus, meiosis results in four different arrangements of these chromosomes. Reprinted with permission of The McGraw-Hill Companies, Inc.,* Principles of Genetics 3/e *by R. Tamarin. (Copyright 1991 The McGraw-Hill Companies, Inc.)*

over process together with random segregation of chromosomes is what produces new combinations of genes in a population. The fertilized egg contains one set of chromosomes delivered by the sperm cell and the other set that was present in the ovum as a result of oogenesis (meiosis leading to egg development).

4.2 POPULATION GENETICS

4.2.1 Hardy–Weinberg Equilibrium

In a population, allele frequencies at any particular locus of interest can be determined by sampling a number of individuals from that population. Allele frequencies may change from generation to generation if the population is subject to certain factors. The following are examples of such factors: (1) if the population is undergoing significant immigration or emigration, or (2) if alleles are subject to mutation, or (3) if individuals with certain genotypes in the population are subject to natural selection, or (4) if there is nonrandom mating, or (5) if the population being studied is relatively small in size, there is a good possibility of genetic drift resulting from sampling accidents. When there is genetic drift, it is possible that allele frequencies will vary significantly over time, and some alleles can disappear completely from the population. Before a locus is used for forensic purposes, it is important to demonstrate that the population is in Hardy–Weinberg equilibrium and that allele frequencies are stable from generation to generation.

Another type of equilibrium is essential if we are to use the product rule to calculate the frequency of an individual's genetic profile. When testing multiple loci, we must be certain that the inheritance of alleles at one locus is not affected by the inheritance of genes at another locus (Sudbury et al., 1993). This could happen, for example, if genes are linked as described above. If genes are located on the same chromosome but far enough apart from each other, they are said to be in linkage equilibrium since they are inherited independently as if they were located on different chromosomes. The inheritance of alleles at one locus must not be affected by the inheritance of alleles at another locus. Of course, loci on different chromosomes are inherited independently of each other as stated earlier in the discussion of Mendel's laws of heredity.

Thus, genes at loci that are used in human identification analysis must be in both Hardy–Weinberg equilibrium and linkage equilibrium if the gene loci are on the same chromosome (Turowska and Sanak, 1995). The product rule can only be applied when alleles at all loci are inherited in a completely independent manner (Weir, 1992, 1993; Weir and Hill, 1993). Most multiplex kits in use today are designed to analyze genes at loci that are each located on a different chromosome. Nevertheless, geneticists have cautioned that where populations of interest have substructure (meaning distinct subpopulations) you can get deviations from Hardy—Weinberg equilibrium. Communities where inbreeding is a factor have genotype frequencies that can differ significantly from the larger population (Balding and Nichols, 1994; Bellamy et al., 1991; Budowle, 1995). This is the reason that the product rule is modified to include a correction factor to compensate for the existence of these subpopulations.

4.2.2 Subpopulations and Substructure

For many years there were both legal and scientific debates as to the significance of the existence of subpopulations and the errors that could be made in statistical calculations if this fact were not recognized. There is a vast literature on the existence of subpopulations within each major ethnic group. It is well known that within subpopulations, individuals tend to mate with others who are within the same ethnic, economic, and social status. Thus they constitute a small group of individuals who are often isolated reproductively (by choice) from others in the larger population. For example, among Caucasians there are Irish, English, Germans, Italians, Greeks, Poles, and so on. Since members of subpopulations tend to socialize with and marry others within that same group, there is reason to believe that gene frequencies for Irish Americans differ from German Americans. If a database consists of gene frequencies for the larger population, the values may be significantly different from gene frequencies in the smaller subgroups (Budowle and Monson 1994). Furthermore, in these subpopulations, individuals may share certain genetic traits (and genotypes) more frequently than expected based on analysis of allele frequencies in the larger population. Thus, in that subgroup, portions of an individual's genetic profile may occur more frequently than one would calculate using allele frequencies derived from the larger population. Statistical calculations for genetic profiles should only be considered as estimates of the actual frequencies. The actual frequency of the profile in question is unknown. However, one can say that the true value lies somewhere in the range between 10-fold greater and 10-fold less than the calculated statistic. For a variety of reasons, we always want to err in favor of the defendant in producing these statistics, but in a case where substructure exists, there is a danger of erring in the opposite direction.

In 1989, the National Research Council (NRC) Committee on DNA Technology in Forensic Science was formed. The committee was chaired by Victor McKusick. In its first report, which was issued in 1992, the NRC made a number of recommendations to improve the quality of DNA-typing information and its presentation in the courtroom. Some of the recommendations included:

1. Studies of the relative frequencies of distinct DNA alleles in 15–20 relatively homogeneous subpopulations.
2. A ceiling principle using as a basis of calculation, the highest allele frequency in any subgroup or 5%, whichever is higher.
3. A more conservative "interim ceiling principle" in calculating the frequency of a matching profile. The analyst should use the highest allele frequency in any subgroup or 10%, whichever is higher, until the ceiling principle can be implemented.
4. Proficiency testing to measure error rates and to help interpret test results.

Both the interim ceiling principle and the ceiling principle itself came under intense attack from those who felt that these recommendations were completely arbitrary and not based in science, and that the resulting statistic was much too conservative and did not reflect what the analysis was actually showing. Not surprisingly, others felt that these recommendations were not conservative enough.

After many database studies performed throughout the world establishing gene frequencies in numerous subpopulations within each of the major ethnic groups, it was found that the differences between these subpopulations was smaller than the differences between the major ethnic groups. As a result, it was decided that allele frequencies determined for the major groups (aggregate databases) should be used to establish the statistics of a genetic match (National Research Council, 1996). On the other hand, it was recommended that because of the existence of such substructure (subpopulations), that a correction factor, theta (θ), be applied in a modified product rule calculation and that the value of this correction factor be empirically determined. Thus for crimes that take place in New York City, the value for the θ factor should be different than for crimes committed in Lancaster, Pennsylvania, where the predominant population is more homogeneous and mostly Amish (of German descent). The θ value used by many crime labs is equal to 0.03, which reflects the existence of substructure typical of a middle- to large-sized city.

In determinations of genotype frequency, the calculation for the frequency of a heterozygote is done using the formula $2pq$. However, the homozygote frequency is calculated as $p^2 + p(1 - p)\theta$, where the value of θ is usually between 0.01 and 0.03 depending on the size of the population and the degree of reproductive isolation. Thus, if we assume that $p = 0.1$ and θ is 0.03, then the homozygote frequency will be $0.01 + (0.1)(0.9)(.03) = 0.01 + 0.0027 = 0.0127$ or 1.27%.

4.3 NEED FOR QUALITY CONTROL AND QUALITY ASSURANCE

As the technology evolved during the early to mid-1990s, with new DNA identification tests being developed, the power of discrimination continued to increase. Initially, laboratories were performing PCR amplifications using the HLA DQA1 and Polymarker kits to test evidence and suspects. If the suspect was not excluded, the statistical calculations for the frequency of a genetic profile were usually in the range of one in hundreds to one in thousands. This left the door open for defense attorneys to argue that there were hundreds of thousands of other individuals who shared the same genetic profile as that of the suspect, and therefore the match had limited significance. On the other side prosecutors could argue that the DNA test had revealed a match between the evidence and suspect and that the jury should consider the significance of this scientific finding.

Laboratories soon were performing DNA analysis with not only the HLA DQA1 and Polymarker kits, but they were also testing the VNTR D1S80 locus. Now, if there was a match, the statistics of a profile ranged from one in hundreds of thousands to one in millions when these tests were used in combination. Nowadays, the statistics associated with a genetic profile are usually based on the analysis of 13 STR loci and are often in the range of one in hundreds of billions to one in trillions or more. No longer can defense attorneys argue that many others in the population share the genetic profile found in both the suspect and in the evidence. Juries understand that when the frequency of a genetic profile is as rare as one in one trillion, that it is essentially an absolute identification. It is precisely because of the tremendous power of the PCR-STR technology that, toward the end of the 1990s, emphasis was placed on *quality control* (QC) and *quality assurance* (QA) (Gill et al., 1991). Failure to perform the DNA tests properly could lead to false exclusions as well as false inclusions, sometimes with tragic consequences. TWGDAM was established to ensure that DNA testing was performed reliably so that jurors could trust the results of DNA analysis.

4.4 SWGDAM (FORMERLY KNOWN AS TWGDAM) STANDARDS

The Technical Working Group on DNA Analysis Methods (TWGDAM) was established by the FBI to provide a forum for forensic scientists to discuss and exchange technical information on DNA testing and to formulate standards and guidelines for forensic DNA laboratories (see Bylaws of SWGDAM, 2003; SWGDAM, 2001; TWGDAM, 1990, 1995). The committee is comprised of scientists from industry, forensic laboratories, and the academic community. The committee has developed guidelines for DNA laboratories, quality assurance guidelines for forensic DNA testing, and guidelines for DNA proficiency testing. The committee changed its name in 1999 to reflect the scientific aspects of its work.

More specifically, the Scientific Working Group on DNA Analysis Methods has been assigned the following responsibilities by the FBI director:

1. To recommend revisions, as necessary, to the Quality Assurance Standards for Forensic DNA Testing Laboratories (effective October 1998) and the Quality Assurance Standards for Convicted Offender DNA Databasing Laboratories (effective April 1999).
2. To serve as a forum to discuss, share, and evaluate forensic biology methods, protocols, training, and research to enhance forensic biology services.
3. When necessary, to recommend and conduct research to develop and/or validate forensic biology methods.

Its membership consists of an executive board composed of a chairman who is appointed by the FBI director, a vice-chairman, executive secretary, and five members. In addition SWGDAM has seven regular members.

The quality assurance standards mentioned in point 1 include educational and training requirements for various categories of forensic laboratory personnel: new employees, technical managers or leaders, examiner/analysts, technicians, laboratory support personnel, and laboratory directors. SWGDAM has addressed the major issues for each of these individuals. They include specific course requirements, in-house laboratory training, and minimal experience needed to perform DNA analysis. The goals of the training are (1) to familiarize the trainee with the general operations of the forensic laboratory, (2) to instruct the trainee on evidence handling in the forensic laboratory, (3) to ensure that a trainee has the formal education and the working knowledge of the fundamental science of forensic DNA analysis, (4) to educate the trainee on the specific knowledge related to the field of forensic DNA analysis, (5) to provide practical instruction to the trainee on analytical procedures used in the laboratory (extraction, quantification, RFLP, PCR, DQA1 and Polymarker, D1S80, STRs, and mtDNA), (6) to learn how to interpret and report analytical results according to the laboratory's policy, and (7) to instruct the trainee on the legal system of his or her own jurisdiction.

4.5 DNA ADVISORY BOARD

The DNA Advisory Board (DAB) was established as a result of the DNA Identification Act of 1994. Various organizations including the National Academy of Sciences submitted nominations to the FBI for appointment to the DAB. The DAB met in 1995 and its first order of business was to develop guidelines for quality assurance in forensic laboratories. As a result, the FBI director issued two sets of guidelines, one for QA standards in forensic DNA testing laboratories (O'Dell, 2003) and the second set of standards for laboratories participating in submission of genetic profiles to the CODIS convicted offender database (DNA Advisory Board Standards, 2000). SWGDAM now has the responsibility for maintaining and revising these standards.

4.6 MITOCHONDRIAL DNA AND Y-CHROMOSOME STR ANALYSIS AND STATISTICAL CALCULATIONS

As described earlier, mitochondrial DNA is found in the form of a single, double-stranded, closed circular molecule. Analysis consists of sequencing the two major hypervariable regions (HV1 and HV2) within the control region (also called the D-loop region). The analysis is performed by determining differences in the sequence of these regions between the DNA under study [exemplar(s) and unknown(s)] and a reference sequence (Cambridge sequence) where each of the nucleotide bases in the mitochondrial genome is known. Since all the variations in the mitochondrial genome are found on the same chromosome and are inherited as a unit, we cannot use the laws of probability or the product rule to calculate the frequency of the determined genetic profile (mitotype). Each variation in sequences is known as a single-nucleotide polymorphism (SNP—pronounced snip). In the case of mtDNA, the complete set of variations is recorded as a mitotype and the counting method is applied. If the database consists of 2000 profiles and the mitotype in question has not been observed previously, then the statistic becomes 1 out of

2001 profiles. It is important to understand that statistics derived by the counting rule are not comparable to the statistics derived from using the product rule. When using the counting rule in expressing the rarity of the mitotype, the significance of the statistic must be understood. The counting method greatly underestimates the frequency of the profile especially when the database is small.

Y-chromosome STR loci have become very important in a number of areas including anthropology, ancestry determinations, and especially in the investigation of rape. There are a number of multiplex kits designed to study the various STR loci of the Y chromosome. These findings are expressed as a haplotype. The Y chromosome has no pairing sister chromosome and thus STR loci do not undergo recombination. (There are homologous regions at the very ends of the X and Y chromosomes, called telomeres, that are capable of recombination.) The haplotype may consist of four or more loci depending on the laboratory's protocols. Regardless of the number of loci examined, since all STRs on the Y chromosome are linked (Fig. 3.41), the counting rule must be used to develop the statistic for haplotype frequency. For both mitochondrial DNA as well as Y-chromosome STR analysis, a finding of an exclusion is absolute and there is no statistic to calculate. Exclusion means that the questioned and exemplar specimens have different origins.

In the case where an exclusion cannot be made, it becomes important to explain to the jury the significance of the determination of a mitotype or Y-STR haplotype. One approach is to provide the jurors with a statistic based on confidence limits. In biological systems, the 95% confidence limit is often used. The calculation of confidence limits depends greatly on the size of the population database. The larger the population, the greater the upper confidence limit and the more significant the statistical frequency. If the mitotype or haplotype observed has never been observed before, then the 95% confidence limit can be calculated as $1 - 0.05^{1/n}$, where $n =$ number of individuals in the specific population database. Thus, if the database consists of 500 individuals, then the 95% confidence limit becomes 0.6%. If $n = 1000$, then the 95% confidence limit becomes 0.3%. If $n = 1500$, the value is 0.2%, and when $n = 2000$, the value reaches 0.0015, which means that you would see the mitotype or haplotype in no more than 0.15% of the specific population. The calculation becomes a bit more complex when the type has been previously observed in the database. The formula for the calculation of the 95% upper confidence limit is now $p + 1.96[(p)(1-p)/n]^{1/2}$, where $p =$ the number of times the type has been seen divided by n, the number of individuals in the database. Thus, if $p = 5/500$, the 95% upper confidence limit becomes 1.87%, and if $p = 2/500$, the upper limit becomes 0.95%, however, if $p = 5/3000$, then the 95% upper limit becomes 0.31%. Another way of explaining the data is that in a population of 1 million Caucasians (or some other ethnic group), one would expect *no more than* 3100 individuals who would express this mitotype or Y-STR haplotype.

4.7 EXPERIMENTAL CONTROLS

As in any experimental situation, the analysis of casework requires the use of controls so that variables can be minimized and the correct interpretation of observations is made. Quality control must be maintained at the crime scene, in the laboratory (facilities, reagents, scales, incubators, analytical equipment), at the multiplex kit manufacturing facility, and during each analysis. Generally, a number of negative and positive controls are included among the test specimens. Examples of negative controls are samples of the substrate adjacent to the questioned (stained) area, an extraction control to demonstrate that the reagents used for extraction of DNA from the sample did not cause any contamination, and the PCR amplification mix-DNA template negative sample.

Each control is important for a different reason. Let us assume that a PCR-STR analysis will include known (K) samples (blood or saliva specimens) and questioned (Q) samples (swabs or a cuttings). The following control specimens should be included and treated in an identical manner as the evidentiary and exemplar samples. Substrate controls should accompany the evidentiary questioned samples (cuttings and swabs). These controls will provide an indication of what if any genetic material was present adjacent to the area sampled. Finding DNA in a substrate control does not necessarily diminish the value of the genetic evidence found in the specimen itself. However, results obtained for the questioned sample must be interpreted in light of the positive substrate control. PCR positive and negative controls should be included to verify that known specimens are providing expected results and that there is no indication of human DNA contamination. The analysts DNA should be tested if the genetic profile is not already known. The failure to obtain the correct results in a positive control is as problematic as finding DNA in a negative amplification control since it may indicate contamination or human error. The multiplex kits generally come with a positive control for testing, but the laboratory can use its own positive control specimen. The negative PCR amplification control contains all reagents and chemicals for amplification except for template DNA. The observation of alleles in a negative control is a good indication of contamination at some point in the procedure. Contamination of a specimen can occur during its collection, transport to the lab, or while it is in the laboratory undergoing analysis. Of course, contamination may also occur after the specimen has been deposited at the crime scene but before it has been collected. Finally, allelic ladders are included in the analysis so that each of the possible alleles at the 13 CODIS loci can be identified. Commercial ladders do not contain every possible allele, however, they do consist of the most common alleles seen at each of the 13 loci. Failure to see an expected allele in a ladder should trigger a review of the ladders and the analysis results. Off-ladder alleles are occasionally observed (see Section 3.3, 4, 3.2.6).

4.8 VALIDATION OF NEW DNA METHODS

Before any new forensic DNA method can be used for casework, it is absolutely necessary that the technique is validated and found to produce reproducible and reliable results (van Oorschot et al., 1996; Budowle et al., 1997; Kline et al., 1997; Wallin et al., 1998; Wilson et al., 1995). Both the Scientific Working Group on DNA Analysis (previously known as the Technical Working Group on

DNA Analysis) and the DNA Advisory Board have developed standards to achieve validation. The new technique must be useful when testing common forensic specimens such as blood, semen, buccal swabs, and hair. Because such specimens can often be found at crime scenes, they should be included in such validation testing. Furthermore, because these substances are sometimes found in limited quantity, the sensitivity of the procedure should be established. This is usually accomplished by performing a calibration or titration analysis to find the minimum amount of material that the test can detect. Testing must include reference samples for which the genetic profile is already known. Repeated testing of the same specimen must produce the same results. Studies should be conducted using mixed specimens in different proportions as well as individual specimens. Furthermore the test results must be the same despite the sample's exposure to various environmental insults (soil, sunlight, high/low temperature, chemicals, humidity, bacteria, etc.) and regardless of what substrate the sample has been deposited upon (natural or synthetic materials, hard or soft, porous or nonporous, etc.). The method must be tested on mock casework to determine its efficacy in producing correct results. Finally, testing should be performed on nonhuman samples. Because primate DNA is so closely related to human DNA, it is not surprising that certain probes and primers produce results with certain animals such as chimpanzees, which are known to share 98.7% of our genome. Samples obtained from other animals should not produce observable results.

4.9 SINGLE-NUCLEOTIDE POLYMORPHISM ANALYSIS

An emerging forensic technique is based on the analysis of single-nucleotide polymorphisms (SNPs), which are present in the human genome with a frequency of approximately one in every kilobase on average (see Table 4.2). Although SNPs occur at a much higher frequency (millions per genome) than STR loci, they are gen-

TABLE 4.2 Advantages and Limitations of Single-Nucleotide Polymorphisms (SNPs) at a Glance

Advantages	Limitations
1. SNPs are found in abundance in mammalian genomes.	1. Most SNPs are biallelic. Even sites with as many as three alleles are individually of limited discriminatory significance.
2. There are many methods available to detect SNPs.	2. The limited number of alleles makes analysis of mixed samples more complicated.
3. The amplification of alleles is not prone to preferential amplification.	
4. Robust multiplex amplification is relatively easy to achieve.	

Source: National Commission on the Future of DNA Evidence, *The Future of Forensic DNA Testing Predictions of the Research and Development Working Group* (2000).

erally less informative. Unlike STRs, which have many alleles at each locus, most SNP sites are biallelic and have only 2. Thus the discrimination potential at each site is low. On the other hand, analysis of 30–40 SNPs can produce data equivalent in discrimination power to the 13 CODIS loci. Until recently, SNPs have been used to map genetic diseases, but there is enormous potential for their use in human identification. SNPs can be detected using a variety of methods including sequencing, chip technology, and real-time PCR. One of the most significant benefits of SNP detection is that PCR products bearing SNPs are small (less than 100 bases) and therefore in highly degraded DNA where even STR DNA cannot be successfully performed, SNP analysis will produce good results.

One of the most exciting aspects of SNP analysis is the ability to determine ethnic origin based on a determination of a SNP profile. Some SNPs are more likely to be found in specific populations. It has been found that some SNPs are "population specific." However, more studies are needed before it can be stated with confidence that a particular DNA specimen originated from an individual who is a member of a particular ethnic/racial group. In the case of a highly decomposed body part, SNP analysis can be *helpful* in identifying the ethnicity of the victim as well as in identifying that individual. Currently SNP analysis is being used to determine one's ancestral composition (see Sections 3.5.7, 4.8, and 7.5.)

4.10 DATABASE SIZE AND COMPOSITION

Occasionally, the question of sampling size is raised. How can you generate statistics for an ethnic group in the range of one in a billion to one in a trillion when your database consists of only 200 people? How many individuals must be sampled for the database to be considered valid? If the database is too small, the observed allele frequencies may differ from the true frequency values due to random sampling. Rare alleles may not even be seen among those individuals tested. There is a way of determining how close gene frequency estimates are to true values. Calculating the standard deviation for the gene frequency estimates accomplishes this. Thus for allele 1 above (p. 156), the frequency is 0.10 and the standard deviation (SD) is calculated as the square root of $p(1-p)$ divided by $2N$, where N is the number of individuals sampled. Thus, $\mathrm{SD} = \sqrt{(0.1)(0.9)/2(100)} = 0.0212$. This means that the true value of the gene frequency for allele 1 is 0.1 ± 0.02. The frequency is thus someplace between 0.08 and 0.12. As the sample size increases, the SD becomes smaller and smaller and relatively rare alleles are observed.

The individuals that are sampled in a specific group (Caucasians, Hispanics, Blacks, Asians) must come from a "genetically homogeneous" group. Thus a Caucasian database *can* include Caucasian individuals who are of Irish, Polish, Italian, and French ancestry since each of these groups has relatively similar gene frequencies (Budowle et al., 2001). However, a Hispanic database that includes East and West Coast Hispanic subpopulations may exhibit deviations from Hardy–Weinberg equilibrium at certain loci. East Coast Hispanics consist of Puerto Ricans, Dominicans, and Cubans for the most part, whereas West Coast Hispanics consist primarily of individuals of Mexican origin. Examination of the genotype frequencies in such an aggregate database reveals that there are fewer heterozygotes than expected based on Hardy–Weinberg equilibrium (Hartl and Clark, 1989). Excess homozygotes are a good indication that the population in question is not homogeneous and as a result fails the test for equilibrium. Despite the differences in gene frequencies that exist in these subpopulations (Lewontin and Hartl, 1991; Kay, 1993; Levy, 1991), they are not larger than the differences between Hispanics and the other major racial groups such as (Caucasians, Blacks and Asians), and therefore many labs will combine all Hispanic subgroups into one large group.

4.11 DNA DATABASES

AN OVERVIEW

As DNA analysis becomes more and more common in criminal investigations, there will come a day when millions upon millions of people will have been profiled. Currently, the DNA profiles of criminals are being retained in an "offender" database so that in the future biological evidence of unknown origin can be compared with DNA profiles in the database. One day every citizen's genetic profile may be stored in a national database. There are many who are concerned about the ramifications of a government agency maintaining such records. It is essential that all DNA data be encrypted and protected from abuse or unauthorized access.

The FBI has selected a number of STR loci for which DNA profiles are being kept in a database (McElfresh ande Kim, 2000). This database is called the Combined DNA Index System, or CODIS. CODIS incorporates data that results from the analysis of 13 STR loci. These loci can be multiplexed in either one or two amplifications. The CODIS loci include those that are also found in the database systems used in Europe and South America, making it compatible with these other databases. Criminals who commit crimes in different parts of the world can be tracked and caught with the help of these systems.

The 13 STR loci are: D3S1358, VWA, FGA, D8S1179, D21S11, D18S51, D5S818, D13S317, D7S820, D16S539, THO1, TPOX, and CSF1PO (see Fig. 4.3). The match probability is in the order of 1 in 6×10^{14}, and typically the frequency of a genetic profile is calculated to be in the range of one in billions to one in trillions. The CODIS STR databases consist of an evidentiary database that contains genetic profiles for evidence found at crime scenes and a convicted offender database that contains genetic profiles of convicted felons.

The 13 loci were selected because of their high degree of polymorphism and their resulting high discrimination potential. Use of these loci ensures that an accurate and reliable match can be established between a genetic profile obtained from crime scene evidence and a profile contained in the database. At least 7 of the 13

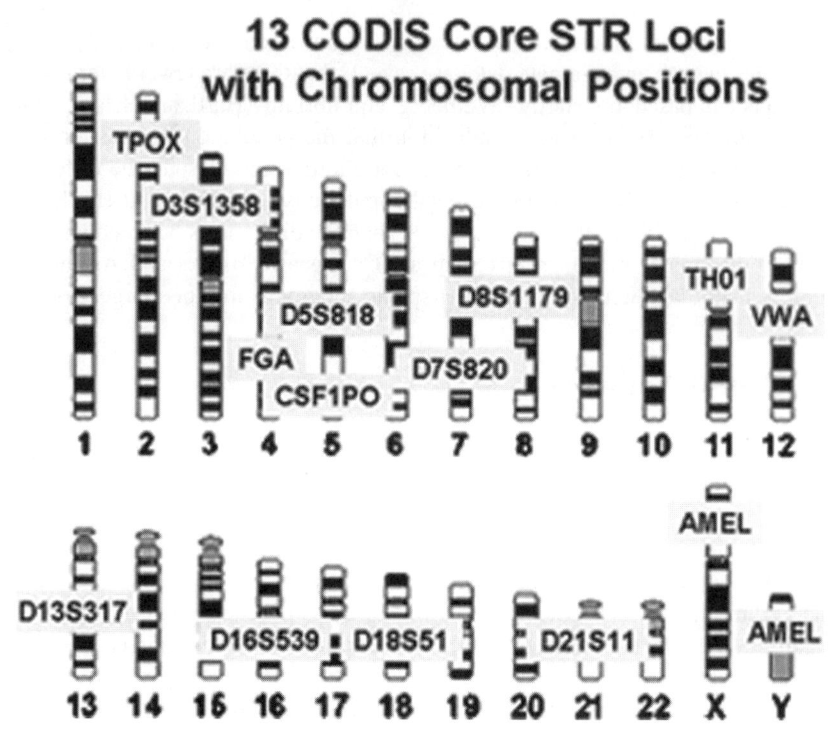

Fig. 4.3 *See color insert. The 13 CODIS loci. Adapted from a power point presentation by Dr. Peter Valone, with his permission.*

CODIS loci are required to identify an individual in the database assuming that the genotype consists of the most common alleles at each locus. Figure 3.39 contains information prepared by the FBI regarding the heterozygosity, number of alleles, and population match probability in a Caucasian American and African American sample for the 13 CODIS loci. Population databases that are maintained by the FBI include Caucasians, African Americans, southeastern and southwestern Hispanics, various Caribbean populations, as well as Japanese, Korean, Chinese, and other Asians (Budowle and Monson, 1993). These databases provide information about allele frequencies at each of the CODIS loci.

There are two searchable CODIS indices: (1) the Convicted Offender Index, which contains profiles of individuals who have been convicted of violent crimes, and (2) the Forensic Index, which contains DNA profiles derived from crime scene evidence. CODIS contains more than just digitized DNA profiles and the names of people who correspond to those profiles. CODIS also contains information about the specimen from which the DNA profile was obtained, the name of the laboratory at which the DNA was analyzed, and the names of the laboratory personnel who handled the specimen. Genetic profile information in the database is protected so that only authorized individuals can have access (Scheck, 1994)

One of the great benefits of having a national DNA database is that serial criminals that cross state lines to commit their crimes can be identified since all states contribute information to the National CODIS database (Dabbs and Cornwell, 1988). Even criminals who cross international borders to commit their crimes will also find no refuge since many of the 13 loci within CODIS overlap with databases maintained by other countries. Groups such as Interpol exchange information between law enforcement agencies in different countries about wanted felons.

According to a report on the Future of Forensic DNA Testing produced by the Department of Justice, by the year 2000, over 100 forensic laboratories were using CODIS. These included state, local, and federal laboratories. The report predicted that by 2005, the CODIS database should have more than one million convicted felon profiles on file. The group that prepared the report also predicted that interstate and international DNA comparisons would be possible and commonplace.

CLOSER LOOK AT DNA DATABANKS—THE NEW YORK STATE DNA DATABASE

New York State is actively using its own databases as well as CODIS to solve previously unsolved cases. Forensic laboratories throughout the state analyze DNA collected from crime scenes and compare the DNA profiles obtained against those contained within the state's database. The laboratories also compare the profiles to those contained in CODIS. New York has decided that persons convicted of a set list of offenses committed on or after the effective date of its DNA law are required to provide a DNA sample that is subsequently included in the state's DNA database. The long list of offenses are provided in Appendix D.

THE ISAAC JONES CASE

When it comes to serial rapists, there is none more infamous than Isaac Jones. Mr. Jones, who is now 42 years old, had been employed as a floor polisher at an office in midtown Manhattan. What he did after hours made him one of the worst predators of women in the history of New York State. Mr. Jones had been implicated in an extraordinary number of rapes, robberies, and violent assaults. Over a period of 5 years the number of his known victims rose to 51. The New York Police Department Street Crime Unit had been hunting for the assailant in the Bronx where most of the crimes had occurred at the time that they had mistakenly shot Mr. Amadou Diallo. Jones had also committed crimes in Manhattan and Westchester, New York. The police apprehended Mr. Jones only after his girlfriend attempted to pawn some of his stolen jewelry. Isaac Jones had been out on the streets after he was paroled for one of his robberies that occurred in 1993. At this time, the State of New York did not take DNA specimens from all felons. If his genetic profile had been in the state's database, he most likely would have been caught after his first assault.

In January 2002, New York State's Division of Criminal Justice Services (DCJS), Office of Justice Systems Analysis, enthusiastically published a research note that profiled the first 100 hits obtained using the New York State DNA Databank. The note was a DCJS study that reviewed the hits and offered recommendations that could be used to improve the database and its usage. Particularly compelling is the fact that, according to the DCJS study, of 104 matches made using the New York State DNA Database (SDIS) over an 18-month period, 33 of the hits or matches were matches in cases in which the police had no suspects and had exhausted every lead that might have helped to crack the cases (McEwen and Reilly, 1994; McEwen, 1995).

THE AROHN KEE CASE

Over a period of 8 years starting in 1991, Arohn Kee was responsible for the murder of 3 teenage girls and the rape of 4 other young women, all in New York City. The victims ranged in age from 13 to 19. The attacks in each case were brutal and vicious, and one of his victims was burned to a crisp on the rooftop of a building. At each of the crimes Mr. Kee left biological evidence. He had been arrested on a petty theft charge, but police had reason to believe that he had committed these heinous crimes. The police obtained a sample of his saliva from a coffee cup that he had discarded. DNA isolated from the saliva linked him to all of these crimes. He was found guilty on 22 counts and was sentenced to serve a term of 400 years without the possibility of parole. Interestingly, Arohn Kee had been arrested in the early 1990s on a felony first-degree burglary charge and convicted. Because New York State did not collect DNA from all felons at that time, his genetic profile had not been included on the state's DNA database. Had his information been included in the SDIS system, it is possible that 1 of the homicides and 2 of the rapes might have been prevented.

4.12 POWER OF DISCRIMINATION

Before DNA analysis became the method of choice, serologists and forensic biologists used cell surface antigens, serum proteins, and various enzymes to study biological evidence. Because these substances are all polymorphic and their inheritance follows Mendel's laws of inheritance, the analyst could compare evidence to a specimen obtained from the suspect to determine if there was a match and therefore an inclusion or if there was an exclusion. Inclusion indicates that the suspect could possibly be the source of the evidence, whereas exclusion is absolute and indicates that the suspect could not possibly have been the source of the evidence. Why are certain polymorphic genetic systems better for human identification than others? The answer lies in the mathematical concept of *power of discrimination.* This is a measure of the ability of the system to identify one particular individual out of a larger number of individuals, a process known as individualization.

The discrimination power of a genetic system is denoted as P_D, whereas the probability of identity is denoted as P_I. The probability of identity, P_I, refers to the ability to determine that two individuals within a population share the same genotype at a specific locus. The relationship between P_D and P_I is given by $P_D = 1 - P_I$. In simple terms, the power to differentiate two people at a particular locus, the power of discrimination, plus the power to identify two people at that same locus, the probability of identity, equals 100%. If the P_I is known, then the P_D can easily be determined using the above formula. The P_I for a single locus can be calculated as the sum of the *genotype frequencies* squared: $P_I = \sum X_i^2$.

If we look at a hypothetical locus with 5 alleles, *a, b, c, d,* and *e*, there would be $5 + 4 + 3 + 2 + 1 = 15$ possible genotypes (in a co-dominant system). Assume the following genotypes and genotype frequencies at locus XYZ:

Genotype	Frequency
aa	.01
ab	.02
ac	.05
ad	.08
ae	.10
bb	.12
bc	.11
bd	.12
be	.12
cc	.10
cd	.05
ce	.04
dd	.05
de	.02
ee	.01

The P_I is calculated as:

$$\begin{aligned} P_I &= (.01)^2 + (.02)^2 + (.05)^2 + (.08)^2 + (.10)^2 + (.12)^2 + (.11)^2 + (.12)^2 \\ &\quad + (.12)^2 + (.10)^2 + (.05)^2 + (.04)^2 + (.05)^2 + (.02)^2 + (.01)^2 \\ &= 0.0918 \end{aligned}$$

Therefore, the P_D is calculated as 0.9082.

The P_I is influenced not only by the degree of polymorphism (number of alleles) but also by the distribution of genotype frequencies. Thus the best polymorphic systems to use for identification purposes are those with a large number of genotypes and with frequencies evenly distributed among the various types. A less useful polymorphic system is one with fewer alleles and with some genotypes in high frequency and with others being very rare.

If one knows the P_I values for N loci (let us assume that $N = 13$), the combined P_I value can be calculated as

$$\text{Combined } P_I = (P_I1)(P_I2)(P_I3)(P_I4)(P_I5)(P_I6)(P_I7) \times (P_I8)(P_I9)(P_I10)(P_I11)(P_I12)(P_I13)$$

This exceedingly small number reflects the probability that two individuals would have the same genotypes at all 13 loci. By deducting this number from 1.0, the resulting number reflects the tremendous power of discrimination. (The closer the P_D approaches 1, the higher the discrimination potential for all 13 loci. Because the P_D is so high when using the 13 CODIS loci, there is no need to analyze any additional loci.

4.13 MIXTURES AND STATISTICS

When analyzing an evidentiary specimen, there are times when the genotype at one or more loci indicates that there is a mixture of two or more individuals. Since a homozygote has only one allele represented twice and a heterozygote has one copy of each distinct allele, finding three or more alleles at a single locus indicates that there is a mixture of DNA from different individuals. Three alleles can be explained by a mixture of DNA originating from three homozygotes or a mixture of DNA from two heterozygotes with one allele in common for both genotypes. Still other possibilities exist to explain the mixture. For example, the observation of three alleles could result from a mixture of DNA from a heterozygote and a homozygote, or a heterozygote and two homozygotes assuming overlapping alleles. The observation of four alleles at one locus can be produced by a mixture of specimens from two to four individuals. The observation of a mixture is not unusual when analyzing sexual assault evidence since the differential lysis procedure is less than 100% efficient at separating DNA from spermatozoa and from epithelial cells in the extracted mixture. The female DNA often "contaminates" the male DNA and can be seen in the male fraction. If a mixture exists, it may be possible to determine if there is a major and minor component based on the peak heights (intensity of fluorescence) observed in the electropherograms. Thus, if a mixture consists of specimens present in the ratio of 3:2, 7:3, or 4:1, the peak heights produced by each genotype will vary significantly in size. If the mixture is in the ratio of 1:1, then the peak heights would all be approximately the same. When samples are mixed in ratios greater than 9:1, the smaller component alleles may be seen in "subthreshold" peak sizes or they may not be seen at all as a result of differential amplification resulting in overwhelming quantities of the alleles of the major component.

4.14 PROBABILITY OF EXCLUSION

There are two ways to handle the statistical calculations for a mixed sample. The analyst may simply indicate the statistics for the major and minor components.

This becomes difficult and even confusing if the major and minor allele peaks are not discernible over all loci tested. The other way to perform the statistical analysis is to calculate the *probability of exclusion,* P_E, which is a very conservative way of dealing with the statistics for a mixture. This calculation treats all peaks as if they have the same peak height or intensity. The P_E describes the frequency of the observed profile that would result from the combined frequencies of all potential donors. In other words, if the alleles seen in the mixture were 8, 9, 10, and 11—even with differences in peak intensities seen—then the P_E calculation would be made by adding together the frequencies of all individuals in the population who could have contributed these four alleles, for example, the frequencies of individuals with genotypes 8,9 and 10,11 plus the frequencies of those with types 8,10 and 9,11, plus the frequency of those individuals with genotypes 8,11 and 9,10 plus the frequencies of 4 homozygotes (8,8 + 9,9 + 10, 10 + 11,11). The resulting statistic is far greater (larger fraction) than that obtained if the individual profiles were discernible and therefore the overall frequency understates the importance of the evidence. For example, a P_E of $1/50$ means that 1 in 50 individuals in the relevant population could be a contributor of one or more alleles in the mixture. All other individuals are excluded as potential donors. This statistic obviously includes *potential* contributors to the mixture who, in fact, did not contribute any alleles but who *could* have (false inclusion). Following the calculation of the P_E, the analyst might indicate that suspect 1 and suspect 2 (assuming 2 suspects) cannot be eliminated as the source of the observed alleles seen in the evidence, however, suspect 3 can be excluded.

4.15 THE LIKELIHOOD RATIO (LR)

When comparing an evidentiary specimen and an exemplar, the analyst is asking the question, "Is there an inclusion, that is, both specimens may have a common origin, or is there an exclusion?" An exclusion would mean that there is no possibility of a common origin between the two specimens. Another way to look at this issue is to hypothesize two different cases. In the first case, the evidence and exemplar specimens come from the same source. In the second case, both specimens have different origins but may appear to be identical as a result of random sampling in the relevant population. Of course, if there is an inclusion over the entire genetic profile, then the analyst calculates the rarity of the profile using the product rule as described in Section 3.2.10. The *likelihood ratio* is a way of determining how likely it is that the observed genetic profile came from the true perpetrator rather than from a random individual. The likelihood ratio test is a statistical test of the goodness-of-fit between two models. A relatively more complex case is compared to a simpler case to see if it fits a particular data set better. In criminalistics, it is expressed as the ratio of the probability that the evidence and exemplar have a common source and therefore have the same genetic profile [which reflects (case 1)] divided by the probability that the evidence and exemplar have different origin but match in all loci tested as a result of random sampling [(case 2)]. The numerator will be 1.0, whereas the denominator is influenced by the statistic produced by the product rule.

$$\text{The likelihood ratio} = \frac{\text{Probability of case 1}}{\text{Probability of case 2}}$$

where case 1 = evidence and exampler have common origin
case 2 = evidence and exampler have different origin

If, for example, we calculate that the frequency of a genetic profile is 1 in 10^9, then the likelihood ratio would be calculated as follows:

$$\text{LR} = \frac{1}{1 \times 10^{-9}} = 10^9$$

What this means is that it is one billion times more likely that case 1 is the correct hypothesis and that it is highly unlikely that the genetic profile of suspect and exemplar matched merely due to random sampling of the relevant population. The likelihood ratio becomes very important when considering the existence of siblings and other close relations who cannot be excluded as contributors of the evidence (Brookfield, 1995b). Their genetic profiles may be identical to the evidence over the number of loci tested and so the calculation of LR will differ from the calculation for a random individual.

4.16 PATERNITY DETERMINATIONS

Currently DNA analysis methods are being used by paternity laboratories to establish whether an alleged father (AF) is the biological father of a child. Before DNA testing became available in the early 1990s, paternity testing was performed by analyzing polymorphic genetic systems such as red blood cell antigens (ABO, Rh, MNSs, Kell, Duffy, and Kidd), red cell enzymes (PGM, phosphoglucomutase; EsD, esterase D; ACP, acid phosphatase; GLO I, glyoxylase; ADA, adenosine deaminase; AK, adenylate kinase), serum proteins (Hp, haptoglobin; Gc, group-specific component; Tf, transferrin), and HLA, human leukocyte antigens (loci A, B, and C) (The D/DR locus is not used in paternity determinations). The HLA system is very polymorphic, and its use extended the capabilities of paternity testing laboratories to both include and exclude an AF. The HLA loci are linked (all on the same chromosome—number 6), and therefore HLA alleles at all loci are inherited as a unit (haplotype). The HLA system follows Mendelian laws of inheritance.

In paternity determinations, specimens are obtained from the "triad" of mother, child, and the alleged father and phenotyped in a number of polymorphic systems. All of the above-mentioned genetic systems are polymorphic, each being composed of a number of alleles for which their frequencies in the relevant population are known. Paternity testing is performed not only to identify the AF as the biological

TABLE 4.3 Probability of Exclusion for Each of the CODIS STR Loci

Locus	Probability of Exclusion
D3S1358	0.6823
VWA	0.6309
FGA	0.7182
THO1	0.5468
TPOX	0.3950
CSF1PO	0.5170
D5S818	0.4768
D13S317	0.5226
D7S820	0.5991
D21S11	0.6980
D18S51	0.7458
D8S1179	0.6096
D16S539	0.5581

Source: Data from Annual Report Summary for Testing in 2001, Prepared by the Parentage Testing Program Unit, American Association of Blood Banks, October 2002.

father but also to protect the wrongfully accused person. This is referred to as the *power of exclusion* (PE).

Each genetic system has a power of exclusion based on the number of alleles in that system, allelic frequencies, and ethnic group examined (See Table 4.3). In paternity evaluations, genetic systems that have higher PE values are preferable over those with lower PE values. For the complete battery of tests a cumulative power of exclusion (CPE) can be calculated by multiplying individual PE values. Ideally, the CPE should be as close to 1.00 as possible, which means that every falsely accused AF would be excluded. The CPE can be calculated using the following formula:

$$\text{CPE} = 1 - (1 - P_1)(1 - P_2)(1 - P_3)(1 - P_4), \ldots, (1 - P_n)$$

where P_1 equals the PE for the first genetic system, P_2 equals the PE for the second system, and so on. CPE values will vary for different ethnic groups since individual allele frequencies will vary with different groups. Paternity laboratories often provide a statistic described as the RMNE, which stands for random man not excluded. This reflects the frequency with which men selected at random from the relevant population (ethnic group) who have the same phenotypes as the AF would not be excluded as the biological father. The RMNE is related to the PE by the formula RMNE $= 1 -$ PE. The RMNE can be calculated for each genetic marker analyzed. Over the total profile the cumulative power of exclusion can be calculated as 1 – the cumulative RMNE value.

4.16.1 Exclusion of the AF as the Biological Father

Nonpaternity can be demonstrated if the phenotypes (or genotypes) of the mother, child, and AF are determined and if the child possesses an allele that must have come from its biological father (because it is not present in the mother) but is lacking in the AF. This allele is known as the obligate allele. Calculation of the paternity index for this genetic system results in a paternity index (PI) value of 0. This is a first-order exclusion. There is another way that the AF can be excluded, but it is less powerful than the example just described. For example, the child can possess an allele that must have come from its biological father (even though the mother has this allele) but it is lacking in the AF. This would be the case if the triad were typed in the ABO blood group system, and the child was an AA, the mother was an AA and the father was a BB. This is a second-order exclusion. Although the first-order exclusion demonstrates a conclusive exclusion, most experts require at least two second-order exclusions before declaring officially that the AF is excluded.

4.16.2 Inclusion of the AF as the Biological Father

The paternity laboratory will analyze the triad's specimens and determine phenotypes (detectable biochemical or immunological traits) for each. The phenotype reflects the genotype since the alleles are co-dominant in most of the systems tested. For each system, what is the probability that the child inherited its paternal allele from the AF as opposed to a random male in the relevant population? The answer to this question depends upon the level of the system's polymorphism and the allelic frequencies in that population (racial group). The probability calculation is based on Bayes theorem in which there are two hypotheses being tested. The first is that the AF is the biological father and the second is that he is not. Unlike criminal matters where the suspect is considered innocent until proven guilty, paternity cases start off with the premise that it is equally likely that hypothesis 1 and 2 are correct. In other words it is just as likely that he is the biological father as not. At each genetic locus a calculation is made to establish the *paternity index* (PI) for that marker. The PI is a ratio of two numbers. The numerator is the probability that the sperm that fertilized the mother's ovum came from the biological father. The denominator is the probability that that sperm came from a random male in the relevant population. The higher the PI the higher the likelihood that the AF is the biological father. Using the systems mentioned earlier, the individual PI values generally range from a fraction to a number less than 10 (although higher values are occasionally found). With the use of HLA loci a PI value can be significantly higher than 10, depending in part on the rarity of the observed haplotypes. The Combined Paternity Index (CPI) is calculated by multiplying individual PI values for each system tested.

$$\text{CPI} = (\text{PI}_1)(\text{PI}_2)(\text{PI}_3)(\text{PI}_4), \ldots, (\text{PI}_n)$$

Thus if you test 6 systems and calculate PI values of 1.3, 1.7, 2, 3.2, 1.1, and 2.7, then the CPI is calculated to be 42.00. The relative chance of paternity is then calculated as follows:

$$\text{CPI}/(\text{CPI} + 1) \times 100\% = 42/43 \times 100\% = 0.9767 \times 100\% = 97.67\%$$

Most scientists who use non-DNA genetic markers will consider any CPI value of 20 or more to be significant since this corresponds to a probability of paternity of 95% or higher. This means that there is a 5% chance that a random person could be the biological father. When the CPI reaches 100 or more, this brings the probability of paternity to the 99% level and provides more confidence for the conclusion of paternity.

Testing DNA has almost completely replaced the genetic testing described above for paternity considerations. DNA is far more stable than proteins, glycoproteins, or sugars and is extraordinarily polymorphic at a large number of loci. DNA testing performed as paternity testing has been used in a number of rape cases where fetal tissue is available to examine or where a child has been born. The premises and concepts remain the same, but the calculations are performed somewhat differently (see Table 4.4).

Occasionally, paternity must be determined in the absence of the mother. For example, if the mother has abandoned the child and cannot be found, a paternity test can be used to verify that an individual is the biological father. There are a number of possible formulas that can be used to calculate the paternity index by examining the alleles shared between child and alleged father. In most cases the formula PI $= 1/4P_r$ (where P_r is the frequency of the shared allele) is acceptable and sufficient (Brenner, 1993).

Table 4.4 can be used to calculate PI values for genotypes found in the triad.

TABLE 4.4 Formulas for Calculating Paternity Index

RMNE and PE[a] Values for VNTR Systems

Band Size Notation	Allele Frequencies for Corresponding Bands
P	p
Q	q
R	r

Combination	M	C	AF	Calculation of Paternity Index	RMNE
1	P	P	P	$1/p$	$p(2-p)$
2	P	P	PQ	$1/2p$	$p(2-p)$
3	P	PQ	PQ	$1/2q$	$q(2-q)$
4	PQ	PQ	PQ	$1/(p+q)$	$2(p+q)-(p+q)^2$
5	PQ	P	P	$1/p$	$p(2-p)$
6	PQ	PQ	P	$1(p+q)$	$2(p+q)-(p+q)^2$
7	PQ	P	PQ	$1/2p$	$p(2-p)$
8	P	PQ	Q	$1/q$	$q(2-q)$
9	PQ	QR	PR (or RS)	$1/r$	$r(2-r)$
10	P	PR	QR	$1/2r$	$r(2-r)$
11	PQ	QR	PR (or RS)	$1/2r$	$r(2-r)$
12	PR	PR	QR	$1/[2(p+r)]$	$2(p+r)-(p+r)^2$
13	PR	QR	QR	$1/2q$	$q(2-q)$
14	PR	R	QR	1/2r	$r(2-r)$
15	absent	PQ	QR	$1/4q$	$2(p+q)-(p+q)^2$
16	absent	PQ	Q	$1/2q$	$2(p+q)-(p+q)^2$
17	absent	PQ	PQ	$(p+q)/4pq$	$2(p+q)-(p+q)^2$
18	absent	Q	QR	$1/2q$	$q(2-q)$
19	absent	Q	Q	$1/q$	$q(2-q)$

[a] PE = 1 − (RMNE).

Source: American Association of Blood Banks.

4.17 LAB ACCREDITATION, CERTIFICATION, REPUTATION, AND FACILITIES

4.17.1 Quality Control

Quality control includes maintaining records for the chemicals, reagents, commercial kits, and other supplies that are purchased by a laboratory for use in testing. Such products are tested to see if they provide reliable results with other reagents prior to their use in casework. They are stored appropriately (dry, frozen, refrigerated, etc.) and any expiration date is noted in a log. All equipment in a laboratory that is used to maintain or provide changing temperatures are calibrated and monitored. Such equipment would include refrigerators, freezers, water baths, thermal cyclers (used for DNA work), and the like. All equipment used for weighing must be calibrated on a regular basis and all records maintained on file. All equipment used for centrifugation must also be calibrated and records kept for inspection. Laboratory procedures are maintained in a protocol manual that is available for inspection.

4.17.2 Quality Assurance

Quality assurance includes continuing education, proficiency testing, analyst certification, and laboratory accreditation. Quality assurance is a way of demonstrating that the analyses are valid and that the results of testing are reliable (Morton and Collins, 1995). SWGDAM (TWGDAM), the DNA Advisory Board (DAB, 1998, 1999), the two NRC reports of 1992 and 1996, and guidelines published by the FBI have all contributed to our understanding of what quality control means in a forensic setting. One aspect of quality assurance is the documentation of a laboratory's quality control program and procedures. For example, the FBI issued Quality Assurance Standards for Convicted Offender DNA Databasing Laboratories. Standard 3.1 states that "The laboratory shall establish and maintain a documented quality system that is appropriate to the testing activities."

3.1.1 The quality manual shall address at a minimum:

(a) Goals and objectives
(b) Organization and management
(c) Personnel qualifications and training
(d) Facilities
(e) Sample control
(f) Validation
(g) Analytical procedures
(h) Calibration and maintenance
(i) Proficiency testing
(j) Corrective action
(k) Documentation
(l) Review
(m) Safety
(n) Audits

Because the technology has become so powerful, quality control and quality assurance are vital since a scientific determination can result in a conviction and lengthy sentence or, in some cases, even a sentence of death. On the other hand, it can be used to exonerate those who have been convicted of violent crimes (Travis and Asplen, 1999). The American Society of Crime Laboratory Directors, Laboratory Accreditation Board, accreditation process monitors a laboratory's compliance with quality assurance and quality control measures.

4.17.3 Proficiency Testing

Proficiency testing is an important component of quality control and quality assurance. It is an outgrowth of the DNA revolution in forensic science. Forensic laboratories employ commercial labs that offer mock casework test trials. The casework is assigned to an analyst who handles it like any other item of evidence. A report is prepared, and the laboratory director can then determine if the results are correct or not. The purpose of proficiency testing is to verify that individual analysts are performing quality work according to the laboratory protocol, reaching the correct conclusion, and reporting it accurately. Failure to do so would result in more training for the analyst. Such proficiency testing can be done in an open or blind fashion. In the former case, the analyst is aware that the casework is not authentic and is being used to test skills and adherence to protocol. In the latter, the analyst is not told the true nature of the casework, and that the results will be used as part of a proficiency examination. Blind proficiency testing is not required presently for forensic laboratories (see Section 3.6.6).

4.17.4 Certification

Certification is a voluntary process of peer review by which a practitioner is recognized as having attained the professional qualifications necessary to practice in one or more disciplines of criminalistics. Certification is offered by the American Board of Criminalistics (ABC), which is composed of regional and national forensic science organizations. Its purposes are to establish professional levels of knowledge, skills, and abilities, to define a mechanism for achieving these levels, to recognize those who have demonstrated attainment of these levels, and to promote growth within the profession. The ABC offers a certificate in criminalistics (general) and certificates in more specific areas such as forensic biology, drug chemistry, fire debris analysis, and trace evidence.

4.17.5 Laboratory Accreditation

In the fall of 1974, crime laboratory directors formed the American Society of Crime Laboratory Directors (ASCLD), which is devoted to the improvement of crime laboratory operations through sound management practices. The members promote and encourage high standards of practice by member laboratories. The Crime Laboratory Accreditation Program of ASCLD (ASCLD/LAB—ASCLD Laboratory

Accreditation Board) is a voluntary program. Any crime lab within or outside the United States may apply for accreditation in order to demonstrate that its management, operations, personnel, procedures, equipment, physical plant, security, and personnel safety procedures meet established standards. Approximately 200 laboratories are now accredited by ASCLD/LAB.

4.18 REVIEWING A DNA REPORT—A SAMPLE RFLP ANALYSIS

Generally, DNA test results are documented in the form of a laboratory report. The laboratory report is a key piece of evidence. We now examine four laboratory reports. The first laboratory report was prepared by the FBI. It is a supplementary report to two previous reports. The first page bears the logo of the lab that prepared the report. This particular report is addressed to "Mr. Kenneth Starr" and identifies him as being located in the "Office of the Independent Counsel."

The first section of the report describes the specimens received by the laboratory. It indicates that the DNA samples were subjected to RFLP examination/analysis. The items deal directly with chain of custody. The report states that it will retain the analyzed specimens and membranes until the submitting agency retrieves them. Also, that the returned samples "should be refrigerated or frozen" and that the samples should be isolated from evidence that has not yet been examined. These statements should be a red flag to an attorney. At trial, the attorney will want to ascertain whether these recommendations were adhered to before the laboratory received the specimens, while the specimens were at the laboratory, and subsequent to their departure from the laboratory, especially if the specimens were later retested.

The second page of the report shows even more key information. The examiner's name appears in the report [it has been redacted (blacked out) to prevent identification]. However, specimen numbers, laboratory numbers, and other internal reference numbers are present. It is important to verify that these numbers are consistent throughout the document. The second page lists the specific genetic (VNTR) loci that were used to test the evidence. These seven loci include D2S44, D17S79, D1S7, D4S139, D10S28, D5S110, and D7S467. The report also indicates that HaeIII was the restriction enzyme used to cut the DNA and that the DNA sample was derived from a semen stain. A review of the facts surrounding this particular report would show that the semen was obtained from the dress of a woman named Monica.

A review of the report shows that two samples were tested. "specimen K39 (Clinton)" and "specimen Q3243-1." Specimen K39 is the biological exemplar of a known individual and specimen Q3243-1 is evidence obtained from the dress. The report concludes that based upon the DNA analysis, "specimen K39 (Clinton) [was] the source of the DNA obtained from specimen Q3243-1, to a reasonable degree of scientific certainty."

It is of note to both prosecutors and defense attorneys that blanket conclusory statements such as that made at the end of this report regarding the source of the DNA, without more authority, should not be allowed to support themselves. This is

FBI LABORATORY

FEDERAL BUREAU OF INVESTIGATION
WASHINGTON, D. C. 20535

Date: August 17, 1998

To: Mr. Kenneth W. Starr
Office of the Independent Counsel
1001 Pennsylvania Avenue, N.W.
Suite 490-North
Washington, D.C. 20004

FBI File No. 29D-OIC-LR-35063

Lab No. 980730002 S BO
980803100 S BO

Reference: Communication dated July 30, 1998 and evidence submitted August 3, 1998

Your No. 29D-OIC-LR-35063

Re: MOZARK;
MC 106

Specimens received: July 30, 1998 and August 3, 1998

Specimens:

This report supplements two FBI Laboratory reports dated August 3, 1998 and August 6, 1998 and contains the results of the DNA-RFLP examinations.

This completes the requested examinations. The submitted items and the probed DNA membranes will be retained until retrieved by a representative of your organization. In addition to the evidence in the case, any remaining processed DNA from specimens examined by DNA analysis is also being returned to you. The processed DNA can be found in a package marked PROCESSED DNA SAMPLES: SHOULD BE REFRIGERATED/FROZEN. It is recommended that these samples be stored in a refrigerator/freezer and isolated from evidence that has not been examined.

This Report Is Furnished For Official Use Only

LABORATORY

FEDERAL BUREAU OF INVESTIGATION
WASHINGTON, D. C. 20535

Report of Examination

Examiner Name:	[redacted]	Date:	08/17/98
Unit:	DNA Analysis 1	Phone No.:	202-324-4409
FBI File No.:	29D-OIC-LR-35063	Lab No.:	980730002 S BO 980803100 S BO

Results of Examinations:

Deoxyribonucleic acid (DNA) profiles for the genetic loci D2S44, D17S79, D1S7, D4S139, D10S28, D5S110 and D7S467 were developed from HaeIII-digested high molecular weight DNA extracted from specimens K39 and Q3243-1(a semen stain removed from specimen Q3243). Based on the results of these seven genetic loci, specimen K39 (CLINTON) is the source of the DNA obtained from specimen Q3243-1, to a reasonable degree of scientific certainty.

No DNA-RFLP examinations were conducted on specimen Q3243-2 (a semen stain removed from specimen Q3243).

BLACK – 1,440,000,000,000
CAUC – 7,870,000,000,000
SEH – 3,140,000,000,000
SEH – 943,000,000,000

DNAU1 - Page 1 of 1

This Report Is Furnished For Official Use Only

mainly because DNA analysis is a science of exclusion, not inclusion. More importantly, it is a science based on laws of probability and statistics. The results of DNA testing should not state that a sample absolutely originated from a particular person unless the statistic developed indicates a rarity that makes it impossible for anybody else to share the observed genetic profile (except for identical twins). DNA analysis is used to determine whether a sample could have possibly originated from a particular person. If the analyst concludes that despite DNA testing, a person can still not be ruled out as the source of the DNA, and that person is a possible source of the DNA, then it should be determined how likely it is that the suspect was the source of the evidence. This is accomplished by thorough statistical calculations that tell us how frequently a particular profile is expected to appear in each of the populations considered.

This report indicates that the DNA profile obtained from the semen found on Monica's dress is expected to be found in only one out of 7,870,000,000,000 Caucasians. Based upon this evidence, it is safe to say that the semen and the exemplar specimen shared a common origin. Population frequencies were provided for persons of various racial groups.

This report contains items that are essential in a scientific report. A report should contain a description of the type of test that was performed. The actual protocol used need not be included in the summary report but should be made available for inspection by the opposing expert. It should describe the samples tested; it should have some information regarding chain of custody and handling of the specimens. The examining agency and examiner should be named. The person requesting the tests should be identified. The report should list the dates the specimens were received by the laboratory and even when the tests were performed (especially if there was a delay in testing). The specific loci tested should be indicated in addition to the type of analysis performed. The results of the tests should be indicated and the analyst's conclusions should be accompanied by supporting statistics. If some of the above information is not presented in the report (i.e., the date that the tests were conducted), it should certainly be documented in the bench notes of the analyst. In that way, the records are preserved and are available for inspection.

4.19 REVIEWING A DNA REPORT—A PCR-BASED DNA EXAMINATION (HLADQA1, PM, D1S80, AND CTT-CSF1PO, TPOX, THO1)

The second DNA report was prepared by the Virginia Department of Criminal Justice, Division of Forensic Science. The report concerns the analysis of DNA obtained from various places on a woman's body, including a bite mark on the woman's right upper back. The DNA taken from the "victim" was compared to a "suspect" named Marvin Albert, who is known to many as a prominent sportscaster.

The first page of the report contains similar information to the RFLP test report described in Section 4.18. The report contains reference numbers and describes the samples that had been tested. It also identifies who had submitted the evidence and

Commonwealth of Virginia
Department of Criminal Justice Services
DIVISION OF FORENSIC SCIENCE

CERTIFICATE OF ANALYSIS

Northern Laboratory
9797 Braddock Road
Suite 200
Fairfax, VA 22032

August 4, 1997

Tel. No.:	(703) 764-4600
Fax:	(703) 764-4633
TDD/Voice:	(804) 786-6352

TO: JANE MORRIS
ARLINGTON COUNTY POLICE DEPARTMENT
1425 N COURTHOUSE ROAD
ARLINGTON VA 22201

FILED
AUG 8

SUPPLEMENTAL REPORT
FS Lab #N97-1490

Your Case #: 970212004

Victim(s): [illegible]

Suspect(s): ALBERT, Marvin

Evidence Submitted By: Jane Morris Date Received: 02/19/97

Item 1 Physical Evidence Recovery Kit from [illegible]

Evidence Submitted By: Jane Morris Date Received: 06/06/97

Item 2 Physical Evidence Recovery Kit from Albert

RESULTS:

This report supplements the test results reported in the Certificate of Analysis for FS Lab #N97-1490 dated April 30, 1997.

Item 1

Spermatozoa were previously identified in extracts of the lips/lip area swabs, the "left upper chest" swabs, the stain from the crotch area of the underpants and the stain from the rear panel of the underpants. Human deoxyribonucleic acid (DNA) was isolated from these samples. In addition, human DNA was isolated from the "bite mark right upper back" swabs and from the blood sample from [illegible].

The sample from the lips/lip area swabs was amplified and typed at the HLA DQA1 locus and the PM system (which includes the LDLR, GYPA, HBGG, D7S8 and GC loci). Refer to the Table for the typing results.

The samples from the "left upper chest" swabs, the stain from the crotch area of the underpants, the stain from the rear panel of the underpants, the "bite mark right upper back" swabs and the blood sample from [illegible] were amplified and typed at the HLA DQA1 locus, the PM system (which includes the LDLR, GYPA, HBGG, D7S8 and GC loci), the CTT system (which includes the CSF1PO, TPOX and TH01 loci) and the D1S80 locus. Refer to the Table for the typing results.

No examination was conducted on the remaining contents of this item at this time.

Item 2

Human DNA was isolated from the blood sample from Albert. This sample was amplified and typed at the HLA DQA1 locus, the PM system (which includes the LDLR, GYPA, HBGG, D7S8 and GC loci), the CTT system (which includes the CSF1PO, TPOX and TH01 loci) and the D1S80 locus. Refer to the Table for the typing results.

No examination was conducted on the remaining contents of this item at this time.

Page 1 of 6

DFS 70-015 (Rev 7/95)

the date on which it was submitted. The report also references other reports that had been prepared previously for the same case.

Page 1 of the report shows that the samples were subjected to analysis at the HLA DQA1 locus and the five loci of the AmpliType Polymarker system (PM), D1S80 which is an AMPFLP, and three STR loci, CSF1PO, TPOX, and THO1, abbreviated as CTT (Budowle et al., 1996, 1997). It should also be noted that separate descriptions appear for each item of evidence and the tests performed on each sample are individually described. This is particularly true for the DNA samples obtained from the suspect as well as those obtained from the victim and her underpants. Page 2 also has a sentence that indicates that an examination was *not* conducted on certain items of evidence. Attorneys on both sides would be wise to either exploit or explain such omissions at trial.

Pages 3 and 4 of the report contain a summary of the test results. These results appear as a table that allows for quick and easy comparison of the results obtained from each item of evidence. In a situation such as this, in which there are multiple samples, a chart is invaluable as it is important to be able to compare the typing results obtained from one DNA sample against those obtained from another sample and in this way determine if there is a match between the two.

Page 5 of the report indicates the laboratory's conclusions and also presents statistics for each genetic profile obtained from the various items of evidence.

In the present case, the DNA profile obtained from the sperm found on the victim's left upper chest and the victims underpants appeared to be consistent with the DNA profile of the suspect, Marvin Albert. The report concludes that "Marvin Albert [could not] be eliminated as a possible contributor of the genetic material isolated from [the] samples." The report then gives the population frequency statistics that support the conclusions with respect to the DNA derived from the bite mark, underpants, and chest. This test is different from the RFLP analysis described in Section 4.18 because it expresses the analyst's conclusions with respect to exclusion, rather than inclusion. If test results and analyst conclusions are not expressed appropriately, then defense counsel may be able to make an argument at trial regarding examiner bias.

The sixth and final page of the report has a section that states that no results were obtained for the analysis of the lip area swabs. The defense attorney can examine the state's expert regarding the failure to obtain a genetic profile from this evidence. The test states that certain DNA samples were insufficient in amount for RFLP analysis. The defense may also wish to note the failure of the testing laboratory to use a more highly resolving DNA test such as sequencing to establish inclusion or exclusion. The laboratory report also states that a separate RFLP report would be prepared at a later date. Both the defense and prosecuting attorneys should obtain a copy of this later report.

Just as in the RFLP report, this report indicates that the samples are being retained at the laboratory until requested. Once again, chain-of-custody issues may be raised, including evidence handling and storage conditions. Information regarding chain of custody may be important if the defense plans to raise a question about contamination of evidence.

Commonwealth of Virginia
Department of Criminal Justice Services
DIVISION OF FORENSIC SCIENCE

CERTIFICATE OF ANALYSIS

Arlington County Police Department
Your Case # 970212004
FS Lab # N97-1490
August 4, 1997

SUMMARY OF HLA DQA1, PM, CTT AND D1S80 TYPING RESULTS

Item	Description	HLA DQA1	LDLR	GYPA	HBGG	D7S8	GC	CSF1PO	TPOX	TH01	D1S80
1	Blood sample from [redacted]	4.1, 4.1	AB	AB	BB	AB	AC	10, 12	10, 10	6, 7	24, 31
2	Blood sample from Marvin Albers	2, 2	AB	AB	BC	AA	BC	10, 11	8, 8	8, 8	24, 24
1	Lips/lip area swabs from [redacted]										
	-non-sperm fraction	4.1, 4.1	AB	AB	BB	AB	AC	---	---	---	---
	-sperm fraction	***	***	***	***	***	***	---	---	---	---
1	"Left upper chest" swabs from [redacted]										
	-non-sperm fraction	2, (4.1)	AB*	AB	BC	A(B)	BC(A)	10, 11	6, (10)	8, 9	---
	-sperm fraction	2, 2	AB	AB	BC	AA	BC	10, 11	8, 8	8, 8	***

Note: Numbers and/or letters in parentheses ( ) represent alleles that are present in lesser intensity than other alleles not in parentheses.
--- Indicates not tested *** Indicates no amplification results

Page 3 of 6

DFS-70 [illegible]

The laboratory notebooks (bench notes) and laboratory testing protocols should be requested, along with the temperature control logs, and documentation of scale and analytical instrument calibrations.

Commonwealth of Virginia
Department of Criminal Justice Services
DIVISION OF FORENSIC SCIENCE

CERTIFICATE OF ANALYSIS

Arlington County Police Department
Your Case # 970212004
FS Lab # N97-1490
August 4, 1997

SUMMARY OF HLA DQA1, PM, CTT AND D1S80 TYPING RESULTS

Item	Description	HLA DQA1	LDLR	GYPA	HBGG	D7S8	GC	CSF1PO	TPOX	THO1	D1S80
1	Crotch area stain from [redacted] underpants										
	-non-sperm fraction	2, (4.1)	AB	AB	BC	A(B)	BC(A)	10, 11, 12	8, 10	8, (6), (7)	24, (31)
	-sperm fraction	2, 2	AB	AB	BC	AA	BC	10, 12	8, 8	9, 9	24, 24
1	Rear panel stain from [redacted] underpants										
	-non-sperm fraction	2, (4.1)	AB	AB	BC	A(B)	BC(A)	10, 11, (12)	8, (10)	8, (6), (7)	24, (31)
	-sperm fraction	2, 2	AB	AB	BC	AA	BC	10, 11	8, 8	8, 8	24, 24
1	"Bite mark right upper back" swabs from [redacted]	2, 2	AB	AB	BC	AA	BC	10, 11	8, 8	8, 9	24, 24

Note: Numbers and/or letters in parentheses ( ) represent alleles that are present in lesser intensity than other alleles not in parentheses.

Page 4 of 6

DFS-75-003 (Rev 7/00)

It is important that one become acquainted with laboratory reports such as those described herein. They are not difficult to understand, and they are indeed crucial to any case involving DNA when used in conjunction with the testimony of an expert witness. These types of report constitute DNA evidence, whether being asserted for purposes of a criminal trial or a paternity test or some other civil litigation.

Commonwealth of Virginia
Department of Criminal Justice Services
DIVISION OF FORENSIC SCIENCE

CERTIFICATE OF ANALYSIS

Arlington County Police Department
Your Case # 970212004
FS Lab # N97-1490
August 4, 1997

CONCLUSIONS:

Based on the above test results, the DNA profile obtained from the sperm fractions o the "left upper chest" swabs (Item 1), the crotch area stain from the underpants (It 1) and the rear panel stain from the underpants (Item 1) as well as the DNA profile obtained from the "bitemark right upper back" swabs (Item 1) is consistent with the DNA profile of Marvin Albert. Therefore, Marvin Albert cannot be eliminated as a possible contributor of the genetic material isolated from these samples.

The probability of randomly selecting an unrelated individual with a matching DNA profile (HLA DQA1, PM, CTT and D1S80) as obtained on the crotch area and rear panel stains from the underpants (Item 1) and the "bite mark right upper back" swabs (Item 1) is approximately:

1 in 2.6 billion in the Caucasian population

1 in 670 million in the Black population

1 in 1.1 billion in the Hispanic population

The probability of randomly selecting an unrelated individual with a matching DNA profile (HLA DQA1, PM and CTT) as obtained on the "left upper chest" swabs (Item 1) approximately:

1 in 310 million in the Caucasian population

1 in 38 million in the Black population

1 in 83 million in the Hispanic population

The DNA profiles obtained from the non-sperm fractions of the "left upper chest" swa (Item 1) and the crotch area and rear panel stains from the underpants (Item 1) are consistent with a mixture of the DNA profiles of [illegible] and Marvin Albert. Therefore, neither [illegible] nor Marvin Albert can be eliminated as possible c contributors of the genetic material isolated from these samples.

No types foreign to [illegible] were detected in the non-sperm fraction of her lips/lip area swabs (Item 1).

Page 5 of 6

4.20 REVIEWING A DNA REPORT—A PCR-STR-BASED DNA EXAMINATION (CODIS LOCI)

The DNA report illustrated in this section was prepared by the the Office of Chief Medical Examiner, Department of Forensic Biology. The report concerns the

Commonwealth of Virginia
Department of Criminal Justice Services
DIVISION OF FORENSIC SCIENCE

CERTIFICATE OF ANALYSIS

Arlington County Police Department
Your Case # 970212004
FS Lab # N97-1490
August 4, 1997

CONCLUSIONS (Continued):

No DNA amplification results were obtained for the sperm fraction from the lips/lip area swabs (Item 1). Therefore, no determination can be made as to the origin of the spermatozoa identified on these swabs.

The genetic material isolated from the lips/lip area swabs (Item 1), the "left upper chest" swabs (Item 1) and the "bite mark right upper back" swabs (Item 1) is insufficient in amount for DNA (RFLP) analysis. The genetic material isolated from the crotch area and rear panel stains from the underpants (Item 1) may be suitable for DNA (RFLP) analysis. The results of this examination will be the subject of a separate forthcoming report.

The evidence will be available at the Northern Laboratory after you have received the results of all requested examinations.

Attest:

I certify that I performed the above analysis or examination as an employee of and in a laboratory operated by the Division of Forensic Science, and that the above is an accurate record of the results of that analysis or examination.

Karen Curtis Ambrozy
Forensic Scientist

KCA

---

analysis of DNA from a suspect and from evidence resulting from a rape. Like the previous reports, identifying information has been redacted. The report mentions the detection of blood on a fitted bed sheet, and semen on a flat bed sheet and a towel, based on the detection of the semen-specific substance, P30. The analysis of all three items of evidence reveals that they share a common origin. The report also indicates that semen was not detected on the fitted sheet or on clothing that the victim wore during the attack. The DNA analysis examined the 13 CODIS loci plus amelogenin (which reveals gender). The flat sheet, Stain 1B1 provides a complete genetic profile for all 13 loci. On the other hand the towel reveals only 6 loci. This is explained by the fact that only the Cofiler kit was used for this item and not the Profiler Plus kit. The first page of the report describes the statistics generated from a 6-locus match and from a 13-locus match. It also provides statistics for each of 4 populations: African Americans, Caucasians, Hispanics, and Asians.

At the time that this report was written, neither the victim nor the suspect had produced a specimen, and, therefore, the evidentiary profile could not be compared to either of the two individuals. However, the profile was sent to the New York State Laboratory in Albany to be entered into the state and national DNA databases. The data was also entered into the local database maintained by the Office of Chief

OFFICE OF CHIEF MEDICAL EXAMINER
Charles S. Hirsch, M.D., *Chief Medical Examiner*
520 First Avenue, New York, NY 10016
Tel: 212.447.2618 · Fax: 212.447.2630 · TTY: 212.447.2086 -http://www.nyc.gov

DEPARTMENT OF FORENSIC BIOLOGY
Robert C. Shaler, Ph.D., *Director*
Howard J. Baum, Ph.D., *Deputy Director*
Marie Samples, M.S., M.Phil., *Assistant Director*
Mechthild Prinz, Ph.D., *Assistant Director*

[redacted] 13, 2002

LABORATORY REPORT

VICTIM: [redacted]	LAB NO: [redacted]
SUSPECT: [redacted]	ARREST NO: K[redacted]
PRECINCT: [redacted]	COMPLAINT NO: [redacted]

SUMMARY OF RESULTS:

Human blood was found on the fitted sheet, item 1A.

Semen* was found on the flat sheet, item 1B, and the towel, based on the presence of P30 antigen.

PCR DNA testing was performed on the fitted sheet, the flat sheet and the towel; the DNA profile of the blood and semen donor were determined. Results indicate that the blood on the fitted sheet and the semen on the flat sheet and the semen on the towel could have all come from the same source. This combination of DNA alleles is expected to be found in approximately:**

6 Loci Match	13 Loci Match
1 in greater than 15,000,000 Blacks	1 in greater than 1 trillion Blacks
1 in greater than 6,800,000 Caucasians	1 in greater than 1 trillion Caucasians
1 in greater than 1,900,000 Hispanics	1 in greater than 1 trillion Hispanics
1 in greater than 10,000,000 Asians	1 in greater than 1 trillion Asians

No semen* was found on fitted sheet 1A, the underwear, t-shirt, or pajama pants .

* Semen has two components: the seminal plasma (which contains the P30 antigen) and spermatozoa. Semen can be identified by detecting either P30 antigen and/or sperm.

**OCME STR database, National Research Council (1996) The Evaluation of Forensic DNA Evidence, Natl. Acad. Press, Washington DC.

Medical Examiner, New York City (see page 2 of report). The third page of the report provides general information about DNA testing and also about how sexual assault evidence is handled. Page 4 lists the 13 loci, the chromosomal locations, the common alleles within each locus and the possible number of genotypes (permutations of the number of common alleles into groups of 2). Page 5 lists results of the analysis. Once again, identifying information has been redacted. It should be

page 2

Further analysis could be done upon submittal of a blood sample from the victim and the suspect. Further analysis will require approximately 60 days.

The DNA results in this case do not match any previous PCR (STR) DNA cases to date.

The DNA results in this case have been entered into the OCME local DNA databank. The DNA results will be entered into the National Combined DNA Indexing System (CODIS).

page 3

EXAMINATIONS:

Blood and other physiological fluids and tissues contain polymorphic ("many forms") genetic markers which can differ from person to person. These genetic markers are inherited, that is, pass from generation to generation and can be used to compare biological samples from different sources. Genetic markers occur because of changes (mutations) that occur in a person's hereditary material, DNA (Deoxyribonucleic Acid).

Alternative forms of DNA are called alleles; they are found at the same location of the DNA (locus, plural loci) on homologous (matching) chromosomes. An individual can have a maximum of two different alleles at a particular locus, one on each homologous chromosome. A group of two alleles from the same locus constitutes a type.

Several different loci may be analyzed simultaneously using a technique known as the polymerase chain reaction (PCR). This technique allows small amounts of DNA to be amplified; after amplification, the alleles present in the sample are identified.

The loci tested may include the short tandem repeat (STR) loci [D3S1358, D16S539, THO1, TPOX, CSF1PO, D7S820, VWA, FGA, D8S1179, D21S11, D18S51, D5S818, D13S317]. The STR loci exhibit length polymorphisms which are variations in the number of core repeats, which are 4 base pairs in length. STR alleles are named according to the number of core repeats present at the locus. Each locus has between 8 and 32 identifiable alleles.

The loci tested may also include the Amelogenin locus, which is located on the chromosomes X and Y, and can be used to determine the sex origin of an unknown sample.

Semen has two components: the seminal plasma (which contains the P30 antigen) and spermatozoa. Semen can be identified by detecting either P30 antigen and/or sperm. To analyze samples which potentially contain a mixture of semen and other body fluids, a technique known as "differential extraction" is used. This technique is designed to physically separate DNA in the epithelial cells from the DNA in the sperm cells. This results in the sample being divided into two "fractions":

> "epithelial cell fraction" -is enriched for DNA from the source of the physical evidence; this is typically DNA from the victim
>
> "sperm cell fraction" - is enriched for DNA from the semen donor(s), if sperm were present

---

noted that this laboratory designates samples that have not been tested with an asterisk (*). A result may also be negative where no alleles have been detected in the amplified product or the result may be designated insufficient, which means that the quantity of sample DNA was too small to amplify successfully so the analysis stopped after the quantification stage. The last page of the report

page 4

Locus	Chromosome	Alleles	Types
D3S1358	3	9, 10, 11, 12, 13, 14, 15, 15.2, 16, 17, 18, 19, 20	91
D16S539	16	5, 8, 9, 10, 11, 12, 13, 14, 15	45
THO1	11	4, 5, 6, 7, 8, 8.3, 9, 9.3, 10, 11	55
TPOX	2	6, 7, 8, 9, 10, 11, 12, 13	36
CSF1PO	5	6, 7, 8, 9, 10, 11, 12, 13, 14, 15	55
D7S820	7	6, 6.3, 7, 8, 9, 10, 11, 12, 13, 14, 15	66
VWA	12	10, 11, 12, 13, 14, 15, 15.2, 16, 17, 18, 19, 20, 21, 22	105
FGA	4	16, 16.2, 17, 17.2, 18, 18.2, 19, 19.2, 20, 20.2, 21, 21.2, 22, 22.2, 23, 23.2, 24, 24.2, 25, 25.2, 26, 26.2 , 27, 27.2, 28, 28.2, 29, 29.2, 30, 30.2, 31, 31.2	528
D8S1179	8	8, 9, 10, 11, 12, 13, 14, 15, 16, 17, 18, 19	78
D21S11	21	24.2, 25, 26, 27, 28, 28.2, 29, 29.2, 29.3, 30, 30.2, 31, 31.2 32, 32.2, 33, 33.1, 33.2, 34, 34.2, 35, 35.2, 36, 38	300
D18S51	18	9, 10, 10.2, 11, 12, 13, 13.2, 14, 14.2, 15, 16, 17, 18, 19, 20 21, 22, 23, 24, 25, 26	231
D5S818	5	7, 8, 9, 10, 11, 12, 13, 14, 15, 16	55
D13S317	13	5, 8, 9, 10, 11, 12, 13, 14, 15	45

indicates item numbers, voucher numbers, the date that the sample was received by the lab, and a description of each item. It also indicates how the specimens were handled after the analysis. Some evidence was retained in the lab while other evidence was returned to the Evidence Unit. Both the analyst and supervisor signed off on the report.

Different laboratories use different formats for their summary reports, some providing more information than others. Great care should be taken in the wording of the report since the report will be offered into evidence and will be the subject of the analyst's testimony. The analyst should testify carefully and with the understanding that the DNA report is a powerful item of evidence. Jurors will be told of a matching profile. Statistics will generally indicate the extreme rarity of that profile. The conclusion of the analyst need not be expressed in absolutes. A conclusion can be explained in terms of probabilities. The conclusion in a report where a match has been obtained should state that one cannot exclude the defendant as the source of the evidence. Although this is a very conservative way of stating the conclusion, it should be remembered that the jury has also been provided with the statistical significance of the match, and thus they are aware of the rarity of the profile.

4.21 REVIEWING A PATERNITY REPORT BASED ON ANALYSIS OF DNA

The DNA report illustrated in this section was prepared by Bio-Synthesis, Inc., Lewisville, Texas. The report indicates that this company is accredited by the American Association of Blood Banks. (The testing group may also be accredited

Table 1- Allelic typing was done with the following results:

ITEM	D3S1358	D16S539	Amel	THO1	TPOX	CSF1PO	D7S820	VWA	FGA	D8S1179	D21S11	D18S51	D5S818	D13S31 7
fitted sheet st. 1A1:	15		XY	9.3	8, 11		8	*	*	*	*	*	*	*
flat sheet st. 1B1:														
sperm fraction	15		XY	9.3	8, 11		NEG	16	**		28, 30	NEG	11	11, 12
epithelial fraction	INS	INS	INS	INS	INS	INS	INS	INS	INS	INS	INS	INS	INS	INS
swab remains fraction	15		XY	9.3	8, 11		8	16	23, 24		28, 30	17, 20	11	11, 12
towel st. 5C:														
sperm fraction	15		XY	9.3	8, 11		NEG	*	*	*	*	*	*	*
epithelial fraction	15		XY	9.3	8, 11		8	*	*	*	*	*	*	*
swab remains fraction	*	*	*	*	*	*	*	*	*	*	*	*	*	*

* = Typing not attempted
NEG = No alleles detected
INS = Insufficient human DNA was detected for this test; therefore, this sample was neither amplified nor typed for this test.

All of the DNA alleles from the blood on the fitted sheet, the semen on the flat sheet and the semen on the towel are the same. Therefore, the blood and the semen could have come from the same source. All of the DNA must have come from a male.

page 6

EVIDENCE RECEIVED:

ITEM	VOUCHER	DATE REC'D	DESCRIPTION
1A	L3	3/26/02	fitted sheet
1B	"		flat sheet
2	"		black underwear
3	"		white t-shirt
4	"		pajama pants
5	"		brown towel

DISPOSITION:

The following items will be retained in the laboratory:

-stains and controls taken from bed sheet 1A, bed sheet 1B and the towel
-DNA extracts for all samples and controls tested

The remainder of the evidence has been returned to the OCME Evidence Unit.

Analyst: Sandra Hayn
Sandra Hayn
Criminalist II

Supervisor: Deborah Briones
Deborah Briones
Criminalist IV

slh:

---

by other agencies.) In addition to various record numbers for internal usage, the report includes the following vital information.

1. Name of mother, child, and alleged father (photo identification and fingerprint is often required by paternity testing labs).
2. Racial group of each of the triad (sometimes child is omitted).
3. The date of the report as well as the dates when specimens were collected and when samples were received by testing lab.
4. Genetic markers (loci) that have been tested. (The report omits the test method, which was PCR-STR analysis.)
5. The alleles observed for each member of the triad at each locus tested.
6. The paternity index for each locus. (Had there not been an exclusion, there would have been a calculation of a cumulative paternity index as well as a probability of paternity calculation expressed as a percentage figure.)

BIO-SYNTHESIS, INC. P.O. Box 28, Lewisville, Texas 75067-0028 Tel.: (800) 227-0627 Fax: (972) 420-0442

The DNA Identity Testing Laboratory is accredited by the American Association of Blood Banks.

Example Paternity Report Indicating Alleged Father is NOT the Biological Father

Accession No: 2002 P10000 **Customer Case No:**

	Name		Race	Specimen Number	Sample Collected	Sample Received
Mother:	Example	Jane	Caucasian	P10000M	7/4/02	7/4/02
Child:	Example	Child 2		P10000C2	7/4/02	7/4/02
Alleged Father:	Sample	Joe	Caucasian	P10000AF	7/4/02	7/4/02

Conclusion:

The alleged father, **Joe Sample**, was excluded as being the biological father of the child, **Child 2 Example**, since he lacks the genetic markers necessary of the biological father. The existence of non-matching markers between alleged father and child results in a Probability of Paternity of **0.00%**

Result:

Genetic Locus	Mother		Child		Alleged Father		Paternity Index
D7S820	12	13	11	12	8	9	0.00
D13S317	11	12	10	12	11	12	0.00
D5S818	11	11	11	11	11	11	2.63
D18S51	13	16	13	14	18	18	0.00
D21S11	31	32.2	29	32.2	28	29	2.41
D8S1179	11	13	10	13	14	16	0.00
FGA	21	22	22	25	20	24	0.00
vWA	15	16	15	19	14	17	0.00
D3S1358	15	17	17	17	16	18	0.00

I, the undersigned Director, have read the foregoing report on DNA analyses conducted on the above named individuals and have verified that the results herein are true and correct. I hereby certify that these analyses have been conducted according to the standard protocols of Biosynthesis, Inc. and are in accordance with AABB/ASHI guidelines.

Sworn and subscribed before me this

__________day of__________,

____

Gregory M. Sawyer, Ph. D.
Director
DNA Identity Testing Laboratory

---

7. The conclusion that the alleged father is (or is not) the biological father.
8. If the alleged father is found to be the biological father, a note of explanation is included. This note might state the following: The alleged father, John Doe, cannot be excluded as the biological father of the child "Little Doe." Based on these data, the probability of paternity is 99.999% as compared to an untested randomly chosen man of the (Caucasian, Black, or Hispanic)

BIO-SYNTHESIS, INC. P.O. Box 28, Lewisville, Texas 75067-0028 Tel.: (800) 227-0627 Fax: (972) 420-0442. The DNA Identity Testing Laboratory is accredited by the American Association of Blood Banks.

Example Paternity Report Indicating Alleged Father IS the Biological Father

Accession No: 2002		**P10000**		**Customer Case No:**		
	Name		**Race**	**Specimen Number**	**Sample Collected**	**Sample Received**
Mother:	Example	Jane	Caucasian	P10000M	7/4/02	7/4/02
Child:	Example	Kid		P10000C	7/4/02	7/4/02
Alleged Father:	Sample	Joe	Caucasian	P10000AF	7/4/02	7/4/02

Conclusion:

The alleged father, **Joe Sample**, cannot be excluded as being the biological father of the child, **Kid Example**. Based on the testing results outlined below, a Probability of Paternity of **99.9994** % (Prior Probability=0.50), reflecting a Combined Paternity of **19,0713**, was obtained, as compared to a random male of the **Caucasian** population. In addition, at least **99.9992** % of the population is excluded from the possibility of being the biological father of the child.

Result:

Genetic Locus	Mother		Child		Alleged Father		Paternity Index
D7S820	9	12	8	9	8	9	3.03
D13S317	11	13	11	12	11	12	1.69
D5S818	11	12	11	11	11	11	2.63
D18S51	12	15	15	18	18	18	16.67
D21S11	29	32.2	29	32.2	28	29	1.79
D8S1179	13	15	13	16	14	16	25.00
FGA	23	24	23	24	20	24	1.83
vWA	17	17	14	17	14	17	4.60
D3S1358	15	17	16	17	16	18	2.25

I, the undersigned Director, have read the foregoing report on DNA analyses conducted on the above named individuals and have verified that the results herein are true and correct. I hereby certify that these analyses have been conducted according to the standard protocols of Biosynthesis, Inc. and are in accordance with AABB/ASHI guidelines.

Sworn and subscribed before me this

________day of________,

____

Gregory M. Sawyer, Ph. D.
Director
DNA Identity Testing Laboratory

population. Prior probability = 0.5. At least 99% of the male population is excluded as the possible father of this child.

The report is notarized and signed by an official of the laboratory. That individual attests that the results are true and correct and that the analysis was conducted according to the company's standard protocols and in accordance with AABB and ASHI (American Society for Histocompatibility and Immunogenetics) guidelines.

BIBLIOGRAPHY

Abrahamson, S.S. Chain of National Commission on the Future of DNA Evidence, The Future of Forensic DNA Testing Predictions of the Research and Development Working Group, National Institute of Justice November 2000 NCJ 183697 Available online: http://www.ojp.usdoj.gov/nij/pubs-sum/183697.htm.

American Society of Crime Laboratory Directors [Online]. Available: http://www.ascld.org/accreditation.html.

National Research Council. *DNA Technology in Forensic Science.*

Committee on DNA Technology in Forensic Science, Board on Biology, Commission on Life Sciences, and the National Research Council. National Academy Press, Washington, D.C., 1992.

National Research Council. *Evaluation of Forensic DNA Evidence.*

Committee on DNA Technology in Forensic Science, Board on Biology, Commission on Life Sciences, and the National Research Council. National Academy Press, Washington, D.C., 1996.

NIST STRbase website: http://www.cstl.nist.gov/div831/strbase.

Scientific Working Group on DNA Analysis Methods. Training Guidelines, *Forensic Science Communications* (2001, October) Available online: http://www.fbi.gov/hq/lab/fsc/backissu/oct2001/kzinski.htm.

Technical Working Group on DNA Analysis Methods (TWGDAM), Statement of the Working Group on Statistical Standards for DNA Analysis. *Crime Lab Digest*, **17**(3), 53–68 (1990).

Weir, B. S. A Bibliography for the Use of DNA in Human Identification. Kluwer Academic, Dordrecht, Netherlands, 1995, pp. 179–213.

REFERENCES

Balding, D. J. and R. A. Nichols. DNA Profile Match Probability Calculation: How to Allow for Population Stratification, Relatedness, Database Selection and Single Bands. *Forensic Sci. Int.*, **64**(2–3), 125–140, (1994).

Bellamy, R. J., C. F. Inglehearn, I. K. Jalili, A. J. Jeffrey, and S. S. Bhattacharya. Increased Band Sharing in DNA Fingerprints of an Inbred Human Population. *Hum. Genet.*, **87**(3), 341–347 (1991).

Brenner, C. H. A Note on Motherless Paternity Case Computation. *Transfusion*, **33**, 51–54 (1993).

Brookfield, J. F. Statistical Issues in DNA Evidence. *Electrophoresis*, **16**(9), 1665–1669, (1995a).

Brookfield, J. F. The Effect of Relatedness on Likelihood Ratios and the Use of Conservative Estimates. *Genetica*, **96**(1–2), 13–19 (1995b).

Budowle, B. The Effects of Inbreeding on DNA Profile Frequency Estimates Using PCR-Based Loci. *Genetica*, **96**(1–2), 21–25 (1995).

Budowle, B. and K. L. Monson. Greater Differences in Forensic DNA Profile Frequencies Estimated from Racial Groups than from Ethnic Groups. *Clin. Chim. Acta*, **228**(1), 3–18 (1994).

Budowle, B. and K. L. Monson. The Forensic Significance of Various Reference Population Databases for Estimating the Rarity of Variable Number of Tandem Repeat (VNTR) Loci Profiles. *Exs*, **67**, 177–191 (1993).

Budowle, B. B., Shea, S. Niezgoda, and R. Charaborty, CODIS STR Loci Data from 41 Sample Populations. *J. Forensic Sci.*, **46**(3), 453–489 (2001).

Budowle, B. T., Moretti, K. Keys, B. Koons, and J. B. Smerick. Validation Studies of the CTT Multiplex System. *J. Forensic Sci.*, **42**, 701–707 (1997).

Budowle, B. B., W. Koons, K. M. Keys, and J. B. Smerick. Methods for Typing the STR Triplex (CSF1PO, TPOX, and HUMTHO1) That Enable Compatibility Among DNA Typing Laboratories. In A. Carracedo, B. Brinkmann, and W. Bar (eds.). *Advances in Forensic Haemogenetics*, Vol. 6. Springer, Berlin, 1996, pp. 107–114.

Bylaws of the Scientific Working Group on DNA Analysis Methods. *Forensic Sci. Commun.*, **5**(2), (2003).

Dabbs, D. and P.D. Cornwell. The Use of DNA Profiling in Linking Serial Murders. *Medico-Legal Bull.* **37**(6), 2–10 (1988).

DNA Advisory Board. Quality Assurance Standards for Forensic DNA Testing Laboratories. *Forensic Sci. Commun.* [Online] (July 2000). Available: www.fbi.gov/hq/lab/fsc/backissue/july2000/codis2a.htm.

DNA Advisory Board. Quality Assurance Standards for Convicted Offender DNA Databasing Laboratories. *Forensic Sci. Commun.* [online] (July 2000). Available: www.fbi.gov/hq/lab/fsc/backissue/july2000/codis1a.htm.

DNA Advisory Board. Quality Assurance Standards for Convicted Offender DNA Databasing Laboratories. *Forensic Sci. Commun.* (1999). [Online]. Available: http://www.fbi.gov/hq/lab/fsc/backissu/july2000/codispre.htm.

DNA Advisory Board. Quality Assurance Standards for Forensic DNA Testing Laboratories. *Forensic Sci. Commun.* (1998). Available: http://www.fbi.gov/hq/lab/fsc/backissu/july2000/codispre.htm.

Parentage and Kinship Analysis. In *Forensic Sciences*, Matthew Bender, Albany NY 2001.

Gill, P. I. W., Evett, S. Woodroffe, J. E. Lygo, E. Millican, and M. Webster. Databases, Quality Control and Interpretation of DNA Profiling in the Home Office Forensic Science Service. *Electrophoresis*, **12**(2–3), 204–209 (1991).

Hartl, D. L. and A. G. Clark. *Principles of Population Genetics*, 2nd ed. Sinauer, Sunderland, MA, 1989.

Kay, D. H., DNA Evidence: Probability, Population, Genetics, and the Courts. *Harv. J. L. & Tech.*, **7**, 101–172 (1993).

Kline, M., D. Duewer, P. Newall, J. Redman, D. Reeder, and M. Richard. Interlaboratory Evaluation of Short Tandem Repeat Triplex CTT. *J. Forensic Sci.*, **42**, 897–906 (1997).

Klitz, W., R. Reynolds, J. Chen, and H. A. Erlich. Analysis of Genotype Frequencies and Interlocus Association for the PM, DQA1, and D1S80 Loci in Four Populations. *J. Forensic Sci.*, **45**(5), 1009–1015 (2000).

Koehler, J. J., K. Chia, and A. Lindsey. The Random Match Probability (RMP) in DNA Evidence: Irrelevant and Prejudicial? *Jurimetics J.*, **35**, 201–219 (1995).

Lander, E. and B. Budowle. DNA Fingerprinting Dispute Laid to Rest. *Nature* **371**: 735–738 (1994).

Lempert, R. After the DNA Wars: Skirmishing with NRC II. *Jurimetrics* **37**, 439–468 (1997).

Levy, H. DNA: Race, Ethnicity, and Statistical Evidence. *New York Law J.*, July 25, 1991 (Outside Counsel Column, at 1).

Lewontin, R. C. and Hartl, D. Population Genetics in Forensic DNA Typing. *Science* **254**, 1745–1750. (1991).

McElfresh, K.C. and Y. K. Kim. Finding the Needle in the Haystack: How Many Loci Does It Take to Find a Single Individual in a DNA Database of 2 Million Individuals, Abstract in 11th Proceedings on Genetic Identification, 2000. Available at: http://www.promega.com/geneticidproc/ussymp11proc/abstracts/McElfresh.pdf.

McEwen, J. E. Forensic DNA Data Banking by State Crime Laboratories. *Am. J. Hum. Genet.*, **56**(6), 1487–1492 (1995).

McEwen, J. E. and P. R. Reilly. A Review of State Legislation on DNA Forensic Data Banking. *Am. J. Hum. Genet.*, **54**(6), 941–958 (1994).

Morton, N. E. and A. E. Collins. Statistical and Genetic Aspects of Quality Control for DNA Identification. *Electrophoresis*, **16**(9), 1670–1677 (1995).

Mueller, L. The Use of DNA Fingerprinting in Forensic Science. *Account. Res.*, **3**, 55–67 (1993).

National Research Council. *Evaluation of Forensic DNA Evidence.*

Committee on DNA Technology in Forensic Science, Board on Biology, Commission on Life Sciences, and the National Research Council. National Academy Press, Washington, D.C., 1996.

National Research Council, *DNA Technology in Forensic Science.*

Committee on DNA Technology in Forensic Science, Board on Biology, Commission on Life Sciences, and the National Research Council. National Academy Press, Washington, D.C., 1992.

O'Dell. S. A. A Quality Assurance System for DNA Testing. *Forensic Sci. J.*, **2**, 1–4 (2003).

Scheck, B. DNA Data Banking: A Cautionary Tale. *Am. J. Hum. Genet.*, **54**(6), 931–933 (1994).

Scientific Working Group on DNA Analysis Methods (SWGDAM): Training Guidelines, January 23, 2001. *Forensic Sci. Commun.*, **3**(4), (2001).

Slimowitz, J. R. and J. E. Cohen. Violations of the Ceiling Principle: Exact Conditions and Statistical Evidence. *Am. J. Hum. Genet.*, **53**(2), 314–323 (1993).

Sudbury, A. W. J. Marinopoulos, and P. Gunn. Assessing the Evidential Value of DNA Profiles Matching without Using the Assumption of Independent Loci. *J. Forensic Sci. Soc.*, **33**(2), 73–82 (1993).

Technical Working Group on DNA Analysis Methods. Guidelines for a Quality Assurance Program for DNA Analysis. *Crime Lab. Digest* **22**(2), 21–43 (1995).

Available online: http://www.fbi.gov/hg/lab/fsc/backissu/oct2001/index.htm.

Technical Working Group on DNA Analysis Methods (TWGDAM). Statement of the Working Group on Statistical Standards for DNA Analysis. *Crime Lab. Digest*, **17** (3), 53–68, (1990).

Travis, J. and C. Asplen. *Postconviction DNA Testing: Recommendations for Handling Requests.* National Commission on the Future of DNA Evidence, September 1999, NCJ 177626.

Turowska, B. and M. Sanak. D1S80 VNTR Locus Genotypes in Population of South Poland; Meta-Analysis Pointer to Genetic Disequilibrium of Human Populations. *Forensic Sci. Int.*, **75**(2–3), 207–216 (1995).

Van Oorschot, R., S. Gutowski, S. Robinson, J. Hedley, and I. Andrew. HUMTHO1 Validation Studies: Effect of Substrate, Environment, and Mixtures. *J. Forensic Sci.*, **41**, 142–145 (1996).

Wallin, J. M., M. R. Buoncristiani, K. D. Lazaruk, N. Fildes, C. L. Holt, and P. S. Walsh. TWGDAM Validation of the AmpF*l*STR™ Blue PCR Amplification Kit for Forensic Casework Analysis. *J. Forensic Sci.*, **43**(4), 854–870 (1998)

Weir, B. S. Forensic Population Genetics and the National Research Council (NRC). *Am. J. Hum. Genet.*, **52**(2), 437–440 (1993).

Weir, B. S. Population Genetics in the Forensic DNA Debate. *Proc. Nat'l Acad. Sci. USA*, **89**(24), 11654–11659 (1992).

Weir, B. S. and W. G. Hill. Population Genetics of DNA Profiles. *J. Forensic Sci. Soc.*, **33**(4), 218–225 (1993).

Wilson, M. R., J. A. DiZinno, D. Polanskey, J. Replogle, and B. Budowle. Validation of Mitochondrial DNA Sequencing for Forensic Casework Analysis. *Int. J. Legal Med.*, **108**, (2), 68–74 (1995).

5

Litigating a DNA Case

5.1 LEGAL THEORY

As mentioned earlier, when discussing the admissibility of DNA evidence, it is important to take note of the standards that govern the admissibility of scientific evidence in general. This section outlines the law that governs the admissibility of scientific evidence and expert witness testimony at trial.

5.1.1 Admissibility of Scientific Evidence: A Primer

Laws and Institutions must go hand in hand with the progress of the human mind.
—President Thomas Jefferson (1743–1826)[1]

Throughout history, advances in science have had a tremendous impact on many areas of the law. Legislators and judges have had to consider how these scientific and technological advances would affect the law of the land.[2] In fact, the impact of science has been dramatic and has resulted in litigation and legislation involving subjects outside of the experience and understanding of laypersons. This section will discuss how the courts have developed rules and doctrines that affect the admissibility of complex scientific evidence that is beyond the scope of most

[1] *See* Tony Lyons, *The Quotable Lawyer* (2002). Globe Pequot Press, Fulford, CT.
[2] *See* Stephen Breyer, *The Interdependence of Science and Law*, Judicature, Vol. 82, No. 1 (1998).

DNA: Forensic and Legal Applications, by Lawrence Kobilinsky, Thomas F. Liotti, and Jamel Oeser-Sweat
ISBN 0-471-41478-6

citizens.[3] It will also introduce the concept of expert witnesses, analyze their role in litigation, and provide an overview of the concerns that judges have when dealing with complex scientific topics and the experts employed to make them more comprehensible.[4]

To understand how and why the court has adopted methods to evaluate expert testimony, it is essential that one have a proper understanding of the admissibility of evidence in general. Fundamentally, for evidence to be admissible in trial, it must be relevant.[5] However, at times, even relevant evidence can be deemed inadmissible and be excluded from the jury's consideration of the question being litigated.

Once evidence is deemed to be admissible, it is presented to the trier of fact. Evidence is presented to the judge and jury through the testimony of a witness.[6] This is done through direct testimony, and cross-examination of the witness.[7] Evidence is never transmitted directly to the trier of fact without any explanation of its significance. The attorney litigating the case must establish the identity and purpose of the evidence. This foundation and purpose is developed through the testimony of the witness. As will be discussed later, the witness is thus the voice of the evidence. The trier of fact can usually associate a piece of evidence with a particular witness or set of witnesses. There is no evidence that cannot be associated with a witness because all evidence is introduced to the judge and jury through the testimony of a witness.

5.1.2 Common Law and The Creation of a Judicial Gatekeeping Function

Many doctrines involving the admissibility of evidence were developed under common law.[8] Courts have developed rules that govern what evidence is admissible and what types of evidence are not to be considered during a trial. When advances in the sciences became the concern of judges evaluating admissibility, judges were forced to develop doctrines governing the admission of scientific evidence as well. Federal courts first took on this task. Their system was subsequently used as a model by the individual states.

[3]*See* Thomas F. Liotti, The Judge's Role as Gatekeeper: The Evolution of the Judicial Gatekeeping Function and Its Impact on the Admission of Expert Testimony, N.Y.B.S.A. Criminal Justice Section Journal, Vol. 8, No. 1 at 66 (2000).

[4]*See* Peter W. Huber, *Galileo's Revenge: Junk Science in the Courtroom 56* (1991). Basic Books, Boulder CO.

[5]*See* infra. *See also* Fed.Rul.Evid. 402.

[6]In order to testify a witness must be qualified. While at common law a multitude of factors that, if present, would disqualify a witness from testifying (make the witness incompetent), the Federal Rules of Evidence is more relaxed. Fed.R.Evid. 601 states in pertinent part that: "[e]very person is competent to be a witness." The rule provides for exceptions but, there are not as many as there were at common law.

[7]After the direct examination of a witness, the opposing party may cross-examine a witness as a matter of right. *See Alford v. United States*, 282 U.S. 686, 691 (1931). It should also be noted that at the federal level, the scope of the cross-examination cannot exceed the scope of the direct examination. *See* Fed.R.Evid. 611(b).

[8]*See Generally*, VII J. Wigmore, Evidence in Trials at Common Law (Chadbourn rev. 1981).

The seminal case in which an Appellate Court developed a doctrine governing the admissibility of scientific evidence was *Frye v. United States*.[9] In this 1923 case, the defendant charged with murder attempted to persuade the trials Court to admit evidence produced by an instrument known as a polygraph, more commonly known as a lie detector. The instrument was designed to monitor a number of body functions such as respiration, blood pressure, and the production of perspiration, which, taken together, would indicate if the individual was telling the truth or lying in responding to specific questions. Frye was convicted of second-degree murder. He then attempted to persuade the Court of Appeals (of The District of Columbia) to admit evidence from a "systolic blood pressure deception test" (*Frye*, at 1013). Frye claimed that the results of the test would demonstrate that he was not guilty of murder.

The Appellate Court decided that it would not allow results derived from the deception test to be admitted as evidence in support of Frye's assertion of innocence. In determining whether results from a systolic blood pressure deception test were admissible and whether experts could testify regarding those results, the Appellate Court concluded that it would only allow such expert testimony if "the matter of inquiry [was] such that inexperienced persons [were] unlikely to prove capable of forming a correct judgment upon it, for the reason that the subject matter so far partakes of science, art, or trade as to require a previous habit or experience or study into it, in order to acquire knowledge of it."[10] The *Frye* Court was aware that some topics were outside the scope of laypeople. Unlike ordinary, nonscientific, evidence that is presented to the jury, the presentation of evidence beyond the scope of the jury was problematic because it was not always comprehensible.[11] The Court put forth a solution to the problem created when a subject matter is beyond the scope of a jury inexperienced in the subject. It stated: "When the question involved does not lie within the range of common experience or common knowledge, but requires special experience or special knowledge, then the opinions of witnesses skilled in that particular science, art, or trade to which the question relates are admissible in evidence."[12] By allowing experts to testify, the Court took steps to ensure that the evidence was not just heard by the jury but that the jury also understood the impact of the information presented to them.[13]

The judges in *Frye* allocated themselves a gatekeeping function in which they could screen out unreliable testimony. The judges stated that "just when a scientific principle or discovery crosses the line between experimental and demonstrable stages is difficult to define."[14] It was left to judges to determine whether or not a

[9]293 F. 1013 (D.C. Cir. 1923).

[10]*Frye*, at 1014, quoting defendant's brief.

[11]All evidence and data is presented to the jury during trial in the form of testimony. Witnesses are the voice of the testimony. The problem that the court contemplated when evidence beyond the scope of laypersons was being presented was that the witness was speaking what might as well have been a foreign language that the jury could not comprehend.

[12]*Frye*, at 1014, quoting defendant's brief.

[13]In many ways the expert witness is analogous to an interpreter.

[14]*Frye*, at 1014.

principle or discovery had achieved the appropriate level of recognition. If the principle or discovery did not rise to the required level, the judges directed that it be kept out of the trial. Making such determinations is essentially the task of the judge who acts as a gatekeeper. The *Frye* Court evaluated whether it should use the results of the deception test and decided that the standard for determining the admissibility of such evidence was whether the principle or discovery was "sufficiently established to have gained general acceptance in the particular field in which it belongs."[15] The *Frye* Court concluded that the "systolic blood pressure deception test" had no such acceptance and rejected the evidence, deeming it inadmissible.

In a two-page decision, the *Frye* Court bestowed upon judges a gatekeeping function that has evolved into an important part of contemporary litigation. Most states adopted their admissibility standards using the *Frye* standard of general acceptance as a guide.[16] In fact, the *Frye* standard of general acceptance stood as the "gold" standard to which federal judges held scientific evidence for decades.

In an effort to institute a uniform, comprehensible system of evidence, the Supreme Court promulgated the Federal Rules of Evidence. Although some scholars hailed the establishment of these rules, others were concerned that the section of the rules pertaining to the admissibility of scientific evidence and testimony made the clear standard put forth in *Frye* lose its clarity. Many saw this lack of harmony between the *Frye* standard and the standard governing the admission of scientific evidence and testimony as a precursor to the elimination of the *Frye* general acceptance standard.

5.1.3 Federal Rules of Evidence and the Expansion of the Judicial Gatekeeping Function

The section of the Federal Rules of Evidence that governs the admission of scientific evidence and testimony does not mirror the common law standard put forth in *Frye*.[17] In the Federal Rules of Evidence, the Supreme Court gave more power to judges, who were performing gatekeeping functions. Their ability to use their discretion to determine whether or not evidence was admissible was expanded. Four rules in particular are of importance in an analysis of the expansion of the judicial gatekeeping function with regard to scientific evidence and testimony: Rule 104(a), Rule 702, Rule 402, and Rule 403.

Rule 104(a) establishes the existence of the judicial gatekeeping function in the Federal Rules of Evidence. Rule 104(a) states in pertinent part that: "preliminary questions concerning the qualification of a person to be a witness, the existence of a privilege, or the admissibility of evidence shall be determined by the court."[18] Rule 702 applies to expert testimony and overlaps the *Frye*

[15] *Frye*, at 1014.
[16] New York adopted the *Frye* standard of admissibility in 1938, *see* infra.
[17] *See Generally*, Fed.Rul.Evid. Article VII.
[18] Fed.Rul.Evid. 104(a).

standard.[19] Rule 702 allows judges to permit expert testimony if it assists the trier of fact in understanding evidence or in determining facts.[20] Taken together, Rules 104(a) and 702 acknowledge the judicial gatekeeping function as set forth in *Frye*. Rule 402 allows judges to admit "all relevant evidence."[21] Rule 401 defines relevant evidence as "evidence having any tendency to make the existence of any fact that is of consequence to the determination of the action more probable or less probable than it would be without the evidence."[22] Rule 402 also states that "evidence which is not relevant is not admissible."[23] Rule 402 contains the level to which evidence must rise to be admissible. Only relevant evidence is admissible.[24] All other evidence is not admitted into evidence. Judges performing the gatekeeping function are filtering out irrelevant evidence and allowing relevant evidence to be considered during trial.[25]

When the Court wrote Rule 403, it took the judicial gatekeeping function to a new level. Rule 403 of the Federal Rules of Evidence bestows enormous power on judges. Rule 403 has no negative language in it. It placed no limits upon judges evaluating the admissibility of evidence. Rule 403 expands their authority. Rule 403 states "although relevant, evidence may be excluded if its probative value is substantially outweighed by the danger of unfair prejudice, confusion of the issues, or misleading the jury, or by considerations of undue delay, waste of time, or needless presentation of cumulative evidence."[26] In one long sentence, the Supreme Court promoted judges from mere gatekeepers to masters of the courtroom. Unlike the power given to judges in *Frye*, which described what they could admit, Rule 403 allows judges to exclude otherwise admissible evidence for such reasons as "undue delay" and "waste of time".[27] The discretionary power of judges was expanded by this express authorization of the exclusion of evidence on these grounds.

After the promulgation of Rule 403 the judicial gatekeeping function had evolved substantially. Legal scholars were intrigued.[28] The Supreme Court bestowed power on judges performing the gatekeeping function without explaining the boundaries of their gift. Rule 403 not only allowed for the exclusion of irrelevant evidence, it allowed for the exclusion of relevant evidence and gave no way to gauge what

[19]Fed.Rul.Evid. 702 states: "If scientific, technical, or other specialized knowledge will assist the trier of fact to understand the evidence or to determine a fact in issue, a witness qualified as an expert by knowledge, skill, experience, training, or education, may testify thereto in the form of an opinion or otherwise."
[20]*See* Fed.Rul.Evid. 702. Fed.Rul.Evid. 702 codifies the gatekeeping function put forth in *Frye*.
[21]Fed.Rul.Evid. 402.
[22]Fed.R.Evid. 401.
[23]Fed.R.Evid. 402.
[24]*See* Fed.R.Evid. 402.
[25]*See United States v. Robinson*, 544 F.2d 611 (2d Cir.1976), rev'd en banc 560 F.2d 507 (2d Cir.1977), cert. Denied 435 U.S. 905 (1978). In *Robinson*, the court considered whether the probative value of evidence at issue outweighed its prejudicial effect. *See also Walker v. United States*, 490 F.2d 683 (8th Cir.1974); *United States v. Cunningham*, 423 F.2d 1269 (4th Cir.1970).
[26]Fed.R.Evid. 403.
[27]Fed.R.Evid. 403.
[28]*See* Dolan, Rule 403: The Prejudice Rule in Evidence, 49 S.Cal.L.Rev. 220 (1976).

could not be excluded. Rule 403 was indeed liberal and left legal scholars wondering how to balance it against the conservative general acceptance standard set forth in *Frye*.[29] It would be a long time before the Supreme Court would give any clear indication of whether the standard governing the admission of scientific evidence found in the Federal Rules of Evidence preempted the *Frye* standard.

Whether the *Frye* general acceptance standard or the liberal standard put forth by the Federal Rules of Evidence was the binding standard was debated among legal scholars for decades. Many argued that the standard contained in the Federal Rules of Evidence preempted that which had been outlined in the *Frye* decision. It was clear that the legal community needed the court to clear up this problem by announcing which standard was proper.

5.1.4 Daubert: The Supreme Court Sets Forth a Standard

The Supreme Court finally addressed the issue of admissibility of scientific evidence in *Daubert v. Merrell Dow Pharmaceuticals, Inc.*[30] In *Daubert*, the Court considered the issue of what standard was proper for use in determining whether expert testimony should be admitted. The Court acknowledged in *Daubert* that "in the 70 years since its formulation in the *Frye* case, the 'general acceptance' test [had] been the dominant standard for determining the admissibility of novel scientific evidence at trial."[31] The Supreme Court ended its silence on this subject in *Daubert* and held that "the *Frye* test was superseded by the adoption of the Federal Rules of Evidence."[32] This announcement ended the debates outlined by the Court in the *Daubert* decision itself regarding the viability of the *Frye* standard. The Court, however, did not stop there.

The Court put forth a new standard on how judges should perform their gatekeeping function. The *Daubert* Court outlined four factors for judges to consider when evaluating whether a theory or technique will assist the trier of fact in a particular case. The Court suggested that judges first determine whether or not the scientific technique has been or can be tested to determine its validity.[33] The second factor that the Court suggested that judges consider is "whether the theory or technique has been subjected to peer review and publication."[34] The Court did, however, note that publication is not always necessary.[35] The third factor the *Daubert*

[29]*See Daubert v. Merrell Dow Pharmaceuticals, Inc.*, 509 U.S. 579, 587, n5 (1994). *See also* Jay P. Kesan, Note, *An Autopsy of Scientific Evidence in a Post-Daubert World*, 84 Geo. L.J. 1985 (1996); C. Robert Showalter, Essay, *Distinguishing Science from Pseudo-Science in Psychiatry: Expert Testimony in the Post-Daubert Era*, 2 Va. J. Soc. Pol'y & L. 211 (1995).
[30]509 U.S. 579, 113 S.Ct. 2786.
[31]*Daubert*, at 585.
[32]*Daubert*, at 587.
[33]*Daubert*, at 593.
[34]*Daubert*, at 593.
[35]*Daubert*, at 593. *See also* Horrobin, *The Philosophical Basis of Peer Review and the Suppression of Innovation*, 263 JAMA 1438 (1990).

Court deemed important to consider is the "known or potential rate of error" of a particular scientific technique.[36] The last factor that the Court required a judge to consider is the degree of acceptance within the scientific community.[37]

The *Daubert* decision provided a roadmap to judges attempting to perform their judicial gatekeeping function.[38] The *Daubert* standard was very liberal, like the Federal Rules, but still incorporated acceptance elements like the more conservative *Frye* standard. However, the *Daubert* decision would not be the end of the Court's evaluation of the judicial gatekeeping function.

DAUBERT FACTORS

1. Has the scientific technique been tested to determine its validity?
2. Has the theory or technique been subjected to peer review and publication?
3. What is the known or potential rate of error of the scientific technique?
4. What is the degree of acceptance within the scientific community?

Courts received an explanation of the judicial gatekeeping function subsequent to the *Daubert* decision. This clarification was necessary because courts encountered growing pains shortly after the *Daubert* decision.[39]

5.1.5 General Electric Company et al. v. Joiner et ux.

Three years after the Supreme Court decided *Daubert* it was forced to again review the standard of admissibility used by a trial court. In *General Electric Co. v. Joiner*,[40] the Supreme Court evaluated the question of admissibility and proper standard of review by an appellate court. In *Joiner*, an electrician working on transformers that were discovered to contain a chemical that had been associated with the development of cancer sued his employer after he was diagnosed with small-cell lung cancer. The electrician "linked his development of cancer to his exposure to PCBs and their derivatives, polychlorinated dibenzofurans (furans) and polychlorinated dibenzodioxins (dioxins)" despite the fact that he was a smoker with a history of cancer in his family.[41] The District Court granted summary judgment for the

[36]*Daubert*, at 594.
[37]*Daubert*, at 594.
[38]*See also* Berger, Procedural Paradigms for Applying the Daubert Test, 78 Minn.L.Rev. 1345 (1994); Symposium, Scientific Evidence after the Death of Frye, 15 Cardozo L.Rev. 1745–2294 (1994).
[39]*See* Joseph A. Saunders, Article, *The Other Side of the "Gatekeeping" Coin*, 18 Whittier L. Rev. 105 (1996).
[40]522 U.S. 136; 118 S. Ct. 512.
[41]*Joiner*, 522 U.S. at 139.

defendant-employer that asserted that there was no admissible scientific evidence presented by Joiner's experts that PCBs promoted Joiner's cancer.[42] The Court of Appeals for the Eleventh Circuit reversed, holding that a particularly stringent standard of review should be applied to the trial judge's exclusion of expert testimony.[43] The Supreme Court reversed the Court of Appeals. The Court noted that *Daubert* did not address the standard of appellate review for evidentiary rulings.[44] The Court went on to elucidate the correct standard of review with consideration of the evolution of the judicial gatekeeping function:

> Thus, while the Federal Rules of Evidence allow district courts to admit a somewhat broader range of scientific testimony than would have been admissible under *Frye*, they leave in place the "gatekeeper" role of the trial judge in screening such evidence. A Court of Appeals applying "abuse of discretion" review to such rulings may not categorically distinguish between rulings allowing expert testimony and rulings which disallow it . . . We likewise reject respondent's argument that because the granting of summary judgment in this case was "outcome determinative," it should have been subjected to a more searching standard of review. On a motion for summary judgment, disputed issues of fact are resolved against the moving party—here, petitioners. But the question of admissibility of expert testimony is not such an issue of fact, and is reviewable under the abuse of discretion standard. We hold that the Court of Appeals erred in its review of the exclusion of Joiner's experts' testimony. In applying an overly "stringent" review to that ruling, it failed to give the trial court the deference that is the hallmark of abuse of discretion review.[45]

The *Joiner* Court went on to note that "it was within the District Court's discretion to conclude that the studies upon which the experts relied were not sufficient, whether individually or in combination, to support their conclusions that Joiner's exposure to PCBs contributed to his cancer"[46] and that the District Court did not abuse its discretion in excluding such expert testimony. The *Joiner* decision expanded the role of the judicial gatekeeping function yet again.[47] By ruling in the manner in which it did, the Supreme Court once again demonstrated that judges have broad discretion while performing their roles as gatekeepers.

5.1.6 Kumho Tire: The Court Continues Its Expansion of the Judicial Gatekeeping Function

Subsequent to the expansion of the judicial gatekeeping function in which trial judges must engage and evaluate when considering scientific evidence, many in the legal community desired the development of a standard to determine the reliability of

[42] *Joiner*, 522 U.S. at 140.
[43] 78 F.3d 524, 529 (1996)
[44] *Joiner*, 522 U.S. at 142.
[45] *Joiner*, 522 U.S. at 142, 143.
[46] *Joiner*, at 147, 148.
[47] *See* Wendy S. Neal, Casenote, *General Electric Co. v. Joiner*, 118 S. Ct. 512 (1997): *The Future of Scientific Evidence in Toxic Tort Litigation*, 67 U. Cin. L. Rev. 881 (1999).

technical but nonscientific expert testimony.[48] In *Kumho Tire Co. v. Carmichael Inc.*,[49] the Supreme Court addressed this issue. In *Kumho Tire*, the survivors of a car accident filed a suit in the U.S. District Court for the Southern District of Alabama, against the tire's manufacturer and distributor. The plaintiffs introduced the testimony of a mechanical engineer who was expert in the area of tire failure. The expert believed that a defect in the tire's manufacture or design caused the blowout. In *Kumho Tire*, the Court had to deal with the admissibility of "testimony of engineers and other experts who are not scientists."[50] The Court held that "*Daubert*'s general holding—setting forth the trial judge's general 'gatekeeping' obligation—applies not only to testimony based on 'scientific' knowledge, but also to testimony based on 'technical' and 'other specialized' knowledge."[51] The Court went on to state:

> We also conclude that a trial court may consider one or more of the more specific factors that *Daubert* mentioned when doing so will help determine that testimony's reliability. But, as the Court stated in *Daubert*, the test of reliability is "flexible," and *Daubert's* list of specific factors neither necessarily nor exclusively applies to all experts or in every case. Rather, the law grants a district court the same broad latitude when it decides how to determine reliability as it enjoys in respect to its ultimate reliability determination.[52]

In *Kumho Tire*, the Court expanded the judicial gatekeeping function. Subsequent to the *Kumho Tire* decision, the Court was allowed to use the *Daubert* standard to determine whether the testimony of a nonscientist was admissible. The *Kumho Tire* Court held that "the trial judge must have considerable leeway in deciding in a particular case how to go about determining whether particular expert testimony is reliable."[53] It is of interest to note that the judicial gatekeeping function was expanded once again in a civil case. It is also of importance to note that the gatekeeping function applies to both criminal and civil cases.

5.1.7 Judicial Gatekeeping Function and Its Evolution in New York State

The *Frye* standard influenced the evolution of the admissibility standards that the states applied. However, most states did not follow the path that the Supreme Court took. At the state level, the admissibility standards differ from the federal

[48]*See* Edward J. Imwinkelried, *The Next Step after Daubert: Developing a Similarly Epistemological Approach to Ensuring the Reliability of Nonscientific Expert Testimony*, 15 Cardozo L. Rev. 2271, 2294 (1994); Kristina L. Needham, Note, *Questioning the Admissibility of Nonscientific Testimony after Daubert: The Need for Increased Judicial Gatekeeping to Ensure the Reliability of All Expert Testimony*, 25 Fordham Urb. L.J. 541 (1998); Jennifer Laser, Note, *Inconsistent Gatekeeping in Federal Courts: Application of Daubert V. Merrell Dow Pharmaceuticals, Inc. to Nonscientific Expert Testimony*, 30 Loy. L.A. L. Rev. 1379 (1997).

[49]119 S. Ct. 1167; 1999 U.S. LEXIS 2189; 143 L. Ed. 2d 238 (1999).

[50]*Kumho Tire*, 119 S. Ct. at 1171.

[51]*Kumho Tire*, 119 S. Ct. at 1171.

[52]*Kumho Tire*, 119 S. Ct. at 1171.

[53]*Kumho Tire*, 119 S. Ct. at 1176.

standard. States across the country have adopted their own admissibility standards. The standards adopted by the individual states tend to be derivatives of one of the three federal standards.[54] Most of them are derivatives of either the relevancy standard or the *Frye* standard. The majority use the *Frye* standard. The *Daubert* standard has been virtually shunned as state courts continue to use the other two standards. The fact that the *Daubert* court did not even use the standard it put forth to make its rulings is arguably one reason the states have not been eager to adopt it.[55] Examination of the admissibility standard that evolved in New York is an excellent case study of state admissibility standards.

In 1938, New York adopted the *Frye* general acceptance standard.[56] After the Supreme Court rendered its decision in *Daubert*, New York reconsidered its admissibility standard. In *People v. Wesley*,[57] the Court decided that it would continue to use the *Frye* general acceptance standard of admissibility.[58] In *Wesley*, the defendant appealed from an order affirming his conviction for murder in the second degree, rape in the first degree, attempted sodomy in the first degree, and burglary in the second degree. DNA profiling evidence was collected from various bloodstained items and the admissibility of that evidence was at issue. The *Wesley* Court held that "since DNA evidence was found to be generally accepted as reliable by the relevant scientific community and since a proper foundation was made at trial, DNA profiling evidence was properly admitted at trial."[59]

Although the *Frye* standard of admissibility is generally used when New York judges perform the gatekeeping function, there have been instances when the Court has considered the criterion put forth by the Supreme Court in *Daubert*. In the matter of *Collins v. Welch*,[60] the Court, while acknowledging that "the *Frye* rule still reigns in New York,"[61] considered the admissibility criterion outlined in the *Daubert* decision and in the Federal Rules of Evidence.[62] Expert witnesses have had a role in our legal process for a long time. As our world becomes more dependent on advancements in technology, the role of the expert witness in aiding judges and juries in understanding concepts beyond the scope of laypersons will expand.[63] It is very probable that the frequency with which experts are used will

[54]Some states have adopted a version of the *Daubert* standard, some a version of the *Frye* standard, and many states have been influenced by the Federal Rules of Evidence.

[55]For a further discussion *see* Barry C. Scheck, *DNA and Daubert*, 15 Cardozo L. Rev. 1959 (1994).

[56]*See People v. Forte*, 279 N.Y. 204, 206; 18 N.E.2d 31, 32 (1938), reh'g denied, 279 N.Y. 788, 18 N.E.2d 870 (1939).

[57]83 N.Y.2d 417, 633 N.E. 2d 451, 454 n2 (NY 1994).

[58]For a discussion of New York law on admissibility, *see also People v. Angelo*, 88 N.Y.2d 217, 222–223, 644 N.Y.S.2d 460, 666 N.E.2d 1333; *People v. Roraback*, 242 A.D.2d 400, 662 N.Y.S.2d 327). In New York, the general acceptance rule is applied to civil actions as well. See *Castrichini v. Rivera*, 175 Misc. 2d 530, 669 N.Y.S.2d 140).

[59]*Wesley*, at 425.

[60]178 Misc.2d 107, 678 N.Y.S.2d 444 (Sup. Ct. Tomkins County 1998).

[61]*Collins* at 109.

[62]*Collins* at 109.

[63]Charles Nesson and John Demers, Symposium, *Gatekeeping: An Enhanced Foundational Approach to Determining the Admissibility of Scientific Evidence*, 49 Hastings L.J. 335 (1998).

increase as well. The criterion used to determine the desired characteristics that an expert should have will change over time. The contemporary shift in what constitutes a desirable witness is just an indicator of what is to come. Experts have been able to use their specialized knowledge to influence the triers of fact in coming to a decision. An understanding of what factors have led us to our current system and way of thinking is essential to any prediction of what the future holds. It seems evident that we are close to seeing radical changes in the law as it relates to expert testimony. Whether the judicial gatekeeping function that has evolved with the role of expert witnesses is an adequate safeguard against the manipulation of the decision processes that decide the outcome of a controversy remains to be seen.

5.2 ADMISSIBILITY OF DNA EVIDENCE

In general, virtually all jurisdictions throughout the United States will admit DNA evidence.[64] However, this statement is limited to DNA evidence that passes muster under the *Daubert* and/or *Frye* standards discussed above. We now provide an overview of the admissibility of polymerase chain reaction–short tandem repeat PCR-STR and mitochondrial DNA (mtDNA) analysis results. There is also a discussion of animal, plant, and viral DNA. Cases in which statistics that govern the interpretation of DNA evidence are also discussed. A list of relevant cases is included in Appendix B. A similar list of relevant statutes has also been prepared and is included herein.

While many of the DNA admissibility issues that arose in past cases involved DNA testing of variable numbers of tandem repeats (VNTRs) using restriction fragment-length polymorphism (RFLP) analysis, the admissibility of RFLP results will not be discussed because such evidence is almost always found to be admissible. The courts of most states have ruled that RFLP analysis testing is admissible under *Daubert* and *Frye*. If cases that discuss the admissibility of RFLP evidence are desired, *The Evaluation of Forensic DNA Evidence*[65] provides a thorough treatment of the admissibility of RFLP in the United States.

5.2.1 PCR-STR DNA Evidence

As described earlier, PCR amplification and subsequent DNA testing of STRs is widely used for forensic identification purposes. PCR-based DNA testing has been found by most courts to be generally accepted in the scientific community. A list of cases in which the admissibility of STR-based testing DNA analysis is evaluated appears as Appendix B. There is usually no dispute about the admissibility of

[64] *See* Thomas F. Liotti, Evidentiary Voir Dire, The Champion, May 2002, pp. 26–28. The Champion is a publication of the National Association of Criminal Defense Lawyers (NACDL).
[65] *See* National Research Council, The Evaluation of Forensic DNA Evidence, 1996, Appendix 6A.

such evidence. Disputes that arise from the admission of PCR-based DNA evidence tend to arise from collateral issues such as chain of custody and contamination.[66]

5.2.2 Mitochondrial DNA

The use of mtDNA test results for the purposes of human identification in legal proceedings has increased in recent years. The admissibility of mtDNA evidence is not as clearly established as is the admissibility of RFLP and PCR-STR-based testing. At the time of this writing, over 16 courts have issued decisions regarding the admissibility of mtDNA testing. A list of many of these decisions is presented in Appendix B. Most of these courts have either ordered mtDNA testing to be performed or ruled that the method was generally accepted in the scientific community.

In one recent case, the trial court placed less emphasis on the testimony of experts who seemed to have a direct financial interest in the admissibility of mtDNA test results because such an admissibility decision would "provide a source of revenue for the companies" as they could provide such testing in future cases.[67] The court in *Holtzer*, however, still concluded that the lower court did not err in holding that mtDNA testing was indeed generally accepted and admissible.[68] In *State v. Scott*, the Tennessee Criminal Appeals Court determined that it was not an error to admit mtDNA without a hearing.[69] The *Daubert* standard is applied in Tennessee.

In *State v. Council*,[70] the Supreme Court of South Carolina held that mtDNA evidence was admissible under the South Carolina Rules of Evidence (SCRE) and under the standard set forth in *State v. Jones*.[71] The standard set forth in *Jones* is more liberal than the *Frye* standard.[72] The Supreme Court of South Carolina has not adopted the *Daubert* standard.[73] The fact that this mtDNA evidence was deemed admissible under a more liberal standard than the *Frye* standard is a fact worth noting. It is also of note that just as in cases involving STR-based DNA evidence, when evaluating the admissibility of mtDNA cases, some courts consider the chain of custody of the evidence at issue in the determination of admissibility.[74]

Not all courts that have heard testimony regarding mtDNA have deemed such evidence to be admissible. In 1998, a Florida court held mtDNA analysis to be

[66]*See People v. Hamilton*, 255 A.D.2d 693, 681 N.Y.S.2d 117 (3d Dept. 1998) (PCR Evidence Admissible under *Frye* subject to appropriate chain-of-custody evidence); *State v. Lyons*, 863 P.2d 1303, 1209, (Or. Ct. App. 1993) ("The potential for contamination presents an 'open field' for cross-examination at trial, but does not indicate that the PCR method is inappropriate for forensic use").

[67]*See People v. Holtzer*, 2003 Mich. App. LEXIS 523 (Mich. Ct. App. Feb. 25, 2003).

[68]*See Id.*

[69]*See State v. Scott*, 1999 Tenn. Crim. App. LEXIS 758, 1999 WL 547460 (Tenn. Crim. App. July 28, 1999), reversed in part on other grounds, 33 S.W.3d 746 (Tenn. 2000).

[70]335 S.C. 1, 515 S.E.2d 508, 1999 S.C. LEXIS 76 (S.C. 1999).

[71]273 S.C. 723, 731, 259 S.E.2d 120, 124 (1979); *See Council*, 335 S.C. at 21.

[72]*See Council*, 335 S.C. at 19.

[73]*Id.*

[74]*See State v. Dean*, 76 S.W.3d 352, 2001 Tenn. Crim. App. LEXIS 791 (Tenn. Crim. App. 2001) (holding that mtDNA was properly admitted and that the chain of custody had been established).

inadmissible because the "database from which the experts wish[ed] to refer [wa]s insufficient to provide reliable statistical conclusions."[75] The court also held that the counting method of quantitative analysis used for mtDNA test results "fail[ed] to provide a meaningful comparison to assist, rather than confuse, the jury."[76] The effect of this case on the admissibility of mtDNA evidence in Florida is unclear as the Court of Appeals of Florida, Second District recently held that a lower court properly admitted evidence of mtDNA analysis at trial.[77] While not all courts have held mtDNA to be admissible, "the case for mtDNA is strengthening with time, not weakening."[78]

5.2.3 Animal DNA

While one usually thinks of human DNA when considering the admissibility of DNA evidence in a legal proceeding, the analysis of DNA from other organisms has been evaluated for admissibility in a number of cases. Individuals committing crimes in homes where pets are present can leave the scene unknowingly carrying on their clothing hair and other biological evidence derived from various house pets such as cats, dogs, and even birds. The Locard principle can explain how such transfers of important evidence can occur. Individuals who profit from sales of animals (or animal parts or their products) on the endangered species list can be brought to justice as a result of animal DNA analysis. Individuals who have purchased a highly prized and/or rare animal and fear that the animal they were given has been switched can determine if they received the same animal that they were expecting through the use of animal DNA testing. Several courts have had occasion to examine issues involving the admissibility of evidence employing animal DNA.[79] The use of animal DNA can be of use to persons involved in a legal proceeding for a variety of reasons. One such reason was made clear in Vermont when the results of DNA analysis conducted on evidence obtained from a deer were used to obtain a search warrant.[80] PCR testing conducted on swine DNA was held to be admissible in Iowa.[81]

5.2.4 Plant and Viral DNA

In addition to animal DNA, courts have had occasion to consider the admissibility of plant and viral DNA as well. An Arizona court considered the admissibility of plant

[75]*See State v. Crow*, No. 96-1156-CFA (Fla. 18th Cir. Ct. May 14, 1998).
[76]*Id.*
[77]*See Magaletti v. State*, 2003 Fla. App. LEXIS 4511, 28 Fla. L. Weekly D 884 (Fla. Dist. Ct. App. 2d Dist. Apr. 4, 2003) (affirming the trial court's determination that mtDNA was admissible).
[78]*See People v. Holtzer*, 2003 Mich. App. LEXIS 523 (Mich. Ct. App. Feb. 25, 2003) (affirming the judgment of the lower court and holding that mtDNA evidence was properly admitted).
[79]*See United States v. Guthrie*, No. 93 6508. 50 F.3d 936, 944 (11th Cir. 1995).
[80]*See State v. Demers*, No. 96–452. 167 Vt. 349, 707 A.2d 276 (Vt., 1997).
[81]*See* U.S. v. Boswell, No. 00–4005. — F.3d —, 2001 WL 1223128 (8th Cir. Iowa 2001).

DNA in *State v. Bogan*.[82] A California court had the unique distinction of finding that a DNA test for HIV was admissible during an attempted murder case.[83]

5.2.5 Statistics

Courts have issued numerous opinions regarding the admissibility of statistical analysis with respect to DNA evidence. In its 1996 report, the National Research Council noted that "the concern that has given courts the most pause in admitting DNA evidence involves the methods for characterizing the implications of an observed degree of similarity in DNA types."[84] Specific methods and concerns regarding statistics have been discussed in an earlier section. One who seeks a more detailed reference describing and discussing statistical issues should see the NRC report described herein as well as the *Reference Manual on Scientific Evidence*.[85]

The decisions issued by courts regarding statistics cover a wide range of issues too diverse to discuss here. The 1996 NRC report lists a host of cases in which courts address concerns related to the presentation of statistics pertaining to DNA evidence at trial. Appendix B, which lists relevant cases involving DNA evidence, also lists some cases regarding statistics. In general, courts are concerned with the way in which analysts describe the frequency with which a particular DNA profile appears in the population at large.[86] In particular, courts tend to scrutinize the method in which the aforementioned frequency is calculated.[87]

The way that the frequency with which a particular DNA profile appears in a particular population is calculated has been argued in the scientific community over the past decade or so. In fact, there have been great debates in the scientific community over the use of statistics in connection with DNA analysis. It is imperative that one who will probably encounter testimony at trial involving statistics conduct research into the particular method of determining the random likelihood of finding a particular DNA profile that will be described.

[82]*See Bogan*, 905 P.2d 515, 519–20 (Ariz. Ct. App. 1995), app. dismissed, 920 P.2d 320 (Ariz. 1996).
[83]*See State v. Schmidt*, N. 99–1412. 771 So.2d 131 (La. App. 3 Cir. 2000) (DNA testing of HIV ruled admissible in an attempted murder case).
[84]*See* The Evaluation of Forensic DNA Evidence, pg.185 (NRC 1996).
[85]*See* David H. Kaye and George F. Sensabaugh, Jr., *Reference Manual on Scientific Evidence* (Federal Judicial Center, 2000).
[86] *See United States v. Yee*, 134 F.R.D. 161, 181 (N.D. Ohio 1991) ("Without the probability assessment, the jury does not know what to make of the fact that the patterns match: the jury does not know whether the patterns are as common as pictures with two eyes, or as unique as the Mona Lisa.").
[87]*See Butler v. State*, No. SC95158. 2002 WL 926283 (Fla., 2002); *People v. Funston*, No. C032472. (Cal.App. 3 Dist. 2002) (holding that error-rate statistics were not admissible); *State v. Belken*, No. 99–2001. 633 N.W.2d 786 (Iowa, 2001); *Arizona v. Garcia*, 197 Ariz. 79, 3 P.3d 999 (Ariz. App. Div. 1999) (*holding that DNA Mixture calculations were admissible under the Frye Standard); State v. Williams*, 574 N.W.2d 293 (Iowa, 1998) (*holding trial court erred in admitting matching evidence without statistics*); *Hull v. State*, 687 So. 2d 708 (Miss. 1996) (holding that trial court erred in admitting matching DNA evidence without statistics); *Nebraska v. Carter*, 524 N.W.2d 763 (Neb. 1994) (the racial group of the perpetrator was unknown and the court held that limiting evidence on statistical frequency to two racial groups was prejudicial).

5.2.6 Paternity

While forensic DNA analysis has had a great impact on the area of criminal law, some areas of civil law have also been greatly affected by the developments in forensic DNA technology that have taken place in recent years. Forensic DNA testing has had a huge impact on the area of family law as well. In particular, DNA testing has been widely employed in determinations of paternity.

Determinations of paternity have a great impact on a variety of legal issues, including issues of inheritance, the eligibility to receive public and private benefits received by legitimate children (e.g., worker's compensation, social security, pensions), as well as the right to use the father's name. The rationale for challenging or seeking to determine paternity can vary. Illegitimacy is one reason to seek a paternity determination. Challenging the paternity of a child born in wedlock may be another. Despite obvious evidence to the contrary, in many jurisdictions, paternity is presumed when a child is born to a woman who is also married. In such cases, it is often presumed that the woman's husband is the father of the child. However, as we all know, this may not always be the case.

The Uniform Parentage Act, which has been enacted by several states, describes relevant concepts in the area of paternity law.[88] The act was recently amended in the year 2000 and the new act is being considered by several states. The new act incorporates much of the old act and incorporates provisions covered by the Uniform Putative and Unknown Fathers Act (1988) and the Uniform Status of Children of Assisted Conception Act (1988). It is important to note that the Uniform Parentage Act is not the law that governs every state. Each state has its own laws governing paternity. However, since many of the laws of the individual states are based upon the Uniform Parentage Act, the act is discussed herein to provide a general overview of the law governing determinations of paternity. Consult the statutes in the state in which your proceeding is taking place in order to determine the law of that particular jurisdiction.

Under the Uniform Parentage Act, paternity is presumed and attributed to a man when: (1) he and the child's mother have attempted to marry legally; (2) he and the child's mother are already married and the child was born during the marriage or within 300 days of its dissolution; (3) if he has acknowledged paternity by filing a written statement; (4) if he has consented to be named as the father on the child's certificate of birth; (5) if he is obligated to support the child by court order or written voluntary statement; or (6) if he receives the minor child in his home and represents the child as his natural child.[89]

Standing is often an issue in legal proceedings that must be addressed. According to Article 6 of the Uniform Parentage Act, the following individuals have standing in a paternity proceeding: (1) the child, (2) the mother of the child, (3) a man whose paternity of the child is to be adjudicated, (4) a support enforcement agency, (5) an authorized adoption agency or licensed child-placing agency, or (6) a representative

[88] *See* Unif. Parentage Act, Article 2, 3 (2000).
[89] *See Id.*

authorized by law to act for an individual who would otherwise be entitled to participate in a proceeding but who is deceased, incapacitated, or a minor, or (7) an intended parent.

Article 5 of the Uniform Parentage Act of 2000 deals with genetic testing. Article 5 governs both individuals who submit to voluntary genetic testing and individuals who are ordered to undergo testing by a court or support enforcement agency.[90] Sworn statements of a party to the proceeding are necessary in order for the court to issue an order for DNA testing.[91] A support enforcement agency may order genetic testing only if there is no presumed, acknowledged, or adjudicated father.[92]

In paternity cases DNA testing must take place in an accredited paternity laboratory. The Uniform Parentage Act specifically states the requirements for DNA testing in paternity cases:

> (a) Genetic testing must be of a type reasonably relied upon by experts in the field of genetic testing and performed in a testing laboratory accredited by:
> (1) the American Association of Blood Banks, or a successor to its functions;
> (2) the American Society for Histocompatibility and Immunogenetics, or a successor to its functions; or
> (3) an accrediting body designated by the federal Secretary of Health and Human Services.
>
> (b) A specimen used in genetic testing may consist of one or more samples, or a combination of samples, of blood, buccal cells, bone, hair, or other body tissue or fluid. The specimen used in the testing need not be of the same kind for each individual undergoing genetic testing.
>
> (c) Based on the ethnic or racial group of an individual, the testing laboratory shall determine the databases from which to select frequencies for use in calculation of the probability of paternity. If there is disagreement as to the testing laboratory's choice, the following rules apply:
> (1) The individual objecting may require the testing laboratory, within 30 days after receipt of the report of the test, to recalculate the probability of paternity using an ethnic or racial group different from that used by the laboratory.
> (2) he individual objecting to the testing laboratory's initial choice shall:
> (A) if the frequencies are not available to the testing laboratory for the ethnic or racial group requested, provide the requested frequencies compiled in a manner recognized by accrediting bodies; or
> (B) engage another testing laboratory to perform the calculations.
> (3) The testing laboratory may use its own statistical estimate if there is a question regarding which ethnic or racial group is appropriate. If available, the testing laboratory shall calculate the frequencies using statistics for any other ethnic or racial group requested.
>
> (d) If, after recalculation using a different ethnic or racial group, genetic testing does not rebuttably identify a man as the father of a child under Section 505, an individual who has been tested may be required to submit to additional genetic testing.[93]

[90] *See* Unif. Parentage Act, Article 5, Section 501 (2000).
[91] *See* Unif. Parentage Act, Article 5, Section 502(a) (2000).
[92] *See* Unif. Parentage Act, Article 5, Section 502(b) (2000).
[93] *See* Unif. Parentage Act, Article 5, Section 503 (2000).

Section 503 is clear with respect to the requirements for laboratory accreditation and the use of genetic databases and population frequency calculation methods.

Situations may present in which the man believed to be the father of a child is not available for testing. Recent disasters such as the World Trade Center attacks demonstrate this reality all too well. The comments with respect to Article 5, Section 502, point out that the alleged father is not the only party who can be tested in order to determine paternity and that a court may make a determination of paternity by testing close relatives of the man believed to be the father of the child. Article 5, Section 508, governs situations in which genetic specimens from the putative father are not available. In such cases, a court may order the following individuals to submit genetic samples: (1) the parents of the putative (possible) father, (2) brothers and sisters of the possible father, (3) other children of the possible father and their mothers, and (4) other relatives of the possible father necessary to complete genetic testing.[94]

Subsequent to DNA paternity testing, the laboratory will create a report of its findings. The Uniform Parentage Act requires that a report of genetic testing must be in a record and signed under penalty of perjury by a designee of the testing laboratory. In legal proceedings a report created pursuant to the Uniform Parentage Act is self-authenticating.[95] With respect to chain-of-custody issues, documentation from the laboratory that conducted the DNA testing of the following information is sufficient to establish a reliable chain of custody. Such documentation allows the results of genetic testing to be admissible without testimony and must contain: (1) the names and photographs of the individuals whose specimens have been taken, (2) the names of the individuals who collected the specimens, (3) the places and dates that the specimens were collected, (4) the names of the individuals who received the specimens in the testing laboratory, and (5) the dates the specimens were received.[96]

After DNA paternity testing has been completed and a report is prepared, some may seek to rebut the analysis or the conclusions of the report. The Uniform Parentage Act declares that a man is rebuttably identified as the father of a child if the DNA test results disclose that: (1) the man has at least a 99% probability of paternity, using a prior probability of 0.50, as calculated by using the combined paternity index combined paternity index (CPI) obtained in the testing, and (2) a CPI of at least 100 to 1.[97] While the Uniform Parentage Act sets these standards, it should be noted that many states have established a presumption of paternity at a lesser probability of paternity.

If one wishes to rebut the results of DNA paternity testing despite a showing by testing done pursuant to the Uniform Parentage Act that there is at least a 99% probability of paternity, and a CPI of at least 100, he has to have new DNA testing performed that conforms to the requirements set forth in the Uniform Parentage Act. These new tests must: (1) exclude the man as the genetic father of the child or (2)

[94] *See* Unif. Parentage Act, Article 5, Section 508(a) (2000).

[95] *See* Unif. Parentage Act, Article 5, Section 504(a) (2000).

[96] *See* Unif. Parentage Act, Article 5, Section 504(b) (2000).

[97] *See* Unif. Parentage Act, Article 5, Section 505(a) (2000).

identify another man as the father of the child.[98] In the unique circumstance in which more than one man is identified by DNA testing as the possible father of the child (i.e., if the putative father is one of a pair of identical twins or has male siblings that share his genotype at many loci), the court can order each of the men to submit to further genetic testing to identify the genetic father.[99] It should be noted that even identical twins can be differentiated by means other than DNA testing.

5.3 LEGAL PRACTICE

5.3.1 Different Stages of a Trial

For one to understand how DNA evidence is used at trial, it is important to understand the trial process itself. This section describes what happens to a suspect who is arrested and is moved through the criminal justice system and trial process. What goes on during various stages of the trial is outlined step by step. Veteran legal practitioners may wish to skip this material and proceed directly to Section 5.4.

5.3.1.1 Arraignment Subsequent to being arrested, a defendant is usually arraigned. An arraignment is a proceeding in which the charges are read to the defendant and the defendant enters a plea of either guilty or not guilty. An arraignment is a major event! Depending on who is being arraigned or what the defendant is being charged with, an arraignment can cause quite a stir. Photographers and reporters might be interested in the arraignment proceeding. An attorney may encounter such persons outside of the courtroom. Often, the defendant's family and friends are present at the arraignment. The attorney may wish to attack the charging instruments or the evidence in the case during the arraignment proceeding.

Bail is usually set at the arraignment proceeding. Bail is based upon whether the judge perceives the defendant to be a flight risk. Bail is a way of compelling the defendant to return to court for future proceedings. Securing the defendant's release is the defense attorney's primary concern at the arraignment. Defendants who are out on bail, rather than incarcerated, can be very helpful during trial preparation. However, securing the defendant's release has an additional purpose. By successfully securing the defendant's release during arraignment, the defense counsel avoids having to deal with the jury's perception of the defendant being surrounded by court officers at trial.

Defense counsel may also be made aware of what evidence has been seized and whether DNA is a part of the case. While DNA is used for the purpose of identification, the defense counsel may also be made aware that there are other identifications of his client, the complainant knowing the defendant or picking him out of a lineup. If blood was recovered from the scene but comparisons to the defendant's blood have

[98]*See* Unif. Parentage Act, Article 5, Section 505(b) (2000).
[99]*See* Unif. Parentage Act, Article 5, Section 505(c) (2000).

not yet been made, then the prosecutor may be requesting that they be allowed to obtain blood samples from the defendant. This is known as a *Schmerber v. California* application.[100] Naturally, once defense counsel is apprised that DNA is a part of the case, it becomes important to ascertain which lab is being used and to obtain its protocols. The history of that lab and its experts may then be determined, and counsel can retain experts familiar with that lab, its experts, and its protocols in order to determine the strength of the prosecution's case and areas where the evidence may have been compromised or contaminated.

5.3.1.2 Grand Jury Subsequent to being arraigned, a defendant may be indicted. An indictment is the instrument that charges a defendant with a crime. Defendants are indicted by the grand jury. The grand jury is composed of a group of citizens brought together to hear the testimony of witnesses and to consider the evidence the prosecution has and to determine if the prosecution has enough evidence to hold the defendant for trial. In most jurisdictions, the defendant has the opportunity to testify at the grand jury proceedings. If, after hearing the testimony put forth, the grand jury believes that the prosecution has a case against the defendant based upon the evidence, it will most likely vote to indict the defendant on the charges. Once the defendant is indicted, the case is prepared for trial. The nonexistence of DNA has a great bearing on whether a defendant should testify. Finding out prior to indictment whether there is DNA evidence is the aim of defense counsel and prosecutors alike.

5.3.1.3 Discovery Discovery is the process by which the prosecution and defense (or the plaintiff and defendant in a civil trial) exchange information relevant to the case. The period during which discovery takes place is one of the most important segments of a legal proceeding. Discovery is not really a phase at trial with a distinct end or beginning. Discovery materials may be requested and/or turned over at any point during a legal proceeding, though some state or federal rules may limit the discovery period and may also require that additional requests be made with the permission of a judge. Each side makes requests for information during discovery.[101] The prosecution has a duty to turn over all evidence that may show the defendant's innocence.[102] This duty exists at all stages of trial and during the pretrial period.[103] However, it should be noted that some state courts

[100]*See Schmerber v. California*, 384 U.S. 757; 86 S. Ct. 1826; 16 L. Ed. 2d 908, 1966 U.S. LEXIS 1129, (1966).

[101]Each state has its own pretrial discovery rules. For an example of state pretrial discovery rules, *see* N.Y.C.P.L. §240.

[102]*See Brady v. Maryland*, 373 US 83, 83 S.Ct. 1194 (1963); *Kyles v. Whitley*, 514 US 419 (1995); *U.S. v. Gil*, 297 F3d 93 (2002); *People v. Ahmed*, 20 N.Y.2d 958, 286 N.Y.S.2d 850 (1967); *People v. Vilardi*, 76 N.Y.2d 67, 556 N.Y.S.2d 518 (1990); *People v. Rosario*, 9 N.Y.2d 286, 213 N.Y.S.2d 448, *cert. Denied*, 368 U.S. 866, 82 S.Ct. 117 (1961); *United States v. Giglio*, 405 U.S. 150; 92 S. Ct. 763; 31 L. Ed. 2d 104; 1972 U.S. LEXIS 83(1972). *See also*, Thomas F. Liotti and H. Raymond Fasano, *Forget Brady and Rosario: Use Foil*, The Attorney of Nassau County, January 1997 at 3 and 17 and Thomas F. Liotti, *The Uneven Playing Field, Part III, or What's on the Discovery Channel*, St. John's Law Review, Vol. 77, No. 1, Winter 2003, pp. 67–74.

[103]*See generally* N.Y.C.P.L. §240.60.

have held that discovery or "Brady" materials should be disclosed to the defendant "at the earliest possible opportunity in advance of trial."[104] Both the prosecution and the defense make discovery requests to one another that are geared toward the discovery of relevant evidence. Such evidence includes police statements,[105] statements by witnesses,[106] information on all prior criminal conduct that the prosecution is aware of and intends to use against the defendant, and DNA test results.[107]

Whether information is "material" to a case and can be considered discoverable pursuant to *Brady* is determined based in large part on case law in which the term material has been explored in great length. Each state may have its own rules. For example, in New York, information is considered material for *Brady* purposes if a reasonable possibility exists that the disclosure would have altered the trial outcome.[108] However, there are some cases that limit this definition. For example, a New York State court has ruled that the failure to perform DNA tests in a case in which favorable results are merely speculative does not violate the spirit of *Brady*.[109]

Whether a case is civil or criminal in nature, it is important to note that both sides have a duty to provide discovery. In a criminal case, the defendant might have a duty to provide reciprocal discovery to the prosecution.[110] When a party fails to respond to a discovery request during the stated time limit or to provide information that such party is required by law to provide, such behavior is called a discovery abuse. Discovery abuses are often a reason to seek intervention from a judge. The judge may have the power to order a party to disclose information or to hand over documents. If the party still refuses, the judge may be able to impose a penalty (sanction) upon the party, or dismiss the case outright, or preclude a party from presenting this information to the jury.

[104] *People v. Bottom*, 76 Misc.2d 525, 351 N.Y.S.2d 328 (Sup.Ct. N.Y. Co. 1974). *See also People v. Hunter*, 126 Misc.2d 13, 480 N.Y.S.2d 1006 (Sup.Ct. N.Y. Co. 1984).

[105] *See People v. Morillo*, 181 A.D.2d 532, 582 N.Y.S.2d 1 (1st Dept. 1992) (holding that the notes of a policeman are discoverable).

[106] *See People v. Boone*, 49 A.D.2d 559, 370 N.Y.S.2d 613 (1st Dept. 1975).

[107] *See People v. Davis*, 196 A.D.2d 597, 601 N.Y.S.2d 174 (2d Dept. 1993), *app. Denied*, 82 N.Y.2d 923, 610 N.Y.S.2d 175 (1994) (holding that statistical data regarding the probability of DNA matching samples is *Brady* material).

[108] *See People v. Vilardi*, 76 N.Y.2d 67, 556 N.Y.S.2d 518 (1990); *People v. Davis*, 81 N.Y.2d 281, 598 N.Y.S.2d 156 (1993); *People v. Scott*, 216 A.D.2d 592, 628 N.Y.S.2d 965 (2d Dept. 1995), *aff'd*, 88 N.Y.2d 888, 644 N.Y.S.2d 913 (1996).

[109] *See People v. Smith*, 204 A.D.2d 140, 612 N.Y.S.2d 12 (1st Dept. 1994), *app. denied*, 84 N.Y.2d 872, 618 N.Y.S.2d 18 (1994). *See also People v. Battee*, 122 A.D.2d 526, 505 N.Y.S.2d 10 (4th Dept. 1986); *People v. Owens*, 108 A.D.2d 1014, 485 N.Y.S.2d 584 (3rd Dept. 1985) (noting that *Brady* material did not include laboratory results that were not conclusively favorable to the defense).

[110] *See* N.Y.C.P.L. §240.45(2). *See also People v. Charron* 198 A.D.2d 722, 604 N.Y.S.2d 311 (3d Depr. 1993), *app. Denied*, 83 N.Y.2d 803, 611 N.Y.S.2d 139 (1994) (holding defendant must provide reciprocal discovery to the prosecution, which includes statements made by a prospective defense witness, other than the defendant, that relates to the subject matter of the witness' testimony; *People v. Figueras*, 199 A.D.2d 409, 606 N.Y.S.2d 237 (2d Dept. 1993) (holding that information concerning the criminal record or any pending criminal action against any witness except the defendant, that the defense team plans to call much be provided to the prosecution).

5.3.1.4 Preparation for Trial—A Lawyer's Guide

> The leading rule for the Lawyer, as for [one] of every other calling, is diligence. Leave nothing for tomorrow which can be done today.
>
> —President Abraham Lincoln (1809–1865)

Being a successful trial lawyer requires devotion to one's clients, an unparalleled work ethic, and love of the profession. The practice of law and preparation for trial require a great deal of time and hard work.[111] Preparation for any trial is exhausting, demanding, detailed work. A DNA trial will require not only legal research but also research of the forensic scientific literature. There are no shortcuts. No matter how long ago you took high school biology, attaining knowledge of DNA is achievable. It simply takes time. This book is geared to shorten your learning curve.

It is important during the course of one's representation of a client that you believe in what you are doing. When times are tough, lawyers need to remember why they embarked on a legal career. The lawyer would like to be recognized as a professional who fights for justice. Never be thwarted by cynics or those who misunderstand the lawyer's high calling.

Every aspect of a lawyer's work involves ethics.[112] Problems may arise multiple times each day; do not wait to be hit over the head with a two-by-four. There is a healthy level of paranoia that all trial lawyers should have. Obtain educated opinions and talk to other lawyers about ethical dilemmas. Write memos to the file. As preparation for trial proceeds, consider the adversary's plight and to what extent professional courtesy will be returned. Hand-to-hand combat with an opponent is a pointless distraction. Making your adversary or the presiding judge an enemy in front of the jury is not helpful.[113] Take the client and the client's case to a jury on its merits. Besides the Canons of Ethics, all lawyers involved in a criminal case should refer to the American Bar Association Standards for Criminal Justice. Whether one is involved in a criminal case or a civil litigation, it is important to be ethical and to follow the governing code of professional responsibility. That means that in a DNA case, a prosecutor is bound to turn over all exculpatory evidence but also evidence of contamination or a failure to follow the chain of custody or protocols.

[111] A great book in which this aspect of the trial lawyer's role is analyzed by Sam Schrager, in *The Trial Lawyer's Art*, Temple University Press, 1999. Schrager uses his introduction to discuss how the trial lawyer's role is that of a storyteller. Also, *see New York Criminal Practice*, 2nd ed. Lawrence N. Gray (Editor in Chief), Supplement, Chapter 11A, 2000. *See also* Thomas F. Liotti, The Lawyer's Bookshelf, *Trying Cases to Win*, by Hon. Herbert Stern, New York Law Journal, January 15, 1992 at 2.

[112] *See* American Bar Ass'n, *Model Rules of Professional Conduct* (1992); Monroe H. Freedman, *Understanding Lawyers' Ethics* (1990); New York State Bar Ass'n, *The Lawyer's Code of Professional Responsibility*.

[113] For more on important considerations before trial, *see* R. Lawrence Dessem, *Pretrial Litigation: Law, Policy & Practice* (1991).

The client's first phone call is the point to begin trial preparation. This could be the beginning of a case that will go to trial. In early client interviews lawyers often ignore asking detailed evidentiary questions such as: Were you injured? Were you cut? What were you wearing? All of the answers to these questions and more may help the practitioner to determine whether it is a DNA case and how best to prepare for it. Prepare the client for that possibility by explaining the costs of legal fees, experts,[114] investigators, and the like. If the client can afford daily transcripts, advise him or her to order them. Tell clients who are faced with serious charges that trial preparation is the most important thing that they will ever do in their lifetime. Some clients say that they cannot afford to go to trial. However, it may well be the case that they cannot afford not to go to trial. It is the lawyer's job to give counsel to the client regarding the best way to proceed. A first step is to determine the existence of DNA evidence and whether it can be or should be rebutted.

Everyone involved with this case is questioning the likelihood that your client committed the crime. You will probably be wondering the same thing. However, there is a lot to consider with regard to asking the ultimate question of the client. There is conflicting advise about whether or not an attorney should ask whether the client actually engaged in the acts he is being accused of committing. An attorney is able to connect the dots. Asking the ultimate question may present ethical problems later. Perjury and suborning perjury are legal, ethical problems that must be considered at the inception of each case. What if the attorney decides to put the client on the stand? What do other defense lawyers and their clients know about the client? What does the prosecution know about the client? Never rely upon the client alone for the facts. Use your best judgment and hire an investigator to collect information about your client and the allegations. On the day of client retention, it might be prudent to ask yourself: "If I had to give my opening statement or summation at this point, what would it be focused on?" It helps the attorney empathize with the client and develop a theory of the case.[115] While polygraph test results are not admissible,[116] you may want your client tested anyway. Positive results may assist you in plea negotiations, but as a general rule you do not want your client, or anyone associated with your client, to blurt out at trial that they passed a polygraph. If there is DNA in the case, then the strategy options are reduced.

It is important to visit the crime scene and conduct your own investigation rather than relying on an investigator to tell you about the crime scene.[117] While no lawyer should make himself or herself into a witness, visiting the crime scene does not do so *per se*. You might consider bringing binoculars, a magnifying glass, small plastic bags, tweezers. Have your investigator take photos, make a video and take the

[114]The costs for DNA and population geneticist experts should be explained in detail.

[115]*See* Thomas F. Liotti, *Avoiding Prosecutions*, N.Y. St. B.J., Feb. 1995, pp. 49–58.

[116]*See Frye v. United States*, 293 F. 1013 (D.C. Cir. 1923).

[117]Criticizing the manner in which the police gathered evidence is an important part of the defense function. *See* H. C. Lee, T.M. Palmbach, and M.T. Miller, *Henry Lee's Crime Scene Handbook*, Academic Press, New York, 2001.

statements of witnesses. Bring a map. Learn the streets, the neighborhood, and so on. Consider what notes and diagrams may be subject to reciprocal discovery.[118] In some cases, it may be useful to have flyers distributed; they may prompt witnesses to come forward. Ads in local newspapers may accomplish the same result. Try to secure copies of all medical and autopsy reports as soon as possible. You may want to have your client physically examined to have blood samples taken or his clothes inspected for forensic evidence. For example, your client tells you that the deceased spit in his face before he killed him. The deceased may have left DNA evidence on your client that could help to corroborate a claim of self-defense.

Visit the library and identify books relevant to your particular case. There have been may books written that can assist an attorney who is preparing to try a case in a particular area of law.[119] Create a list of books that may be of value or assistance to you. Keep the list that you have prepared current as your case progresses. As the trial date approaches, the ability to quickly locate these texts can be of great assistance to the attorney and the client. Such texts may also assist in the preparation of motions.

The media is always a factor to consider. The first thing an attorney must consider at the outset of a case is whether the case could be considered a high-profile case. Is it newsworthy? If there is a news angle, the attorney should know what that is. What, if anything, should counselors be saying to the media? If possible, secure the client's permission to speak to the media before doing so. What is one's adversary saying to the media? What do the Canons of Ethics permit? Ask these questions: "Am I legally permitted to comment? If so, what may I say? Is a comment needed? Will it help or hurt my client? How far away is trial? Is there any likelihood that my comments will prejudice potential jurors?" If the prosecutor says that my client's blood or DNA was found at the scene, what if anything should be said to counter that?

Counsel may prepare a news release for interested media; in that way there are fewer chances of making an erroneous or potentially harmful statement.

Talk to your adversaries. Talk to the police, the prosecutor, and the medical examiner if one is involved in the case. Prepare a witness and exhibit list for your opponents. Try to talk to the adversary's witnesses.

Setting up a folder and loose-leaf system for each witness is advisable. Information collected about a witness is placed in the file, and the loose-leaf contains the witness's prior statements or testimony. The file will contain the *Rosario*,[120]

[119]A well-composed and researched book that may be of assistance to a lawyer who is preparing for a criminal trial in the State of New York is Ronald E. Cohen and James C. Neely, *Criminal Trial Advocacy* (8th ed., 1999). The aforementioned text is a comprehensive handbook prepared by the Appellate Div-

[120]*See People v. Rosario*, 9 N.Y.2d 286, 213 N.Y.S.2d 462 (1st Dep't 1989), which required prosecutors to make available to the defendant written or recorded statements made by those the prosecutor plans to call as witnesses at trial.

Giglio,[121] and *Brady*[122] material. The main file consists of pleadings, reports, and correspondence—all highlighted in yellow and indexed with Post-its. Given the opportunity, go and watch your opponent at trial in another case or police witnesses as they testify, particularly if it is a DNA case. Find out what you can about them. Transcripts of their prior testimony, vitae, Department of Motor Vehicle reports, plaintiff/defendant civil action indexes, credit checks, and judgment indexes will be useful.

Consider the charges and evidence against your client. Review the complaint or indictment for deficiencies. Do the charges mirror the statute? What are the elements? What must be proven? What evidence does the prosecution have?

5.3.1.5 Jury Selection/Voir Dire When a jury is being selected, the attorneys on both sides examine or question each perspective juror. This examination is often referred to as voir dire. Voir dire is a very important part of the trial process. During voir dire, the jury is chosen. It is important that each attorney be able to contribute to the decision of who will sit on the jury. Attorneys on both sides should take great care to ensure that people who are predisposed or who may not return a favorable ruling are eliminated from the jury pool. Questions about a juror's knowledge of DNA and respect for science should be on the list of questions. The following section will explore the voir dire process and relevant considerations in greater depth.

5.3.1.5.1 Profiling Jurors

> Wooten's strategy was to pick one member [of the jury] who was strong and intelligent and one who, in his opinion, wasn't. You tried to present your case in story form to the juror who was not intelligent, whereas you argued the contradictions before the one who was.
>
> —Norman Mailer
> *The Executioner's Song* (1990)[123]

If the DNA evidence is strong and you are trying to refute it, then you may want jurors who have limited knowledge of science. If the evidence is weak, then you may want jurors to uphold it. Therefore, you may wish to have sophisticated persons who have a grasp of the relevant scientific principles on the jury. If you are seeking to highlight contamination or misconduct, you want to choose jurors who are very independent with a strong streak of rebelliousness in their personalities. When doing this, do not make the mistake of subscribing to various stereotypes. Pay close attention to the clues that people present. For example, in order to succeed in certain occupational areas, the ability to grasp complex information is crucial.

[121]*See United States v. Giglio*, 405 U.S. 150; 92 S. Ct. 763; 31 L. Ed. 2d 104; 1972 U.S. LEXIS 83 (1972) (prosecutor's promises made to his witness in return for testimony are material for Brady purposes).
[122]*See Brady v. Maryland*, 373 U.S. 83 (1963), in which the Supreme Court held that it is a violation of due process for the prosecution to suppress evidence that could be favorable to a defendant.
[123]This quote and many others of practical value can be found in *The Quotable Lawyer* (Globe Pequot Press Guilford, CT, Tony Lyons ed., 2002).

A person may not look or act like he or she has such a skill. Remember, it is jury duty. Jury selection can at times be just as adversarial as a trial. However, the aforementioned adversarial tension between the lawyers and jurors is just not as apparent as it can be between the attorneys during a trial. Rare and golden is the citizen who dutifully and without any reluctance puts his or her job and life on "hold" in order to serve his or her country as a juror to ensure that a defendant receives a fair trial. This must be kept in mind when interpreting the behavior of potential jurors.

5.3.1.6 Quickie Voir Dire[124]: Making the Most of Fifteen Minutes

Thomas F. Liotti and Ann Cole[125]

For art and science can not exist but in minutely organized particulars
—William Blake, *Jerusalem* (1818–1820) To the Public

Voir dire is the blending of art and science. During a well-executed voir dire, the trial attorney uses his or her presentation skills to glean as much information as possible from prospective jurors while planting the seeds of his case in the juror's minds. Lawyers must be able to search for sources of bias either for or against the client. They must know when to stop asking questions of jurors believed to be favorable, how to rehabilitate a juror the prosecution seems to have successfully removed for cause, and how to use a juror who will surely be excused for cause as a foil to educate the rest of the panel. As with most aspects of trial work, only practice and experience can help an attorney turn voir dire into an art form. This discussion endeavors to provide a useful framework for lawyers preparing for jury selection.

The late Cat E. Bennett was a pioneer in the field of jury selection,[126] and all who have followed have tried to build upon her work. Much of what we write here applies to both civil and criminal trials. Ever since the Court of Appeals in *People v. Jean* [127] approved time limits for voir dire of 15 minutes in the first round and 10 minutes in each round thereafter, we have been forced to focus on what is really

[124]*See also* Paul M. Lisnek, *The Hidden Jury and Other Secrets Lawyers Use to Win* (Sourcebooks, Inc., Naperville, Ill., 2003).
[125]Ann Cole is a jury and trial consultant in New York. She has served as a consultant to CBS on the O.J. Simpson case and has been retained or court appointed for the defense in a number of important cases including the Oklahoma City bombing case. Together, Tom and Ann have picked a great many juries. Tom and Ann are grateful to Jason Spector, a former law clerk in Mr. Liotti's office, for his assistance in the research and writing of this portion of the book, which was also published as an article, and Kathleen Murphy of Ann's office for her assistance. This article was previously published as follows: *New York State Bar Association Criminal Justice Section Journal*, Summer, 2000, Vol. 8 No. 1 at 20 and *New York State Bar Association Journal*, September, 2000, Vol. 72, No. 7 at 39. This article was also printed in the *New York Law Journal* on June 22, 2000 at 1, 2 and 3 and reprinted for the *New York Law Journal* CLE Seminars at their request for their *Advanced Medical Malpractice Techniques* seminar on October 2–3, 2000 at the Helmsley Hotel, New York City.
[126]See Bennett and Hirschhorn, *Bennett's Guide to Jury Selection and Trial Dynamics in Civil and Criminal Litigation*, West Publishing Eagan, MN, 1993, 1995.
[127]*See Jean*, 75 N.Y. 2d 744 (1989).

important and eliminate unnecessary questions. This discussion is written in an effort to help you to hit the ground running.

Do your warm-up exercises before you begin voir dire. We recommend that you practice public speaking, preferably on controversial subjects so that you lose a sense of self-consciousness and become more confident. Practice engaging your audience, asking them questions and studying their reactions. Another good exercise is to engage strangers in conversation. Observe how long it takes you to get them talking or answering your questions. The ride up to your office on the elevator is an opportune time for such endeavors. Remember what you said that elicited their responses. You might want to adopt some of those questions or comments for your voir dire.

5.3.1.6.1 Study the Case and Define the Objective Even before specific voir dire questions are contemplated, read everything you can about the case and visit the scene. In order to persuasively argue the defense case, you must be able to paint a word picture of your case. You must first clearly define and distill the key issues and the key facts. It is at this point where you may need to determine your prospects for essential nullification. If the law is bad for you, determine if there is something about the facts or about the conduct of your adversaries that will make the jury angry, so angry that they may overlook the letter of the law or purposely disregard it, providing your client with a more equitable "street justice." You can never start too soon to get the jury angry with your opponent. And, of course, if the judge is not the fair and impartial arbiter that he or she should be, you may have to get the jury angry at your judge as well, especially since he or she will make the legal rulings and charge the jury with the specifics of the law. If a judge is late, get to the courtroom early. Get set up and be ready to go. Keep looking at your watch, shaking your head. Jurors are taking time off from work. They already think that the legal system is a big time waster. Their time is valuable. You care about their time. Does the judge? Does your adversary? Stand in the well of the courtroom looking like you are ready. Point to that courtroom clock. Point back toward the judge's chambers. Shake your head some more. You can reinforce this later on when you ask questions about their work, their time, and whether they will be able to give their full attention to the case. Start with the basic premise that jurors do not want to be there, and they view you as being a part of the rude system that has interrupted their lives. You have to show them that you care about their time, that you are not a part of the system, and that you too resent having your time wasted and so does your client.

5.3.1.6.2 Focusing on the Goal Lines Do you want to get an acquittal? Get divorced? Win a monetary award? Do you want punitive damages? When combined with the facts of your case, each of these objectives will lead you to search for a particular type of person with particular biases. For example, a plaintiff's attorney will not want a doctor on the jury of a medical malpractice case if the issue is a close judgment call. But, what if there was particularly egregious behavior on the part of the defendant doctor? You might very well want to have more competent colleagues who will come down on the defendant very hard.

5.3.1.6.3 Profile Establish a profile ahead of time to determine what kind of jurors you want. Very often I want antiestablishment, antiauthoritarian jurors. We are always mindful of *Batson v. Kentucky*.[128] See, also, *People v. Blunt*,[129] *People v. Kern*,[130] and *People v. Garcia*[131] and deliberately try to not, even inadvertently, discriminate on the basis of race, gender, or national origin. Thus, we do not establish a "cognizable group" that we favor or to which we are opposed. If a prospective juror is an immigrant, we want to know what hurdles she encountered in her assimilation. We are interested in their attitudes, not the color of their skin. Every juror must be considered as an individual. Voir dire would be unnecessary if we simply generalized about people or categorized them on the basis of their skin color. For example, generally we like jurors who are angry. A judge once asked us about a cognizable group against which he thought we might be discriminating. In that case we assured the judge, in a jocular way, that our only cognizable group was to find the dumbest people that we could find. He assured us that it would not be a problem in that particular county.

5.3.1.6.4 Body Language Before you engage each prospective juror in conversation, take a good look at each of them. What are they reading, if anything? What does it mean if a prospective juror is reading the *New York Times* as compared to the New York *Post*? How are they dressed? It is said that you can tell everything about a man's wealth by his shoes. Particularly for women, do they have a modern or traditional haircut? Do they wear a lot of accessories? A word of caution, always keep in mind that a prospective juror who works or does not work full time might not be dressed as they normally would. Therefore, what does it say if a businessman wears a suit and tie for jury duty?

How does each potential juror carry himself? Does this tell you if he or she will be a leader or a follower? Are jurors likely to be like Henry Fonda in *Twelve Angry Men*? If necessary, will the juror stick to his position even if he is only one who feels a certain way? Do not be afraid to ask him. This may be your hold-out juror. A mistrial or hung jury may be a victory in your case. Which jurors like you and your client? Which smile at you and laugh at your jokes? Can you make eye contact with each juror you are willing to sit? Does each look at your client? Or, do any sit with cold looks on their faces.

5.3.1.6.5 Background Information and Preparation In all likelihood, the court where your case will be tried will have booklets that they give out to jurors. Get familiar with them. Also, most courts now show a video to potential jurors to prepare for jury duty. Take a look at it. You can ask questions about the jurors understanding of it, which will help you to glean information about intelligence, their level of information retention and ability to communicate information to others,

[128]*See Kentucky*, 476 U.S. 79 (1986).
[129]*See Blunt*, 162 A.D. 2d 86 (1990).
[130]*See Kern*, 75 N.Y. 2d 638 (1990).
[131]*See Garcia*, 217 A.D. 2d 119.

all critical aspects of deliberation while creating a bond with the prospective jurors. Keep an eye out for monuments or other items around the courthouse that may be a source of initial pleasantries. For example, on the wall of a courthouse in Mineola, New York, the following words are etched: "Justice, God's Idea, Man's Ideal." Ask potential jurors what do those words mean to them?

Talk to the commissioner of jurors, as he or she possesses a wealth of knowledge about the jury pool or typical selection procedures. You want to know more about jury service than the judge or your adversary. Find out when and how much jurors get paid? For how long must their employers pay them? Look at the statutes that allow jurors to be excused for health, hardship, or other reasons. You want to empathize with jurors and show them that you care about them. To our surprise, on a recent case, we learned from the commissioner of jurors that although our jury selection was occurring midweek, we would nonetheless be getting a fresh batch of jurors, not rejected or recycled from other cases. Often when you begin jury selection in the middle of a week you will be faced with jurors who have been "rejected" during voir dire in another matter. You must inquire whether this is the first panel each juror has been on during this jury service. If it is not, inquire about the effects of the previous rejection and its possible impact on your case.

5.3.1.6.6 Questionnaire The Office of Court Administration typically has a one-page questionnaire that jurors complete. Prepare blank questionnaires ahead of time so you can listen more than you write. Take careful notes on your prepared form. The questionnaire tells of any prior jury service. If potential jurors deliberated to verdict, you may not want them. Why? Conviction rates are high. If they were on a criminal case that went to verdict, they may have voted to convict. You have to ask if they were satisfied with their jury service. Were they fair to both sides? If they say yes, then you may have a juror who voted to convict. If the juror, on the other hand, tells you about how she thought the police were shady or that she distrusted them, that gives you a different perspective on a particular juror.

5.3.1.6.7 Motion in Limine; Requests for Additional Challenges and More Time for Jury Selection Before voir dire you must determine what you want to (and likely can) limit in the way of proof against your client. To limit the proof, you must also curtail your voir dire, limiting your questions only to those topic areas that will be important at trial. If your case is complex or has received media coverage, you may want extra challenges or more time for voir dire. Make a written motion to that effect. Each judge will treat this request differently, so you must ask around if you lack personal familiarity. If the judge will not expand your time, consider posing particular questions to the entire panel, and follow-up with focused questions to those that respond. Then ask the rest of the panel how they feel about their follow-up responses.

5.3.1.6.8 Requests for Voir Dire by the Court There are questions that you may not want to ask; appearance is everything. Therefore, you should request, in writing, that the court ask those questions for you. For example, the typical question: Is there any physical, emotional, or mental reason why you cannot serve as a juror? You may not want to ask the question that causes the juror to admit to a bladder problem or that they were sexually abused as a child.

5.3.1.6.9 Voir Dire This is not a science. Are you a good judge of character? The lawyer should concentrate on asking the questions while a juror consultant should make the notes, preparing the basis for challenges, both for cause and peremptories. Since you have unlimited cause challenges, the more jurors you can remove for cause the better. You will have to determine what will impede their ability to be fair and impartial, then ask those jurors questions designed to elicit the desired response. You want to reveal some of the worst parts of your case to the jury, but not all in the first round. You introduce the case incrementally and in pieces. If you give them too much in the beginning, then you may shock otherwise good jurors into disqualification. To be granted the challenge for cause, the attorney must get the juror to say he or she cannot be fair. It needs to come from the juror's mouths, not just the attorney's. In the first part of voir dire you introduce the subject matter. Ask jurors: "Is there anything about that subject that offends you"? "Look at my client, look him in the eyes, now in your heart of hearts, tell me if you presume him innocent?" "Nowhere in the United States Constitution does it say that the people are entitled to a fair trial. Would you be satisfied with someone of your like mind presiding as a juror in your own trial or that of a member of your family or a close friend?" "Is this a level playing field or is it already slanted against my client before we even start?" "How many of you would require my client to take the stand and testify in his own behalf?" "How many of you would require that he vindicate himself?"

Voir dire means to "speak the truth." How forward and rude. We barely know these prospective jurors and we are asking them to speak the truth about their private thoughts and experiences. We have only 1.5 minutes per juror to put the juror at ease, get acquainted, and try to find out what is really on his or her mind. Then we ask the jurors to do all the talking. It is not a perfect system, but it is all we have. Ask the jurors to talk to you, any time you get an opportunity to know them better, ask. "Talk to me now." "Help me out here." "Can you be fair to my client?" "Now can you follow this case wherever it may lead?" "Can you be a trier of fact?" "Can you critique the evidence, evaluate it, turn it inside-out in searching for the truth, in trying to find out what happened?" In a criminal case you might ask, "are you afraid of the police?" "You will be forced to look at police misconduct in this case. Can you do that?" "I am telling you up front that my client allegedly made a confession. The police will say he gave himself up. But, in looking at that statement, will you go beyond it? Will you have the courage to determine whether my client was overreached by the police?"

In a civil case, you might say, "you will hear evidence in this case about my client's injuries and damages. I will be asking you for compensatory and punitive damages. Do you agree that people deserve to be compensated for their injuries resulting from negligence?"

You must advocate for your client every moment you spend in front of the jury. In the end, the jury you select will decide your client's fate. Therefore, you must invest some time and resources into preparing your jury selection methodology. Proper preparation will allow you to maximize your time and efforts in front of the jury.

5.3.1.6.10 DNA Voir Dire If there is DNA evidence in your case, then you may want to bring it out during voir dire. For example: "Now ladies and gentlemen, there is DNA evidence in this case which the prosecution says links my client to the crime. How many of you think that if that is so, that there is no need for a trial? How many of you feel my client must be guilty? Do you believe that labs can make mistakes, that evidence may be contaminated, or in some cases planted? Can you accept that a defendant may be framed with DNA evidence?[132] As far as you are concerned, is science infallible? When can science be incorrect?

5.3.1.7 Opening Statements The opening statement is the attorney's first opportunity to tell the jury his or her side of the story. The prosecution should let the jury know what to expect during the trial. The defense attorney should tell the jury why his or her client should be found innocent and how he or she will be persuading the jury that such is the case at trial. Though the jury got to meet the attorneys during voir dire, this is really the first time the jury gets to see each attorney working as a trial lawyer. It is important that you make a good impression on the jury.

Opening statements also serve another important purpose. This is one of the limited times during the course of the trial where the lawyer can speak directly to the jury and, thereby, influence the thought processes of the jurors. Each attorney's theory of the case should be brought out during the opening statement. Do not leave it to the individual jurors to figure out your theory of the case. Let the jury know up front. If the crime committed upsets you, let the jury see that such is the case. Do not be afraid to show emotion! As long as the emotion is genuine, the jury will appreciate it. An honest admission as to the complexity of DNA and the difficulties encountered in understanding it may help the attorney and client to identify with the jury.

The opening statement should be practiced before it is presented to the jury. It should be reasoned and sound, and it should not appear to be rehearsed or forced. At no point should an attorney fail to deliver an opening statement. Such a failure on the part of either the prosecutor or the defense attorney communicates a bad message to the jury. It makes it seem as if you have nothing to say or are afraid to say what you are thinking. If a side has nothing to say, the jury will have no choice but to base its decision entirely on what the other side has to say. Do not allow this to happen. Additionally, by not delivering an opening statement, the jurors may conclude that the attorney has failed to prepare for the trial. As the old adage goes, "When you fail to prepare, you prepare to fail!" The attorney who awaits the presentment of proof before attacking the other side's case and evidence is offering

[132]*See* Thomas F. Liotti, *Trying Cases by Overcoming Horrific Facts*, The Attorney of Nassau County, Part I, July, 2003 at 7, 13 and 14 and Part II, August, 2003 at 10, 12, 13 and 14.

too little, too late. If DNA itself were enough to solve the crime, we would not need trials.

When delivering the opening statement, pay attention to the jurors. They do not want to hear you lecture them, but they will be very interested in hearing what you have to say to them. While they are listening to you, you have the opportunity to see who is or is not paying attention. Does anyone look confused? If a juror appears to be confused during the opening statement, which should be a clear and concise statement of a party's theory of the case, he or she may also be confused during the testimony of an expert witness who is presenting technical information about DNA testing protocols. Know your audience.

Prosecutors and plaintiff's attorneys tell you what they are going to prove. The late Edward Bennett Williams, a renowned trial lawyer often said that his approach, as a defense counsel, was to "seize the burden," or tell jurors what he would prove or disprove. Instead of telling jurors that his client did not have to testify or prove his innocence, he told them his version of the evidence, without apologies or reservations. His opening, he realized, had to be as forceful as that of the prosecution. He would confront the worst evidence against his client by sharing it with the jurors in voir dire and in his opening. He would thereby defuse the prosecutor's strongest evidence.[133] In all likelihood, if Williams ever had a DNA case, he would tell jurors about it during the voir dire and during his opening. Sooner or later you have to confront the weaknesses in your cases and the strengths of the opposition's. Why not defuse the worst of it by revealing it to the jury before your adversary does.

5.3.1.8 Direct Examination

> We Lawyers are always curious, always inquisitive, always picking up odds and ends for our patchwork minds, since there is no knowing when and where they may fit into some corner.
>
> —Charles Dickens
> *Little Dorritt* (1857)

The direct examination is the plaintiff's or prosecution's time during the trial to tell their side's story through witnesses familiar with the events that lead to the necessity for a legal proceeding. Here we explore basic concepts that are important to direct examination in general. Later we cover the admissibility of DNA evidence at trial, which is most often done in criminal trials on direct examination by the prosecutor.

A good direct examination[134] is one that is clear and coherent. Witnesses must be prepared and the attorney on direct must be focused and aware of the value of each witness. Each witness has a purpose in a legal proceeding. Each witness represents a piece of the puzzle that the attorney is assembling for the jurors so that, in the end,

[133]See, Hon. Herbert Stern, *Trying Cases to Win*, Wiley, New York, 1991.

[134]*See* Thomas F. Liotti, *The Art of Direct Examination*, Nassau Lawyer, June, 2002 at 3, 17 and 20.

they can see the big picture. Because the witness is the voice of the evidence, it is important during direct examination that the witness be the center of attention. Memorable, convincing witnesses can win a case!

Successful trial practice requires skills in all parts of the trial. While direct examination may not be viewed as the most glamorous part of the trial, the sound rendition of its mechanics is essential. During the twentieth century, oral argument and cross-examination were considered to be of extreme importance to lawyers. As a result, less was written about direct examination techniques as cross-examination of witnesses and the chance to trip up a witness on cross-examination were viewed as the exciting portions of trial practice. However, every good trial attorney knows that cases are won and lost during direct examination and on the strength of the plaintiff's or prosecutor's case in chief.[135]

The direct examiner must know that which he must prove in order to establish a *prima facie* case. The direct examiner should begin at what is alleged in the summons and complaint, the petition, or the indictment. Does that first document set forth in sufficient detail what must be proven or have elements been omitted or not been sufficiently pled?

Has the draftsperson specified dates, times, places, actors, parties, and the elements of a cognizable cause of action or charge? Look to the complaint or indictment. Does it set forth subject matter and *in personam* jurisdiction? How will that be proven? Do the dates fall within the statute of limitations? How will the dates be proven? If the pleadings are inadequate, do they have to be amended? Can they be? Will counsel have to make a motion at the end of the case to conform the pleadings to the proof? What must be proven during the trial so that the motion may be properly made at the end of the case?

Review the Pattern Jury Instructions, Criminal Jury Instructions, or Judge Leonard Sand's[136] text on federal jury charges concerning the anticipated charge to the jury. Try to secure a copy of your judge's charge pretrial. What does the charge say must be proven? It is a rare judge who will hold a charging conference or ask for requests to charge before the start of a trial. But, the direct examiner needs to review charges used by the judge in similar cases. There are key words that you will want witnesses to use during their testimony and that you as the trial lawyer will want to use during your opening and summation that are lifted or even parroted from the charge. Jurors will identify with witnesses and their testimony if a judge used the same words that they had used during their testimony. The elements will be subliminally proven in the minds of jurors if the key words of proof are repeated again and again, during the opening, during testimony, in the objections, in the closing argument, and in the charge.

The attorney responsible for direct examination must also consider what witnesses are needed to flesh out the case? Do they have to be subpoenaed? What elements of proof will be presented through each of them? You should prepare a witness list with a proffer as to the testimony of each witness.

[135]*See* Thomas A. Mauet, *Trial Techniques*, (Aspen 6th ed. 2002).
[136]*See* 3 L. Sand, et al., *Modern Federal Jury Instructions* (1992).

What exhibits will be offered and what demonstrative evidence will be used? Through which witnesses will they be presented or refuted? Develop a checklist of proof that must be satisfied.

You may wish to assign an investigator to research the background of your witnesses. What are their financial circumstances? What possible motives do they have for testifying? Have they testified previously or given any statements in connection with the case? If an expert, where have they testified; what were they paid; what is in their curriculum vitae that might help, and what cases have they helped to win or lose? Get copies of their transcripts[137] from prior cases in which they testified. If the witness testified in a grand jury, determine whether a transcript is available. Find out if the witness made any prior statement to the police, investigators, or anyone else. Is there a supporting deposition? An F.B.I. 302? (An FBI 302 is a transcript of an interview conducted by an FBI agent and a defendant, witness, informant, etc.) Did your investigator tape record any conversations with witnesses or take any statements from them?

5.3.1.8.1 Presentation of Your Witnesses You want witnesses to look good in front of the judge, to evoke sympathy, to wear earth tones, and to make eye contact with all of the jurors. Take them over to the courtroom and give them the feel of it. Let them watch other witnesses testify at the courthouse, on Court T.V., or at the movies. Watching courtroom dramas on video is fun, and you can critique a witness's performance for them. They then practice their testimony at least once everyday. Witnesses tend to be self-conscious in front of juries. They often speak into their chests, muffle their voices, cover their mouths, and look away from the jury. Experts should be advised to look at the examiner as the questions are asked and then turn to the jurors to answer them.

Make sure the courtroom has microphones and that your witnesses know how to use one. The ideal witness not only knows the science and technology of forensic DNA analysis but can communicate the basics in simple terms to the jury. At the same time, the witness should appear confident, use good diction, and project testimony with a strong voice and good image that makes him or her credible to the jury. In a high-profile case, it would be beneficial to use a mock jury and to videotape the witnesses so that they can see and hear their own performances as you critique the delivery of their testimony.

5.3.1.8.2 Tell a Story The expert witness should deliver narrative testimony, almost without interruption. The lawyer is superfluous except to keep the witness on track. If the witness gets lost, they should remember where they are chronologically. The objective is to tell a story that will keep the jurors keenly interested. The witness must be persuasive, which means that they must be authoritative factually and scientifically precise. The expert should not hesitate to request water, pause, or to take a break in testimony, if any of these become necessary.

[137]*See* Thomas F. Liotti and Peter B. Skelos, *Credibility Questions for Prosecution Expert on Syndromes in Cases of Rape, Child Abuse*, Outside Counsel, New York Law Journal, April 20, 1989 at 1, 2 and 7.

5.3.1.8.3 Best Cross-Examination Is Direct Examination A skillful trial attorney will anticipate the toughest cross-examination questions and steal the thunder of his adversary by asking those on direct. While ordinarily you never ask a why question on cross, you may try to do so on direct. On direct, the witness is yours. He or she is not hostile.

Examples

Direct: You gave a confession to the police. Why?
Direct: Why did you stab your husband?
Direct: Why did you call the plaintiff a thief?

5.3.1.8.4 Leading the Horse to Water While you are not permitted to ask any leading questions on direct, you can always debate what a leading question is. Does the judge and your adversary know it when they see or hear it? Test the waters! Pushing the envelope with a leading question controls the witness, helps with recall, and gives the impact of the question and answer, a double bounce of information to the jurors. Leading questions are used to telegraph answers to the witness.

A modified version of leading questions may be used when, in lieu of the full question, the examiner prepares a well-coached witness to listen for key words or clues. The witness has gone over her testimony so often that when the examiners evoke a key word or clue word or phrase, it should trigger a prepared response from the witness. Plan the direct so that it is bullet proof. This will help you to make out your case and win (almost) every time.

5.3.1.9 Cross-Examination Francis Wellman was one of the great nineteenth-century trial lawyers. According to statements made in his book, he became widely known for his spectacular coups in the cross-examination of witnesses.[138] It is said that he often emphasized that he depended not on trickery but, rather, on hard work and methodical preparation. Pretrial preparation is the key to success if one would like to deliver a strong cross-examination. The attorney should have an idea of what every witness will testify about before the witness opens his or her mouth. Pretrial discovery should be used as a device in which the attorney learns as much about the other side's theory of the case as possible. During pretrial discovery, the attorney should be gathering information that will show not only what the witness will say on the stand but also what the witness will try not to say!

It is not necessary to cross-examine every witness. In fact, it may be damaging to do so. Some witnesses are introduced only to establish some technical fact or collateral matter. If the witness has not hurt your case and offers nothing by way of being able to create reasonable doubt in the minds of the jurors nor can his testimony strengthen your case, the witness should probably not be cross-examined. You should not cross-examine a witness if there is no reason to do so.

[138]*See* Francis L. Wellman, *The Art of Cross Examination*, 4th ed., Simon & Schuster, New York, 1903, 1997.

There are two basic purposes of cross-examination. The first purpose of cross-examination is to elicit favorable testimony that will help you advance your theory of the case.[139] The second purpose of cross-examination is to conduct a "destructive cross."[140] A destructive cross-examination is one in which the witness is discredited because testimony favorable to the cross-examining attorney's side has been elicited from the witness. Mauet points out that destructive cross-examination is particularly effective when a witness has what seems to be a great deal of credibility giving testimony on direct examination.[141] Eliciting favorable statements and admissions from such a witness is even more damaging due to the fact that the witness seems credible to the jury. Such statements and admissions will carry more weight when the witness seems to be trustworthy and reliable.

Cross-examination should have a structure. The cross-examination should not be random and unorganized. Cross-examination may sharpen or dilute the evidence. Cross examination may put jurors to sleep or cause them to question the credibility of DNA witnesses. For example, in the O.J. Simpson case, Barry Scheck cross-examined Mr. Fung on evidence collection for eight grueling days. Scheck succeeded in demonstrating the mishandling of evidence by the Los Angeles Police Department. Peter Neufeld did likewise with other witnesses, confusing the evidence by an adroit command of the science. He and Scheck both asked "complicated" questions that ultimately caused the jury to believe that the evidence was contaminated or, worse, "planted." In the end of his cross-examination, Mr. Fung came off the stand feeling like he had been run over by a tank. Some may think that Scheck overdid his examination, but apparently the jury listened attentively.

Generally, it is not a good idea to repeat the direct examination on cross-examination since you are reinforcing proof. Such an approach is only effective where the testimony seems memorized and repeating it will demonstrate such to the jury. One example of using this method effectively took place during a cross-examination that took place at a trial subsequent to a huge fire at a New York City factory in the early 1900s in which over 100 people were killed. During the trial, a witness who seemed to know little English was asked to repeat her testimony again and again. The witness proceeded to give her horrible and moving recount of the events that lead up to the deaths of so many people again and again, exactly the same way, every pause, every word, the same. By the time the witness was finished the last time, her credibility was shot. If you cannot accomplish such a feat on cross-examination, then you should not allow the witness to repeat what was stated on direct.

The strongest points should be put forth at the beginning and end of the cross-examination.[142] The jurors will remember the first things they hear. They will also remember what they hear last. However, one should avoid cross-examinations that follow a predictable outline. The order of the cross-examination should not be

[139]*See* Thomas A. Mauet, *Trial Techniques*, pg. 218 (Aspen, 6ed. 2002).
[140]*Id.*
[141]*Id.*
[142]*Id.* at pg. 219.

predictable. The attorney must be able to get the witness to make certain statements or points without realizing he or she is doing so. By skipping around or varying the order in which points are addressed, the witness will be less likely to anticipate the purpose of the questions being asked.[143] There is a fine line, however, between skipping around and varying the order in such a way that the jury becomes confused and misses the point.

Keep control of the witness during cross-examination. Never argue with the witness. If the witness is giving you a hard time, ask the court to address it. This does two things. First, it should fix the problem. Second, by having the court address the problem, the jurors will not believe that you are engaging in some lawyer trick and may legitimately believe the witness was engaging in some sort of misconduct when the court intercedes to remedy the problem. Ask questions in such a way that the witness's testimony is controlled. Never ask questions that ask "what," "how," or "why" or questions to which you do not know the answer.[144] All questions on cross-examination should be leading questions if possible. Leading questions are those to which one must answer either yes or no. This method allows the attorney to tailor the testimony of the witness. The attorney should ask questions that make a statement of fact in such a way that the witness must either agree or disagree by answering yes or no.[145]

The attorney should be the center of attention on cross-examination. On direct, the witness tells the story. On cross-examination, the attorney is telling the story, shaking things up, and making clarifications. During the cross-examination, the attorney must be effective, but if he or she is too effective, it may lessen the impact of his or her cross. This is particularly true when the attorney asks "one question too many."[146] The attorney should only establish his or her points on cross-examination. He or she should draw the dots but must be careful not to connect them. Allow the jury to do so. Better yet, save the big questions for closing arguments, where they can be posed aloud without a witness who may respond with a damaging answer. This can be accomplished by suggesting points on cross, without asking the questions that will serve to clarify these points. This method helps to build the foundation for statements during closing that address the ultimate point, despite the fact that such a point was not addressed during the cross-examination. If you have built the foundation to ask questions during closing argument that leave doubt in the minds of jurors, you have been effective on cross-examination!

The length or brevity of the cross-examination will depend on your objectives. If it is to confirm the direct, then you may keep it short. If it is to confuse the direct or take it into a different direction by developing your own theories or turning the witness into your expert, then it may take longer. In the final analysis, the jury will tell you how effective you have been by their verdict.

[143] *Id.*
[144] *Id.* at pg. 221.
[145] *Id.* at pg. 225.
[146] *Id.* at pg. 227.

5.3.1.10 The Art of Objecting[147] Objecting is an art! Be proactive. Objections should be made whenever appropriate. There are multiple reasons to make timely objections. One reason to object is to ensure that the adversary does not obtain information to which he or she is not entitled. One may object in order to ensure that the defendant is not prejudiced by inadmissible evidence; to ensure that the other side behaves properly; and to create a good record so that the issue in controversy can be raised on appeal. Failure to raise certain objections in a timely manner may eliminate the defendant's standing to appeal at a later date. Counsel should be aware of whether he is in a jurisdiction in which failing to join in the objection of a co-defendant will lead to a waiver of the objection on the part of his client.

Objections should be as specific as possible. Frivolous objections should be avoided at all costs! However, it is sometimes possible to object and fail to state a reason. This is sometimes necessary when an attorney knows that something is improper but he cannot put his finger on exactly what it is. In some cases, the attorney may get lucky and have the objection sustained without having to justify the objection. Counsel should stand when making objections. Do not interrupt the person asking the question but make sure to object before the witness begins to answer the question. If it is necessary, make sure that you move to strike any testimony subsequent to objecting so that testimony offered after an impermissible question is not considered by the jury. Ask the judge for a corrective instruction. Remember that objections can be made at all points in the trial, including closing arguments. Be prepared to object at closing if necessary. While some consider this impolite, it is the job of an attorney to protect the rights of his or her client at all stages of the trial. Make sure the judge actually rules on every objection. Make sure that you consult a book on trial techniques and making objections before the trial. Practice making objections by watching trials and silently making objections when one is called for. Take note of the objections that are being made.

Legitimate reasons to raise an objection include:

- Immaterial or irrelevant response.
- Collateral matter.
- Argumentative.
- Asked and answered.
- Badgering.
- Answer is beyond the scope of the witness's personal knowledge.
- Cumulative.
- Compound question.
- Assumes facts not in evidence.
- Improper impeachment.

[147]This segment of the book is taken from an article on the same subject. *See* Thomas F. Liotti, *Helpful Practice Hints—The Art of Objecting*, the New York State Bar Journal, July/August, 1998, Vol. 70, No. 5 and the Mouthpiece, Vol. MII, No. 1, January/February, 1998 at 17 and 18.

- Irrelevant.
- Leading question.
- Witness is not an expert.
- Misquoting or mischaracterization of witness testimony.
- Improper opinion evidence.
- Prior bad acts.
- Narrative answer.
- Question is too general, not specific enough.
- Best evidence.
- Unresponsive.
- Speculation.
- Ultimate issue.
- Impermissible parole evidence.
- Witness is not competent to testify.
- Witness does not understand what truth is or is to young to give testimony.
- Communications are privileged.
- Hearsay.
- Risk of prejudice outweighs probative value.

Make sure that you take the time to practice making objections. It can make all the difference at trial. If you and your experts have thoroughly reviewed the evidence, including exemplars that the people intend to use, you may be in a position to either object to them or use them as your own exhibits. In a DNA case you want to get access to whatever the prosecution will be using in the way of exhibits as early as possible. You want to determine whether the exhibits are necessary or misleading.

5.3.1.11 Closing Arguments Closing arguments take place at the end of a trial or legal proceeding. Closing arguments are the last time the attorney gets to communicate directly with the jury before they deliberate and make their decision.[148] Closing arguments should be the first thing considered and written by the attorney at the beginning of the case.[149] The attorney should then try to use the closing arguments as a guide when calling witnesses and attempting to prove the defense side of the case.

There are a number of items to consider when deciding what to tell the jury during closing arguments. The opening statement and the closing argument are the attorney's only real opportunity to testify during the trial. During closing arguments, the attorney should inform the jury how the facts support his or her theory of

[148]*See* Mauet, *infra*, at pg 361.
[149]*See* Roger Haydock and John Sonsteng, *Trial: Advocacy Before Judges, Jurors and Arbitrators*, 2ed., West Publishing, Eagan, MN 1999.

the case. If there are questions that were not asked but for which the foundation was laid, the attorney should pose those questions to the jury during closing arguments. This concept was described in more detail when we discussed cross-examination. During the trial, the attorney should keep a notebook in which he or she makes note of what each witness said, the evidence introduced through each witness, and any other relevant things that come up during trial. The attorney should review this trial notebook prior to making the closing arguments.

Another item to be considered when drafting closing arguments is the jury instructions or charge.[150] The *jury instructions* are the judge's instructions to the jury that explain the law that applies in a particular case. The jury instructions explain the law as it should be applied. Each attorney will draft recommended jury instructions and will submit those instructions to the judge. Many jurisdictions have *pattern jury instructions* that are published and available as a guide to assist attorneys in drafting instructions specific to a particular case. The attorney should be aware of the possible instructions the judge might give if instructions are given to the jury *after* closing arguments in the jurisdiction in which the case is being tried. If the instructions are given *before* closing arguments, the attorney should be sure to draft closing arguments that paint his or her side in a favorable way in light of the instructions given.

During the closing arguments, the attorney must outline and reiterate the strengths of his or her side. The counselor must also address the weaknesses of the case. Try not to allow the adversary to discuss these weaknesses. Anticipate what your adversary will say in his or her closing arguments. Addressing weaknesses can help the attorney seem more credible. Let the jury know that you are not asking them to ignore the weaknesses of the case, (you should tell them that you know that there are weaknesses) but convince them that those weaknesses are outweighed by the strengths of your case.

During the closing arguments, the attorney should consider who has the burden of proof in the case.[151] If you have the burden of proof, you should explain to the jury how the burden has been met.

The attorney's appearance and mood are important. Show the jury that you believe in your client, in your client's version of the events, and in your case. Argue your side persuasively and effectively. Do not read from a script. Even if your closing arguments have been drafted in intricate detail, reading from a script should be avoided. Instead, the attorney should prepare a list of key words and topics that must be addressed during the closing arguments.[152]

The attorney should not use closing arguments merely to remind the jury what they heard during the trial. The jury members were present at trial. They know what took place. Be creative. Tell a story. Use exhibits. Use magic if you have that sort of talent. There is an old tale that describes a teaching method utilized

[150]*See* Mauet, *infra*, at pg 361.

[151]*See* Roger Haydock and John Sonsteng, *Trial: Advocacy Before Judges, Jurors and Arbitrators*, 2ed., West Publishing, Eagan MN 1999.

[152]*Id.* at pg. 591.

by Socrates in which it is said that Socrates would slap a student every time he mentioned an important point. Every time you make an important point, do something to make that point stand out for the jury. Try and liven up your closing. The jury will appreciate you for making their time in the courtroom more exciting. However, do not forget to argue your client's version of the facts. Rehearse your examination of witnesses. Obsessive, tireless preparation is the key to success.

Finally, make sure that you make eye contact with the jury. Don't lecture the jurors. Talk to them. They are people who are taking time away from their lives and families to perform a very important function. They are probably taking their role very seriously. They are watching you carefully. Try to make sure they like you, or if your personality is one that is beyond repair, try to make sure that they can at least tolerate you. If your personality is totally repugnant and you are aware of this fact, ask the jury not to hold your weaknesses, character flaws, and/or behavior against your client.[153] However you decide to persuade the jury that you have made a more persuasive case, make sure you always do your best to protect the interests of those you represent.

One final point, review what props or exhibits your adversary has and consider using some as your own. Have all your exhibits and charts set up for summation. As the jurors file into the courtroom, they should see an impressive array of exhibits set up before them. You are now in center stage.

5.4 DNA FOR DEFENSE ATTORNEYS—CONTESTING DNA EVIDENCE

Forensic DNA analysis has changed dramatically since the mid-1980s. The technology has evolved from a technique that was often contested at *Frye* hearings to a different kind of technique, PCR-STR analysis, that has been deemed admissible in virtually every state. The defense counsel should obtain an expert who can help explain how the procedure is performed. The attorney should consider the possibility of a challenge to the way the procedure was followed in the particular case at hand. Did the analyst perform the test as it is described in the laboratory's protocol? Were there any shortcuts taken? The prosecution expert should be examined thoroughly to determine if he or she has all the educational background and the expertise necessary to perform the test and do the statistical analysis required to come to some scientifically valid conclusion. The attorney should consider how to challenge items of evidence that contain mixtures of DNA from two or more sources. The attorney should also consider what analyses were not done that could have been done. Such testing might provide additional information to assist the jurors in the decision-making process. The chain of custody of the evidence should be carefully examined, and it should be determined if the evidence had

[153] *See* Thomas F. Liotti, *Highest Verdicts of 2002, Getting Jurors to Like You*, New York Law Journal (a special section and edition) March 17, 2003 aqt 51, 53, 59 and S13, and The Nassau Lawyer, March, 2003 at 3 and 20.

been collected, packaged, documented, and transported to the laboratory properly. The issue of possible contamination can be raised. Contamination could have occurred from the moment the evidence was deposited at the crime scene to the time the analysis is performed in the laboratory. The DNA profile of the analyst should be made available to eliminate this potential source of contamination. The attorney may also want to determine if DNA sequencing or Y-STR analysis was done, and if not, then why not. Lastly, the defense counsel may be interested in pursuing the scientific literature that describes variations of specific alleles resulting from single base changes in one (or more) of the tandem repeats. Thus, you can imagine that an allele with a specific length or size is detected and becomes part of an individual's genotype at a given locus. The literature may describe the existence of two different alleles (each allele has a different sequence as a result of a single base mutation). Since these alleles share the same size but have different sequence, an apparent inclusion might actually be an exclusion. To determine this a more detailed analysis such as DNA sequencing would have to be conducted. Finally, it is in the best interest of the defense to convey to the jurors the complexity of the science. Jurors who do not understand the technology may doubt that the test is meaningful. This may also contribute to doubt about the prosecutor's case.

5.5 DNA FOR PROSECUTORS

Prosecutors have scientists and a criminalistics laboratory at their disposal to perform complete analyses of evidentiary items that contain biological specimens. Where possible the prosecutor should use an accredited crime lab and certified examiners. The prosecutor should review the summary reports of the laboratory and ask appropriate questions about the conclusions drawn. The prosecutor should familiarize him or herself with the tests that have been used in the case and should ask the criminalist if other tests have been performed to confirm the findings. If they have not, he should find out why they were deemed to be unnecessary. The prosecutor should review the expert's credentials and prepare the expert for direct testimony. Models, graphics, charts, or any other teaching aids should be made available. The material should be presented in such a way as to not confuse the jurors who have little experience with DNA other than reading about it in newspapers or seeing it presented in a TV trial. Jurors are not selected based on their educational experience or level attained in school, and most will not have taken university physics, biochemistry, molecular biology, population genetics, or statistics. In fact, if a potential juror had any or all of these courses, he or she would no doubt be subjected to voir dire and asked to return to the jury pool room. The expert should have a thorough knowledge of the techniques employed in the case at hand, even if that person was not the analyst that performed the testing. The significance of the DNA match between the evidence and defendant is described in the statistical analysis. This number is very important, and it must be explained clearly to the jury that a person's genetic profile is virtually unique. Furthermore, find another person with that same profile, one would have to test different ethnic groups and analyze millions, billions, or trillions of individuals. The jury should then be told that the world's population is no more than 5.5 billion people, making the profile essentially an absolute identification. It is also important to explain to the jury how one can develop such extraordinary large numbers (i.e., one out of trillions) when the individual databases from which the numbers are derived consist of only several hundred individuals.

5.6 DNA FOR JUDGES

Judges have extraordinary responsibility in deciding on the admissibility of scientific evidence and in accepting the credentials of expert witnesses, thereby allowing them to testify at trial. Science and technology are constantly changing and advancing, making it hard even for scientists to keep up with the latest improvements. It is therefore difficult for a judge to hear experts talk about scientific techniques and procedures and make decisions that may bias the case against the defendant. It is important for judges to receive at least fundamental training in forensic DNA analysis. Because it is used so extensively in trials where the defendant is accused of assault, burglary, rape, or homicide, it has become more likely than not that DNA evidence will be part of the trial. The judge must maintain an open mind and make decisions based on the merits of each case since each case is different from all others. It is untrue that DNA analysis is the same in every case. For example, some DNA analyses can produce incomplete profiles where only a small number of loci produce results, some may include genotypes at specific loci that cannot be replicated, some electropherograms indicate the presence of artifacts while others do not, some alleles are present in very small amount (low peak height) and could be interpreted as a stutter peak (artifact) rather than as an authentic allele. Extra alleles may indicate contamination, a mixture, or there may be some other explanation. Judges should not hesitate to work with "neutral" scientists to learn more about the scientific foundation for DNA analysis, or about the techniques employed, or if additional information is needed on how electropherograms should be interpreted.

6

DNA Evidence at Trial

6.1 ATTACKING AND DEFENDING DNA EVIDENCE

The first thing that every person who is involved in a trial must have is a theory of the case. If a crime has been committed, there is a person or persons responsible for the crime as well as a victim. There is a crime scene and there is physical evidence that can be found on either the suspect, the victim, or at the crime scene.[1] This evidence will be analyzed and eventually presented at trial. Physical or biological evidence is introduced to help support a theory that is being offered to explain the events leading up to or during the crime. DNA evidence is a brought before the court because it can provide important information for the jury to hear and evaluate.

6.1.1 Theory of the Case/Plan of Attack

Before trial, it is essential for both sides to formulate a plan of attack. This plan of attack is sometimes referred to as the theory of the case. If one side does not have a plan of attack, more likely than not, the trial is lost before it even starts. The prosecutor should have a theory of how the crime was committed. In law school, students learn the mnemonic of MOM, which stands for *means, opportunity, and motive.* These are three essential elements that prosecutors must prove. If the prosecutor is seeking to introduce DNA evidence, this evidence should advance this theory

[1] *See* H. C. Lee, T. M. Palmbach, and M. T. Miller, *Henry Lee's Crime Scene Handbook*, Academic Press NY (2001).

DNA: Forensic and Legal Applications, by Lawrence Kobilinsky, Thomas F. Liotti, and Jamel Oeser-Sweat
ISBN 0-471-41478-6

in some way. The prosecutor should be using the results of DNA testing to prove that the defendant is guilty of the crime that he or she is being charged with committing.

The defense counsel, on the other hand, should have a different plan of attack. The defense attorney should formulate a theory of the case that tends to mitigate the effect of the DNA testing results on the jury. The defense may offer a variety of alternative theories where appropriate. These theories may include among other things, the possibility that the evidence was contaminated, the possibility that there was a laboratory error (mishandling of samples or error in test interpretation) or that improper laboratory procedures were followed, that there is some problem with the tests employed (i.e., insufficient number of loci producing typing results), and the possibility that adverse environmental conditions could have had some adverse effect on the DNA being tested. Regardless of the specifics of the theory, there must be a strategy or plan of attack. If the evidence cannot be successfully challenged, then perhaps another explanation for its discovery at the crime scene should be developed.

6.1.2 What is Required for DNA Test Results to be Admitted into Evidence?

It is important to understand how DNA test results are properly presented and successfully admitted into evidence. The presentation of DNA evidence includes three aspects that the court must consider in determining admissibility. The first involves the admissibility of the DNA test results in the form of a summary report. The second involves the offering of the statistical significance of the DNA test results. The third addresses the analysts conclusions with respect to the determined genetic profile and his interpretation of the statistical frequency data in relevant populations within the major racial groups.

For DNA test results to be considered by a jury in their proper context, all three of these issues must be addressed. If the party seeking to have DNA evidence considered fails to fulfill any of these three components, then the DNA testing results cannot be properly evaluated by the jurors. In the next section, these stages are considered and explained in greater detail.

6.2 DNA FOR THE PROSECUTOR OR THOSE WHO SEEK TO ADMIT DNA EVIDENCE

As explained in the previous section, the party who seeks to have DNA evidence admitted has a serious task to accomplish. Irrespective of the admissibility of DNA evidence in the jurisdiction in which a trial is taking place, the procedural admissibility of such evidence is a separate matter and is really considered on a case-by-case basis. Just because DNA evidence is deemed by the state to be admissible does not mean that such evidence will be admitted at any one trial. It is imperative that the party seeking to introduce DNA test results have the evidence considered in the context of the frequency with which such a DNA profile occurs in the population. It is also imperative that the prosecutor or party seeking to introduce such evidence properly prepare the expert to describe to the jury the proper inferences that should be drawn.

6.2.1 Effective Admission of DNA Evidence Takes Place in Three Stages

This section explains the procedure for having DNA test results admitted into evidence. As stated, for DNA evidence to be evaluated in the proper context, three stages of admission must take place. The first stage is completed when the DNA test results in the form of a summary report are admitted into evidence. The second stage is completed when statistics that describe the frequency with which a DNA profile of the type at issue occurs in the population are admitted into evidence. The third stage is completed when an expert has described how the DNA test results should be considered in light of the relevant population genetics or frequencies of occurrence.

It is important for each of these stages to be successfully completed. If the DNA test results are admitted but somehow the other party successfully stops the relevant statistics from being admitted, there is no context for the evaluation of the DNA evidence. The jury simply learns that DNA testing was conducted. The next sections explain each stage of admission and give some advice on how to successfully have DNA evidence admitted and considered at trial.

6.2.1.1 Admission of DNA Test Results At the first stage of admission, there are several goals. The first goal is to ensure that the expert witness is properly qualified as an expert in DNA evidence. This is important for a number of reasons. If the witness you are using to introduce the DNA test results is not qualified as an expert, he or she will be limited as to what he or she can say on the stand and will not be qualified to draw certain types of conclusions.

In addition to having the witness qualified as an expert in DNA evidence, it is also important to have the witness qualified as an expert in statistics as they are applied to DNA analysis. At the very least, he or she should be qualified as an expert who is able to interpret the results of DNA analysis and testing. It should be remembered that the witness may not be the same individual who performed the analysis. When the witness has been deemed qualified, the foundations have been laid for the first and second stages of the admission of the DNA test results and their associated statistics. To have the witness qualified as an expert in DNA analysis and interpretation, a number of steps must be taken. The witness must be called and questioned. It is important to show why the witness should be labeled an expert. Federal Rule of Evidence 702 defines an expert as one who possesses relevant "knowledge, skill, experience, training, or education."

The witness can be offered as an expert by introducing him to the court and to the jury. He should be questioned regarding his education, training, and experience. Once the witness is shown to possess expertise in the area of DNA analysis, the attorney should ask the court to rule that the witness is indeed an expert and can provide testimony. "Your Honor, may the witness be qualified as an expert in the field of DNA analysis?" You can also put your adversary on the defensive by asking if he will stipulate to the witness's qualifications and, if not, then hammering those credentials in front of the jury almost *ad nauseum.*

Once the witness is qualified as an expert, he or she should be questioned about the DNA tests performed and the results of those tests. The relevant laboratory techniques should be mentioned. Any accreditation and quality control measures utilized should be explored. The chain of custody of the DNA evidence should also be described, if relevant. The expert should be examined as to every aspect of the testing that was done and the results of that testing, as well as any other factors that may be relevant. These factors include training of laboratory technicians, the equipment used and the condition thereof, quality control for reagents used, and the reasons the specific DNA tests were performed. If there is any weakness in the evidence, these weaknesses should be exposed by the party offering the evidence and explained. If the jury first learns of these weaknesses or any discrepancies during cross examination, the jury may assign less significance to the laboratory's findings.

It is also important to show why the witness can be trusted. If the witness seems aware of the limitations of his testing methodology but is still able to draw conclusions that are adverse to the defendant, this may have a significant impact on the jury's decision at the end of the trial. Unlike other witnesses, the expert may render an opinion about his findings. Such an opinion should be based in science, and he may be asked to explain why he has reached a certain opinion. The jury will most likely concede that the expert has vast experience in the area of DNA evidence or else he would not be testifying. They will most likely be impressed with his or her educational credentials. If the expert has "hands-on" experience, such as time spent as a laboratory technician or scientist, and supervisory or teaching experience, these credentials should be emphasized. If the expert is a professional witness who testifies frequently, this should be addressed in such a way that the jury sees such activities as powerful and persuasive. There is a big difference in the level of credibility assigned to one viewed as an expert "for sale" and a witness who is painted as an expert in his or her field whose knowledge is so great that he or she is frequently called upon to educate jurors so that they can make informed decisions at trials in which DNA evidence is being offered. Do not let the other attorney paint your expert as a "hired gun." By all means, convey to the jury that the witness is an expert with no bias one way or the other. That he is knowledgeable and can perform the scientific analysis correctly. That he is qualified to write a report and to testify to people who are in search of truth but who have little or no experience with DNA and human identification.

When the expert witness has been introduced, has been qualified by the court as an expert, and he or she has testified as to the DNA testing performed and the circumstances surrounding such testing, stage 1 of admission is complete. The foundation for stage 2 of admission has also been laid.

6.2.1.2 Admission of Statistics that Govern DNA Results In stage 2, testimony concerning population genetics or statistics that govern the interpretation of DNA evidence is proffered. It is essential that the witness testifying as to statistics that govern the interpretation of DNA test results is qualified by the court to do so. This qualification process takes place in stage 1 of admission.

During stage 2 it is important to describe how DNA testing results are interpreted. The jury will not understand how conclusions (proffered during stage 3 of admission) are drawn if they do not understand the process that was used to interpret the DNA testing. This is perhaps the stage where the jury is most likely to "tune out." Many people have an aversion to math. The analysis that is employed to interpret the meaning of DNA test results is number driven. It is a numbers game. Even if you like numbers, do not make the mistake of assuming the jurors will. Use demonstrative evidence during this stage. Do your best to simplify the concepts presented. You should practice going through this portion of the direct examination with the expert witness before trial. Try to stress the fact that the expert must assume the role of teacher. Make sure you work hard not to lose the jurors during this testimony. If you find that you as the attorney cannot follow the expert's explanation of the statistics and or population genetics, then chances are the jury will not either and your strategy must be reassessed. Assure the jurors that you understand that the concepts can be complicated but that they are manageable. However, do this in such a way that you do not insult the intelligence of the jury.

Once stage 2 of admissibility is completed, the jury will be familiar with the techniques used to analyze DNA test results. It is then time to move on to stage 3 of admission, which deals with the conclusions drawn when the results of DNA tests are considered in light of the population genetics and statistics that govern the frequency with which certain DNA profiles appear in various populations. If stage 2 is successfully completed, the jurors will not simply accept the expert's conclusions in stage 3 of admission. The jurors will be sufficiently informed that they will be able to draw their own conclusions and will less likely be persuaded by defense attacks on the expert's conclusions.

6.2.1.3 Admission of an Explanation of the DNA Results

6.2.1.3.1 Stage 1: Admission of DNA Test Results

6.2.1.3.1.1 USE OF AN EXPERT WITNESS It is important to use an expert witness to introduce DNA evidence for the reasons described earlier in this text. Great care should be taken when choosing the person who will testify about the DNA testing performed and the results thereof. This section describes in greater detail how one introduces DNA test results into evidence.

6.2.1.3.1.2 SELECTION AND PREPARATION OF EXPERT WITNESS The attorney offering DNA test results into evidence does not always have total control of the identity of the expert witness who will testify at trial. Generally, the expert is an employee of the laboratory that performed the tests. This being the case, the attorney must work with the expert who has been chosen to testify. However, the attorney has a great deal of control over the testimony preparation of this expert witness. It has often been said that "preparation beats talent 100% of the time." This being the case, it is important for the attorney introducing DNA evidence not to dwell on the shortcomings of the witness if they exist and cannot be changed. For example, a witness who speaks in a monotone voice will most likely not be transformed into a "silver-tongued orator" or a charismatic hero before the trial commences nor is it the attorney's job to see that such a transformation takes place. Instead, the attorney seeking to introduce DNA evidence at trial should work with the expert witness to make sure that he or she knows the case, that he or she knows the evidence, and, most importantly, that he or she can adequately and effectively convey the information relevant to the case to the jury in a manner that they can comprehend. Although the witness will not be testifying about rocket science, for some jurors, it might as well be just that because, for some of the jurors, things are about to go way beyond their ability to understand.

The attorney seeking to introduce DNA evidence should understand a number of things. The time spent preparing the expert is not just for the benefit of the expert. It is also for the benefit of the attorney. This is the attorney's time to ask questions. The expert witness and the attorney will be working as a team at trial. The role of each person must be defined. The attorney should understand the evidence. The expert should understand the legal process. The attorney must stress that the expert will be telling the story. Though the attorney will be guiding the expert through direct examination, the expert witness is not simply a puppet. If the expert drops the ball, the attorney or prosecutor might not be able to "lead" the expert back on track. This is because leading questions are not allowed during direct examination. This being the case, the expert must be prepared as to the theory of the case and the way in which the story should be told. The attorney, on the other hand, should be comfortable enough with the evidence that he or she can effectively guide the expert through the examination and assist the expert if he or she is not on track. In other words, practice makes perfect. Prepare the expert for what he or she should expect during cross-examination. Do not go easy on your expert since the defense attorney will not be as nice to the expert as you. Prepare the expert for the types of issues that might come up on cross-examination. Practice by cross-examining your expert. Make sure that by the end of your practice session, you are convinced that the expert's opinion is sound. If you are not convinced, the jury probably will not be convinced either.

Decide what exhibits will be used at trial. Formulate how you will explain the evidence in a manner in which the jury will understand. The day of trial should not be the first time you have exposed the expert to your style of direct examination. Practice both direct and cross-examination in the way you will at trial. There should be no surprises the day that the expert testifies. Make sure the expert knows what treatises are prominent in the field. If you intend to use any of these works, review them with the expert before trial. Practice in a setting that creates the feel of a courtroom. If possible, use a videotape recorder to record the expert's testimony. Work with the expert to strengthen both his testimony and your examination. Help the expert to become an effective storyteller. If possible, help the expert bring the evidence to life. Most people do not remember complex numbers and technical information, but they do remember people. The jury will associate the DNA evidence with the expert witness who is describing it. Make their experience a positive and memorable one.

Ask the expert what issues might come up. The worst time to find out about the huge laboratory contamination scandal that was in the newspaper is during the defense attorney's cross-examination. Do not assume anything. Ask the expert to brief you on anything that might be relevant for the defense to attack. Be sure to ask about the expert's credentials, experience, and any skeletons in the expert's closet. Do an Internet search on your expert. During high-profile cases, it might be wise to hire an investigator to look into your expert's educational experience and training. Verify the expert's credentials. Sometimes credentials become greatly exaggerated. Attendance at a workshop is not the same as attendance in a credit-bearing academic course. Ask others about the expert. Did the expert fail a proficiency test? Was the expert's testimony ever called into question? The defense attorney will.

6.2.1.3.1.3 INTRODUCING THE EXPERT A jury always asks three basic questions whenever a new witness appears. These questions are "Who is she, why is she here, and can I trust her?"[2] According to Mauet, these questions must be resolved quickly, usually in the first minute.

The above being the case, you must introduce the witness. Get his or her name and purpose on the table and in the minds of the jurors:

Example

Q Can you state your name for the record?
A My Name is Dr. Mary Clemens.
Q What is your occupation Dr. Clemens?
A I am a scientist at the Ivy DNA laboratory.
Q How long have you been at the laboratory?
A I have been working at the laboratory for approximately 21 years.
Q Do you specialize in a particular field?
A I specialize in forensic DNA science and population genetics.
Q What is forensic DNA science?
A Forensic DNA science is a branch of science concerned with the identification of individuals who match DNA profiles found at a particular place, such as a crime scene.
Q What is population genetics?
A In a nutshell, population genetics is the branch of science in which statistics that explain the frequency with which a particular DNA pattern appears in a given population are studied and analyzed.

The expert should be introduced to the jury, including all of his or her relevant credentials. The attorney should then ask the court to qualify the witness as an expert in the conduct of DNA analysis and in the discipline of population genetics, or at the

[2]*See* Thomas A. Mauet, *Trial Techniques*, 4th ed., Aspen publishers, N.Y. (1996).

very least in the interpretation of DNA test results. This will be explained in greater detail in the next section. Next, have the witness explain to the jury why she is in court today:

Example

Q Dr. Clemens, did I ask you to examine any DNA evidence recently?

A Yes, you asked me to examine DNA evidence found in the car where the corpse of Jane Doe was discovered.

Q Did you examine this evidence?

A Yes, I examined the evidence through the use of DNA testing methods and prepared a report that I forwarded to you.

Q Dr. Clemens, are you prepared to tell us about the tests that you performed, about your findings, about your conclusions, and the basis for drawing such conclusions?

A Yes.

If the expert herself did not perform the testing, make sure he or she says so:

Example

Q Did you personally perform tests and analysis on the DNA evidence from the crime scene?

A No. I did not personally perform the tests but I did supervise the testing of the evidence by a laboratory technician who is employed by our lab and I saw the results of these tests. The technician was trained and qualified to perform such testing.

Q So what did you do?

A I supervised the testing and analysis of the DNA and participated in the process of determining which DNA tests should be employed given the amount of DNA obtained from the crime scene. I subsequently worked with my assistants to prepare a report on our findings, which I sent to you.

Q Can you tell the jury what your duties at the laboratory are?

A I am responsible for, among other things, supervising all DNA testing that is conducted at our laboratory, for implementing quality control measures, for ensuring that our technicians are adequately trained and certified, for ensuring that the laboratory is accredited, for ensuring that the equipment is functioning properly and that appropriate procedures are being followed, for preparing reports regarding the laboratory findings, and I periodically testify at criminal and civil trials about the results of DNA analysis performed by our laboratory.

Once the witness has been qualified as an expert and the jury understands his or her occupation and purpose for being in court, it is important to address any "trust issues" that may be present. For instance, in the example above the expert stated that she periodically testifies at criminal and civil trials. The other side may attempt to use this fact to paint the expert as a "hired gun" or a scientist who testifies professionally. The defense attorney will make it seem as if you have paid the expert witness to develop opinions favorable to the prosecution. Such a characterization can have an adverse impact on the jury's perception of the expert. Make sure that you make it clear that this is not the case.

Example

Q Dr. Clemens, you stated earlier that you periodically testify at criminal and civil trials. Are you compensated for such appearances?

A Yes.

Q Does your compensation include an amount for time spent in court?

A Yes.

Q Tell us how you are being compensated

A I charge $300 per hour for any consulting work that I do, irrespective of the nature of the work or who has hired me in this capacity. This fee is charged irrespective of what my opinion and conclusions are. My conclusions are formed independent of my compensation.

Q Are you being paid in this case

A Yes.

Q What are you being paid?

A I am being paid $300 per hour. I have spent about 10 hours working on this case to date, excluding today. This time includes the time I spent drafting the DNA testing summary report that I sent to your office.

Q Are you being paid for your time here today?

A Yes.

Q What are you being paid?

A I am being paid $300 dollars per hour, which is, as I stated earlier, my normal hourly rate.

Notice the use of the word compensation in the example above. It was not used by accident. It is important to stress that the expert is not paid, but rather, the expert is a professional who is compensated for his or her time. This distinction between being paid off and being compensated for one's time is important. Understanding this distinction is key to helping the expert gain the trust of the jury.

It is likely that the defense will attempt to attack the expert on the grounds that he or she testifies professionally. This type of witness is often referred to as a "professional testifier" or a "hired gun." It is important to anticipate this type of attack and mitigate the affect of such an approach.

Example

Q Dr. Clemens, do you perform a great deal of consulting work?
A I consult with attorneys on approximately 50 cases per year.
Q In what kind of cases do you testify?
A As I stated earlier, I testify in both criminal and civil cases.
Q Who are you usually retained by?
A The majority of my clients are prosecutors or state agencies, though I have been retained by private individuals in various civil matters.
Q Do you earn a significant portion of your yearly income by doing this consulting work and testifying at criminal and civil proceedings?
A No. I am employed to run the DNA laboratory and I am also an associate professor of biology at Ivy University. However, when necessary, I am called upon to help explain forensic DNA analysis performed by Ivy Laboratory to attorneys, jurors, and others.
Q You work for Ivy Laboratory, right?
A Yes.
Q They send you out to testify and they get paid for your testimony, right?
A Yes.
Q They would not send you out or get paid if your testimony was not favorable to the prosecution, right?

The fact that the expert has testified in other matters has another useful advantage to the person calling this witness. It can actually help strengthen the jury's perception of the witness if presented in the right fashion. If the expert has testified in other civil and criminal trials, it is likely that the witness has been qualified as an expert by other courts. Make the court and the jury aware of this fact.

Example

Q Dr. Clemens, have you been qualified as an expert by any other courts?
A Yes, I have been qualificd as an expert by other courts.
Q How many times have you been qualified as an expert by other courts?
A Over 50 times. I have been qualified as an expert by state and federal courts in four different states.
Q In what areas have you been qualified as an expert?
A I have been qualified as an expert in DNA analysis and in population genetics.

The next section discusses methods that can be employed to introduce the expert's credentials and experience.

6.2.1.3.1.4 INTRODUCING THE EXPERT'S CREDENTIALS (EDUCATION, TRAINING, AND EXPERIENCE) In his treatise on trial techniques, Thomas Mauet teaches that the objective in presenting the expert's education, experience, and training is to make the expert seem impressive and, at the same time, to make the expert seem likeable. Mauet insinuates that there is some tension between these two goals. This is mainly because when you bring out the expert's credentials, it is easy for the jurors to perceive the expert witness as pompous. This can be prevented by using leading questions to bring out the most impressive credentials so that the expert simply agrees with the attorney conducting the examination rather than making statements that can be perceived as pompous. Such leading questions are acceptable according to Mauet because "these are preliminary matters not in dispute."[3] However, he cautions that despite the above, some judges may sustain objections to the use of leading questions for foundational testimony.[4]

Mauet suggests that the attorney questioning an expert witness on direct bring out the following credentials during the examination:

1. Undergraduate and graduate education and degrees
2. Licenses and certifications
3. Teaching and publications
4. Positions held in important professional associations
5. Public offices held
6. Previous experience as an expert witness
7. Any other accomplishments that have a direct bearing on the witness's expertise

You should ask the expert for his resume or curriculum vitae before you go to court. Become familiar with this document. You may even wish to introduce it into evidence. If you are not able to get the document into evidence, make sure to get the expert to discuss his more impressive credentials in court.

6.2.1.3.1.5 QUALIFYING WITNESS AS EXPERT IN DNA EVIDENCE Once the witness's credentials have been established, it is important to make sure that the witness is qualified as an expert by the court.

Example

Q	Your honor, at this time I would like to ask the court to recognize Dr. Clemens as an expert in DNA analysis.
Judge	Any objections?
A	No objections at this time.
Judge	The court recognizes Dr. Clemens as an expert in DNA analysis. You may continue your examination counselor.

Here you are looking to qualify the witness as an expert in two areas. The first area in which you want the witness qualified is DNA analysis. The second area is DNA

[3] *See* Mauet, at 285.
[4] *See Id.*

statistics or the interpretation of DNA test results. Remember, just because one is an expert in DNA techniques does not necessarily mean that that same person is an expert in population genetics or is expert in the methods of determining the frequency with which DNA profiles appear in a given racial population.

If the defense attorney wishes to challenge the expert witness's qualification, he may ask to voir dire the witness. Some practitioners do this as a matter of course, to preserve issues for appeal. Others simply refrain from launching such challenges. You should be prepared for such a challenge either way.

6.2.1.3.1.6 QUALIFYING WITNESS AS EXPERT IN DNA STATISTICS While the defense attorney may be less inclined to object to the witness's qualification as an expert in the area of DNA analysis, he may seek to ensure that the witness is not qualified in the areas of population genetics, statistics, or the interpretation of DNA testing with respect to the frequencies with which various DNA profiles appear in the population at large. The attorney must appreciate the fact that these are two different areas of science and if the defense attorney is successful in preventing the witness from being qualified in the areas of population genetics or DNA statistics, it may be necessary to call another witness who is qualified to testify as to these areas.

Some attorneys may seek to just abstain from seeking both qualifications in the hopes that the defense attorney will recognize the fact that there are two levels of analysis with respect to DNA testing, namely the laboratory testing and the interpretation of that testing. This inaction may lead the defense attorney to point out the witness's lack of qualification:

Example

Defense	Objection! Judge, the witness is now giving statistical information. He has not been qualified as an expert in statistics or population genetics or anything of that sort.
Prosecutor	Your honor, the witness has been qualified as an expert in DNA analysis. He should be qualified to speak on these matters. At this time, I would like to also offer the witness as an expert witness based on his qualifications in the area of DNA statistics.
Defense	Your Honor, an expert as to DNA testing and comparing samples and so forth is one thing. However, if he is going to testify about population genetics or statistics, that is a far different matter. These are two totally distinct areas. If you are going to consider this request, I would like permission to voir dire the witness.

As the above example demonstrates, if you do not seek to have the witness qualified in the area of DNA statistics, the defense attorney will probably object to testimony involving DNA profile random-match statistics. It is important to have the witness qualified as an expert in both areas. Having the witness qualified as an expert in population genetics or in the statistics that govern the frequency with which certain DNA profiles occur in the population at large will also lay the foundation for testimony in this area, which is stage 2 of DNA admissibility as described herein. Aside from requesting an evidentiary voir dire on credentials where you can inquire as to what qualifies the witness to describe herself as a "scientist," you may wish to explore her understanding of statistics, population genetics, and the mathematical formulas and computer programs used to formulate the numbers about which she wishes to testify.

6.2.1.3.1.7 USING THE EXPERT TO DESCRIBE DNA EVIDENCE TECHNIQUES Once the witness is qualified as an expert in DNA analysis, the witness must be questioned about the testing procedures that took place with respect to the evidence and the conditions and circumstances surrounding that testing.

The following items should be addressed on direct with respect to the testing of DNA of unknown origin obtained from a crime scene (note that these items apply to the direct examination of a DNA expert in general and are not limited to crime scene evidence):

1. Basic principles of genetics and a description of DNA
2. A description of the DNA analysis and tests employed including the advantages and limitations of these tests
3. Quality control and quality assurance standards generally and in the expert's laboratory
4. Proficiency testing and audit information
5. Equipment employed, condition of equipment and controls
6. Any safeguards in place to prevent contamination and errors
7. Retesting procedures and information
8. Laboratory technician qualifications and training information
9. Laboratory accreditation information
10. A checklist of protocols. Who oversees compliance? What warnings and cautionary instructions are given in the protocols of which counsel should secure copies in order to review with a defense expert for compliance?

This list is not comprehensive. It is important to consider additional issues that may be unique to a particular case. However, this list contains the basic items that should come out during stage 1 of admissibility, namely, direct examination with respect to the testing of DNA evidence.

Describing the basic principles of genetics to the jury is essential. An attorney may consult biology and molecular biology texts for diagrams and charts.

However, you may find that the best charts and demonstrative evidence can be found from outlines sold at graduate and medical school book stores that are used by students to study for examinations. Often these works take a "for dummies" approach to key concepts that can generally be difficult to comprehend. The Internet can also be a great resource for materials that can be used to prepare demonstrative evidence to be used during direct examination or to acquire information about labs and experts.

When describing the DNA tests employed, it is important to have the expert discuss both the advantages and disadvantages of these tests. The rationale for the use of a particular DNA test should also be explained by the expert. Do not ignore the disadvantages of the DNA test employed. It is probably better for you to have the expert point these disadvantages out for a number of reasons. The first is that the defense attorney will probably point them out. The second is because if the jury sees that the expert acknowledges these disadvantages but is still able to draw his or her conclusions, the jury may trust those conclusions more. It is almost certain that the conclusions drawn by the expert will seem less authoritative if the jurors learn of the disadvantages of the test from the other side or the other side's expert witness. This work includes charts that outline the advantages and disadvantages of various types of DNA testing. These charts (National Commission on the Future of DNA evidence, The Future of Forensic DNA Testing. Predictions of the Research and Development Working Group, National Institute of Justice, November 2000 Available online: http://www.ojp.usdoj.gov/nij/pubs-sum/183697.htm) were created from a work published by the U.S. Department of Justice. Check and see if any of these charts will be of any help to your particular situation. Chances are, they will be. Use these charts as a starting point. At the very least, the information contained therein should be addressed during direct examination.

The quality control and quality assurance standards used by the laboratory should be described to the jury. The warnings contained in the protocols regarding contamination may also be considered. You may have the expert testify as to the fact that reagents and equipment are properly maintained and monitored.[5] Section three of the National Research Council report describes standards of laboratory performance. This report was heavily consulted during the writing of this chapter. If the quality assurance and control programs used by the laboratory are widely accepted by the forensic-science community, such acceptance should be explained to the jurors. The fact that failure to meet the requirements of a quality control and assurance program will lead to the loss of accreditation or licensing should also be explained.

During direct examination of the DNA expert, proficiency testing and audit information should be described if applicable. Proficiency tests are simply tests of specimens submitted to the laboratory in the same manner as evidence samples. Audits are simply independent reviews of laboratory operations to ensure that the

[5] *See* National Research Council, *The Evaluation of Forensic DNA Evidence*, 1996.

laboratory is operating at a desired level of performance. As described in section three of the National Research Council report, "[p]roficiency-testing and audits are key assessment mechanisms in any program for critical self-evaluation of laboratory performance.[6]

The equipment used and the condition of the equipment should be described. The quality control procedures used by the laboratory conducting DNA testing to ensure that equipment is properly functioning should be described and explained by the expert. According to the NRC report, "failure of the standards and controls [in place] to behave as expected in a test signals a problem with the analytical system and might disqualify test results."[7] Make sure that any issues that may be present at the laboratory with respect to laboratory equipment as described in the media or in laboratory logs that were turned over to the defense during discovery are addressed.

Contamination is probably the most important concept to address during direct. It is probably the DNA term with which the jurors will be most familiar. Even if contamination had no effect on DNA test results, it is most likely that jurors will trust the findings of DNA test results less if it is established that the DNA samples were contaminated at some point. The O.J. Simpson defense team used the phrase "compromised, contaminated, and corrupted" to describe the DNA evidence in that case. It conveyed a vivid impression to the jurors. Contamination takes place when a foreign material is mixed with an evidence sample.[8] There are different stages in the case at which contamination can take place. Address them all with the relevant witnesses. Contamination can take place at the crime scene, it can take place during transit, during storage of the samples, and even in the laboratory. It can be purposeful, or inadvertent. Irrespective of why contamination takes place, once it is established or even deemed possible, the expert's opinion can seem as tainted as the evidence itself. Make sure the expert takes great care to explain why he or she believes that the evidence was never contaminated. He or she should describe the appropriate safeguards and controls that the laboratory uses to ensure that contamination does not take place and to detect such contamination when it does occur. This explanation should take place with regard to potential inadvertent contamination by laboratory personnel and others, as well as possible contamination during the process of PCR if such testing was employed.

Information should be conveyed during the direct examination regarding the steps taken to prevent errors in the laboratory and elsewhere. Retesting procedures and information should also be explained. Procedures for maintaining reagents and equipment should be described as well as procedures for handling and preserving the integrity of evidence, for maintaining laboratory safety, and laboratory security. The more safeguards described to the jurors, the more trustworthy the DNA testing process will seem to jurors.

[6]*Id.* at 78.
[7]*See Id.* at 82.
[8]*See Id.* at 82.

Practical examples are useful.

Q Doctor, mucous contains DNA, correct?
A Yes, it can.
Q If I am working in a lab and handling evidence but I am not wearing gloves or a mask, and if I sneeze, that could run the risk of contaminating evidence, correct?
A Yes

Handing the evidence packages to the witness, you may ask him to open them and explore the following line of questioning.

Q A moment ago, I asked you to examine this evidence, isn't that so?
A Yes.
Q You took those plastic packages and opened them up correct?
A Yes.
Q As you did that, did you run the risk of contaminating the evidence?
A Well it has already been tested.
Q I understand, but you would never make that careless mistake in the laboratory, correct?
A Yes.
Q Even though you made it here?
A Yes.
Q By the way, do you think that your fingerprints are now on the plastic evidence bags?
A Possibly.
Q But not before today?
A I do not know.
Q Since you handled the packages, is it not likely that your DNA is now on the packages as well.
A Possibly.
Q And the reason you answered possibly and not definitely is that you had not tested it for your DNA, correct?

A portion of the direct examination should focus on the training and qualifications of laboratory technicians who work at the laboratory. Steps taken to ensure that laboratory personnel understand the "principles, use, and limitations of methods and procedures applied to the tests performed" should be explained during the direct examination.[9] The education, and experience of the laboratory personnel should also be included in the expert's testimony.

Information regarding the laboratory conducting the DNA testing and its accreditation must be explained in detail. The fact that the DNA laboratory would not be

[9] *See Id.* at 76.

allowed to operate without proper accreditation must be explained to jurors by the expert. It is important for the jurors to see that the laboratory is a safe place in which testing is taken seriously and performed by qualified personnel in a clean and professional environment. The expert must come across like a trained professional and not like the "nutty professor."

6.2.1.3.2 Stage 2: Admission of Statistics that Govern DNA Results To begin the completion of stage 2 of admissibility, the attorney must make sure the witness was qualified in stage 1 as an expert in population genetics or, at the very least, the witness was recognized as one who is qualified to testify regarding the interpretation of the results of DNA testing. If the witness was not so qualified, it may be necessary to call another witness qualified in the areas of population genetics or DNA statistics to testify as to the frequency with which particular DNA profiles appear in a given population. Your witness may come across as a professional testifier and hired gun. Therefore, you must bring out the expert's academic and scientific credentials.

6.2.1.3.1.2.1 INTRODUCING STATISTICS AND MEANING The expert should be questioned on direct examination about the frequency with which a particular DNA profile appears in a given population. As stated earlier, this is the portion of the examination in which you are most likely to lose jurors. Take great care to simplify the concepts as much as possible and to use demonstrative evidence and other techniques geared toward keeping the attention of the judge and jurors.

The following factors should be addressed during this stage of admissibility:

1. General population genetics principles
2. Information about searchable DNA databases and their use if relevant
3. General concepts surrounding statistics databases used in the calculation of profile frequency and how calculations for all relevant racial groups were conducted
4. Information about how relatives can be relevant to calculations or tests employed
5. Potential for analyst bias and safety measures in place
6. Strengths and limitations of calculation measures employed

This list is not comprehensive. The attorney seeking to have DNA statistics admitted into evidence should work with the expert witness to refine this list. Each individual case is unique and there is no absolute formula for success in this regard. However, the concepts above are relevant to the understanding of how the conclusions proffered in stage 3 of admissibility were drawn and should be understood for a number of reasons. The most important reason is because, as stated earlier, if the jury understands the method used to analyze the results of DNA testing, the calculations that govern determinations of the frequency with which a given DNA profile appears in a particular racial population, how DNA databases work, and how calculations are manipulated to account for various factors, the more likely they are to be capable of following the expert's rationale for drawing his conclusions in stage 3 of admissibility. If the expert simply spoon-feeds his opinion to an overwhelmed jury, it is not certain how this will affect the decision made by the jurors. However, an informed jury that understands the calculations, or can at the very least follow the rationale for the calculations, is more likely to trust the conclusions of the expert.

First and foremost, the expert should be questioned regarding general population genetics principles. The jury should be made to understand that testing DNA samples is simply the first step in the process of DNA analysis. The second step comprises the use of statistics to determine how common a particular DNA profile is in a given population. The expert should explain how the calculation of the frequency with which the DNA sample found at the crime scene was made.

If a database was used to find the suspect, the concept of a DNA database should be explained to the jury. The fact that the DNA sample of unknown origin found at the crime scene matched a DNA profile in the database should be explained. Any calculations that are relevant to the use of DNA databases should also be explained. If the calculation method employed is widely used or accepted, the jury should be made aware of such information.

The expert should be questioned regarding general concepts that govern the use of statistics databases. Methods used to calculate the profile frequency should be explained. The fact that frequencies were calculated for each of the possible racial populations that the suspect may belong to should be described. Why each racial group must be addressed separately should be explained to the jury. If the person who may have contributed the DNA sample is from a group for which no adequate database exists, the expert should address this fact and explain to the jury how groups that are closely related were chosen or, if relevant, the testimony of a physical anthropologist should be sought.[10]

There are various instances in which relatives can be relevant to the analysis of a DNA sample obtained from a crime scene or some other relevant source. For example, The National Research Council states the following as Recommendation 4.4 of its report on *The Evaluation of Forensic DNA Evidence*:

> If the possible contributors of the evidence sample include relatives of the suspect, DNA profiles of those relatives should be obtained. If these profiles cannot be obtained, the probability of finding the evidentiary profile in those relatives should be calculated.[11]

This recommendation should be addressed by the expert, if relevant. However, there is at least one other area in which an attorney may want to focus the questions during direct examination of an expert, namely, the impact of relatives on frequency calculations. When mitochondrial DNA analysis is employed, there are a number of concerns with regard to relatives.

[10]*See The Evaluation of Forensic DNA Evidence*, p. 123.
[11]*Id. The Evaluation of Forensic DNA Evidence*, p. 6.

Mitochondrial DNA is maternally transmitted. "Since mitochondrial DNA is always transmitted through the female, all the children of one woman have identical mitochondrial DNA as identical twins."[12] This is a significant fact when evaluating the potential source of a particular DNA sample. The NRC report points out that "[a] disadvantage for forensic use [of mitochondrial DNA] is that siblings cannot be distinguished, nor can other maternally related relatives, such as cousins related through sisters."[13] The NRC study points out the fact that mitochondrial DNA analysis can be combined with data from analysis of nuclear DNA.[14]

The potential for analyst bias should be addressed at some point during the direct examination. In publicized cases, this can be a great area of attack for a defense team. Standard questions regarding the potential for bias should be asked of the expert. Ask what procedures the laboratory has in place to limit the possibility for analyst bias. Ask how the laboratory flags such cases. In addition to these questions, the attorney should make an investigation of her own. Read the DNA reports prepared by the lab. Is there a good argument for examiner bias? You should read the section herein on reviewing a DNA report for a good example of a case in which examiner bias is possibly present. If your laboratory report contains any conclusion that states that the suspect is the source of the DNA sample, you have a potentially serious case of examiner bias. Why? This is because DNA is a science of exclusion, not inclusion. The current state of DNA analysis is such that it can only be used to exclude persons as the source of a sample. DNA analysis cannot be used to make an absolute conclusion regarding the source of a DNA sample. If this was done, then the conclusion flies in the face of justice and is incorrect. The defense attorney will recognize this fact if she is competent and will seize upon the opportunity to show that the conclusion is tainted. Still not sure about what constitutes possible examiner bias? Read the laboratory reports from the Marvin Albert case and the Bill Clinton investigation in Chapter 4. Notice the conclusions in the Marvin Albert case are exclusionary. They point out that Marvin Albert could not be excluded as the potential source of the DNA sample and subsequently provided statistics regarding the frequency in which such a DNA profile can be found in the population at large. In the Bill Clinton report, conclusions were drawn regarding the source of the DNA sample. Could this be because of the donor and his high profile? Who knows! The point is, you should not have to ask.

A strong case can be made for examiner bias because of the way the report is structured and worded. If your report is similar, address it. You can't change the report, however, you have the expert on the stand. If necessary, address the report and its wording and clean the situation up on direct. Clear the situation up before the defense attorney has an opportunity to turn it into a scandal and proof of a "witch hunt" during cross-examination. If the case is high profile, if there are strong politics surrounding the case, or there is pressure whether actual or perceived to get a conviction in the case, look to and eliminate the issue of examiner bias.

[12] *See The Evaluation of Forensic DNA Evidence*, p. 72.
[13] *Id.* at 73.
[14] *Id.*

Finally, the strengths and limitations of calculation measures employed in determining the frequency with which the DNA profile appears in the population at large should be explained. This is the catch-all provision of stage 2 admissibility. Any relevant information regarding population genetics and calculations used in your particular case should be addressed here. Was there a deviation from the norm in your case? Did you employ novel calculation methods? Is it necessary to convey information that will serve to support the conclusions that will be described during stage 3 of admissibility? If there is anything that you have conceived that has not been addressed herein, which must come before conclusions are drawn by the expert, this is the time to get that information into evidence.

6.2.1.3.3 Stage 3: Admission of an Explanation of DNA Results Stage 3 of admissibility is the most important. Stage 3 is the stage in which the DNA expert gives his opinion regarding the evidence and its impact. With regard to scientific evidence such as DNA, laypersons are not allowed to make conclusions regarding the impact of such evidence. However, experts are allowed to draw such conclusions. This is the reason reaching stage 3 of admissibility is crucial! The attorney would like the expert to be able to get on the stand and to testify regarding his or her conclusions with respect to the impact of the evidence being presented.

6.2.1.3.3.1 INTRODUCTION OF EXPERT'S OPINION REGARDING THE MEANING OF DNA EVIDENCE WHEN VIEWED IN LIGHT OF THE RELEVANT STATISTICS Once stages 1 and 2 of admissibility have been completed, the expert should be questioned regarding his or her opinions with respect to the DNA evidence being presented. At this point, he or she has been certified by the court to be an expert. He has explained the testing performed on samples discovered at the crime scene and that done on samples obtained from the suspect. He or she should have also already described the statistics that govern the appearance of various DNA profiles in the population at large.

It is during stage 3 that the expert should be questioned regarding his or her opinion with respect to whether the defendant committed the crime in question, given the results of the DNA testing performed. As stated above, while a normal lay witness would not be allowed to give such testimony, an expert may draw conclusions and give his or her opinion with regard to these matters. If you are seeking to have DNA evidence that implicates a particular defendant introduced at trial, irrespective of how incriminating the evidence is, make it a point to have the expert witness tell the jury that, in his or her opinion, the profile matches that of the defendant. Do not simply leave it to the jurors to make this match. This should be done for a number of reasons. First, you have taken great pains to introduce the expert, to bring out his credentials, and to build his credibility in front of the jury. Take advantage of your hard work! Let him or her help you by stating a favorable opinion. Second, after hearing the testimony with regard to the evidence, the jurors have probably already drawn the conclusion that the defendant's DNA profile matches the DNA profile found at the crime scene. The expert's disclosure of his or her opinion in such a case will help to strengthen the opinion of the jurors.

An example of the language that the attorney should consider using follows:

Q Now Doctor, to a reasonable degree of scientific certainty, do you have an opinion concerning the DNA testing and analysis that your laboratory performed?

A Yes.

Q What is it?

A To a reasonable degree of scientific certainty, it is my opinion that the laboratory testing and analysis confirm beyond equivocation or question that the samples recovered match[15] those of the defendant's DNA.

Q Now Doctor, can you tell us how you arrived at that conclusion?

A Yes, certainly .... In looking at chart 1, you will observe ...

6.2.1.3.3.2 USE OF DEMONSTRATIVE EVIDENCE—SPARE THE JURY While this discussion is being presented here, as with many of the concepts discussed herein, it applies to all stages at which DNA testing is being presented to the jury. When one is attempting to present highly technical concepts to anyone, whether the audience be a group of highly interested molecular biologists or a group of people who just have no interest in anything being presented, one should do so in as simple and entertaining a manner as possible. It is for that very reason that demonstrative evidence is so important. Demonstrative evidence should be used wherever possible to help keep the audience, in particular, the jurors, engaged. Keeping the jurors interested and happy is the key! Remember, you can give the most eloquent direct examination ever contemplated, but it is wasted if the audience has not paid attention or is not able to comprehend the information being presented!

Demonstrative evidence is different from what is normally referred to as evidence. Evidence is the term usually utilized to describe some article that has come from a crime scene or that has played some part in the series of events that led to the litigation in which such evidence is being used. Demonstrative evidence on the other hand is usually some sort of chart, graph, or illustration used to assist the jury or judge in understanding a particular proposition or occurrence of events. Demonstrative evidence is usually created by the parties to the litigation, as opposed to being collected from a crime scene or being borne of the events that lead to the litigation in which it is being utilized. A chart summarizing a financial transaction, a graphic showing sites on DNA that were examined, a map showing the area surrounding a crime scene and a video showing images relevant to the case are all examples of demonstrative evidence.

Demonstrative evidence is a very important part of proving one's case to the trier of fact (jury or judge). A witness's testimony can be compelling. Each word uttered by a witness helps to paint a picture in the mind of the individual

[15] The word "match" is key. It will be used throughout the case particularly in summation with respect to each DNA marker. It has devastating implication. For example: "This sample was taken from the crime scene and it 'matches' the defendant's genetic marker."

jurors. However, words are but imperfect vessels of expression. The problem with words is that they allow someone to visualize what you want them to see but they see that picture the way they want to see it. If the juror has never experienced the type of scenario being described or has never been to the neighborhood in which it took place, he or she might not be capable of painting an accurate picture that will compel him or her to subscribe to the clients version of the story. An attorney does not want to place her client's fate on the life experience and understanding of each member of the jury. The attorney must take into account that the jurors might not have the life experience, viewpoint, or frame of reference necessary to understand the events being described by the witness. A lawyer has a duty to zealously advocate for her client. One way to do so is to eliminate the potential for misunderstanding or misinterpretation of the testimony by the jury. This can be accomplished by using demonstrative evidence. An attorney should never rely on the pictures painted by the jurors based on their experiences to be representative of the client's version of events. The attorney must show the jurors and judge his own pictures and thus eliminate free-thinking among the jurors. When a client's fate rests in the hands of a juror, the attorney wants to take every step necessary to ensure that the jury has a proper understanding of the version of the story most beneficial to the client. Failure to take such steps could very well serve to allow the adversary to paint his version of the events that took place in the minds of the jurors.

There is no single source for demonstrative evidence. It can come from textbooks, journals, symposia, seminars, or lectures. It can even be the product of the attorney, client, or some other third party. One is only limited by his or her imagination! As long as the demonstrative evidence has probative value and conforms to the law of the jurisdiction in which the case is being tried, there is no limit on what can be done. Computer programs that allow users to create charts and graphics have become very popular in recent years. These programs can be utilized to create professional looking graphs and charts that can be utilized as demonstrative evidence during the trial. In addition, programs such as Microsoft Powerpoint can be used to create presentations and slides that can be displayed in the courtroom using a laptop and projector. When a client does not have a great deal of money to spend, such slides can be printed out on transparencies and a simple overhead projector can be used to display the demonstrative evidence. Charts can be created on graphic programs and then taken to a local photocopy shop for magnification. Given the state of technology today, it is relatively easy to have professional quality charts and graphs created.

An attorney should request permission from the judge to use demonstrative evidence and try to clear such evidence with his or her adversary in advance. If possible, the adversary should be made aware of and have a chance to examine such evidence. The likelihood of getting permission to use demonstrative evidence decreases when an attorney fails to inform the court and adversary of her intent to use such evidence. The evidence will most likely be admitted if it has probative value, little prejudicial effect, and if it assists the witness with testimony being given. Demonstrative evidence should assist the jury in understanding the evidence being presented. It should be noted by the attorney intending to use demonstrative evidence to present her case that while such items are often admitted into evidence and allowed to be part of the record during the course of the testimony, some judges will not allow the jury to use such evidence when they deliberate. This is not the case with traditional evidence borne of the events leading to the litigation. The jury is usually permitted to examine such evidence during deliberations.

The following is an example of how demonstrative evidence may be presented at trial:

Example

Q The chart to your left is marked for identification as people's exhibit Y. Do you recognize it?

A Yes.

Q What is people's exhibit Y, Mr. Castro?

A It is a blueprint or picture of the different rooms in my house on 105th Street.

Q Is this blueprint or picture a fair and accurate depiction of the house that you built on 105th Street?

A Yes.

Q Would this blueprint or picture assist you in explaining where you found your deceased spouse on the night in question?

A Yes, it would

Q Your Honor, at this time I offer people's exhibit Y as illustrative evidence[16] and ask that the witness be allowed to use it during his testimony.

However you decide to present your evidence to the jury, make sure you keep them interested and engaged!

6.2.1.3.3.3 ATTACKING POTENTIAL DEFENSE ARGUMENTS USING THE EXPERT

There are a host of arguments that the defense team will make during trial. Many of these arguments will center around the use of the expert in particular. Others will involve other topics. The following discussion explores some of the arguments that can be used against the expert being used at trial. A good lawyer is a prepared lawyer. He or she should prepare for any arguments that can be anticipated and even for those that cannot be anticipated. In addition to the arguments presented here, a lawyer who seeks to have DNA evidence admitted at trial should also read the section herein on attacking or mitigating the effects of DNA evidence.

[16]The court may not allow the exhibit to be received in evidence so the next best way to share the exhibit with the jury is to have it marked for identification, and the expert can then use it as an aid in testifying. By itself, it may not be evidence.

6.2.1.3.3.3.1 Professional Witness If the expert being used has testified in a number of trials, the defense team may attempt to use this against the prosecution The defense attorney will attempt to accomplish this by pointing out to the jury the fact that the expert testifies often and usually for the prosecution. The defense attorney may also point out that the expert only testifies for one side. He or she may attempt to create the inference that this is the case because the money is coming from testifying in favor of the government. In addition, if the expert testifies with such frequency that he or she has no time to do any of the things the prosecutor used to bolster the expert's credentials, namely, to teach classes or to engage in research, this might be pointed out by the defense. If you are a prosecutor, anticipate these attacks and dispel them before they are made. Chances are, these arguments will come out if they are applicable, and the failure of the prosecutor to point them out first will decrease the expert's credibility before the jury!

6.2.1.3.3.3.2 Expert Paid by Prosecution/Plaintiff The attorney who is seeking to admit the results of DNA evidence should anticipate the attacks described above If the expert is being paid, such should be mentioned by the attorney paying the expert, and not by the other side. It should be made clear that the expert is not being paid for his or her opinion but, rather, that the expert is being compensated for the time spent in and out of court and has only been compensated for his or her time when reviewing the file, writing reports, testing, and traveling. The attorney who is making payment must make it clear that the compensation is for time and not for a particular opinion. Payment is made irrespective of the jury's verdict or the outcome. The fees paid are consistent with those paid to other experts. This statement about fees should come from the mouth of the expert if possible!

6.3 DNA FOR THE DEFENSE OR THOSE WHO SEEK TO MITIGATE THE EFFECT OF DNA EVIDENCE

> The most beautiful words in the English Language are "not guilty."
>
> —Maxim Gorky (1868–1936)

This section is geared toward those who seek to mitigate the effect of DNA evidence. It is also of relevance to those who seek to have DNA test results admitted into evidence as one who is attempting to do so must be aware of the potential arguments against the admission of such evidence.

The ability to mitigate the effect of DNA evidence is important. If the trier of fact distrusts the DNA test results being presented, they will have less impact on the decision ultimately rendered by the trier of fact at the conclusion of the proceeding. This section describes several routes of attack that can be used to mitigate the effect of DNA test results at trial.

6.3.1 Preventing the Admission of DNA Evidence in Part or in Its Entirety

As mentioned in Section 6.1.2, there are three stages that must be completed for DNA evidence to be properly admitted and received by a jury in such a manner that such evidence can be most effective in influencing the trier of fact. If the party who opposes the introduction of DNA evidence can block successfully one of the stages of admissibility, the effect of the DNA evidence on the trier of fact may be lessened.

In particular, if the opponent of the introduction of DNA evidence can somehow have the DNA test results precluded in their entirety, he or she may have a better chance of being successful. This is rarely possible. However, in some jurisdictions, when DNA evidence is being offered in a postconviction proceeding and there is no statute governing the admissibility of DNA evidence, such may be possible. At trial, however, most jurisdictions allow the DNA test results to be admitted into evidence. Therefore, one who seeks to mitigate the effect of such results must chip away at the reliability of such results and the analysis being offered by pointing out every factor that will put the accuracy and reliability of such results in doubt.

6.3.1.1 Preventing Admission at One of the Three Stages To mitigate the effects of DNA evidence, once should prevent the successful completion of each stage of admissibility. There are several ways to accomplish this as are described in the following sections.

6.3.1.1.1 Admission of DNA Test Results The fist stage of admissibility is the admission of DNA test results. This is a stage at which attack is crucial in order to mitigate the effects of DNA test results on the decision of the trier of fact. If somehow the DNA test results themselves are not admitted, the DNA evidence will not be a factor in the trial. More likely than not, however, one will be limited to attempting to limit the admission of the results or to attacking the results themselves or the testing that led to the results. Each of these attacks is powerful in its own right and can potentially mitigate the effect of DNA evidence.

6.3.1.1.1.1 ROUTES OF ATTACK

6.3.1.1.1.1.1 New Type of DNA Test One route of attack is the objection to DNA evidence where a new type of DNA test has been performed The phrase "new DNA test" specifically pertains to a DNA testing procedure that has not yet been accepted by the highest court of the jurisdiction in which the proceeding is taking place. For example, while nuclear DNA testing protocols such as RFLP were deemed generally accepted for the purposes of forensic identification, there was a time when mitochondrial DNA testing protocols were not. It had to be made clear that these were two distinct methods with distinct implications. Different methods were used to test each type of DNA, and different statistical databases were used to draw conclusions regarding the frequency with which a particular DNA profile appeared in the population at large or, more specifically, in various ethnic populations. The point is, when the DNA testing performed is new or has not been previously deemed admissible in the jurisdiction in which the proceeding is taking place, it is essential that the attorney opposing the admission of such evidence ask for a *Daubert* or *Frye* hearing at which the admissibility of such evidence can be decided. At such a hearing, the attorney who opposes the introduction of DNA evidence must make a case against the admissibility of such evidence. Remember, just because DNA testing has taken place does not mean that the results of such testing are reliable, generally accepted, or admissible!

In such a hearing, the attorney opposing the admission of the results of the new DNA testing method may wish to call an expert of his own. An Internet search should be performed as well as library research on the limitations of the test in question. The medical journals should be searched for limitations on such testing. If the test is performed using patented technology, the patents should be consulted. Often later patents will point out the limitations of a particular technology. Ask yourself, does some patent point out the limitations of the DNA testing protocol in your case? Are there journal articles that point out the limitations of this type of testing? Are the statistics that the proponent of the evidence is relying on generally accepted? Is there debate in the scientific community with regard to the use of those statistics? Answer these questions for yourself. Even if you cannot come up with answers, you must require answers of the proponent of the new testing method!

6.3.1.1.1.1.2 Expert Not Qualified to Testify as to DNA Results The expert's qualifications should almost never be accepted without question It is imperative that the party in opposition to the admission of DNA test results conduct its own background check on the expert. This can be accomplished by conducting a search on the Internet using various search engines. For instance, google.com and dogpile.com are two Internet search engines that one may use to find information on the expert. Lexis and Westlaw should be used to find news articles that mention the expert. A search of medline or other scientific databases may yield journals in which the expert's work has been criticized.

The basic premise of this point of attack is that the opponent of the DNA evidence should object to the qualification of the expert. Try to limit the scope of the witness's expertise as much as possible. If the witness is a biologist with little experience with DNA try to get the court to limit the witness's recognized expertise to general biological principles. While this may be hard to do, it is worth a try.

6.3.1.1.1.1.3 Laboratory Not Accredited The attorney who seeks to exclude or mitigate the effects of DNA test results should make sure to do research on the laboratory that performed the DNA testing in his or her particular case Perform a search on the Internet as well as a search of news articles on Lexis and Westlaw. Is there any negative information in your search results that can be used at trial? Recently, there have been a host of laboratories at which negative practices have been discovered. Make sure the laboratory in your case has a clean record. If it does not, make every attempt you can to let the trier of fact know this.

Is the laboratory that performed the DNA testing in your case accredited? Who has accredited the laboratory? Is this entity's endorsement recognized in the forensic community? The Laboratory Accreditation Board of the American Association of Crime Laboratory Directors (ASCLAD-LAB) is one of the most recognized accrediting agencies. Is the laboratory accredited by ASCLAD-LAB? Has the laboratory successfully maintained the proper quality control and quality assurance standards of the accrediting agency? Does the lab have an ongoing proficiency testing program? Does it maintain documentation of all of its protocols? Do news reports describe any accreditation problems? If there are, the trier of fact should know of them.

6.3.1.1.1.4 Testing Not Performed by Certified Technicians If the DNA testing in your case was not performed by certified technicians, such a fact is relevant The American Board of Criminalistics offers certification following an examination of one's general knowledge of forensic science as well as skills and knowledge in a specific discipline. The 1992 report of the National Research Council explained that "individual analysts [should] have education, training, and experience commensurate with the analysis performed and the testimony provided.[17] In addition, the report stated that each analyst should have "a thorough understanding of the principles, use, and limitations of methods and procedures applied to the tests performed."[18]

How does one ascertain if the DNA testing in a particular case was performed by certified technicians? ASCLAD-LAB requires that the laboratories that it certifies document their laboratory operations, including proficiency testing, the training and experience of personnel, internal and external audits, and the individual laboratory's procedures for handling samples. The attorney should request this information in discovery requests.

6.3.1.1.1.5 Lack of Discovery Material or Notice with Respect to the Admission of DNA Evidence (including attacks using Federal Rules of Evidence 403) If material that is important to the defense has not been turned over by the other side, an objection should be made It is important to make sure that all evidence that may be relevant to a determination of one's guilt or innocence is turned over and evaluated by the party seeking to mitigate the effect of DNA evidence on the decision of the trier of facts in a particular case.

6.3.1.1.1.6 Improperly Obtained DNA Evidence: "Fruit of Poisonous Tree" As with any evidence, defense counsel should evaluate whether the DNA evidence in his or her case can be suppressed A motion to suppress evidence can be made for a number of reasons. One reason is that in certain jurisdictions, evidence that was

[17] *See* National Research Council. Evaluation of Forensic DNA Evidence. Committee on DNA Technology in Forensic Science, Board on Biology, Commission on Life Sciences, and The National Research Council, National Academy Press, Washington D.C. pp. 104–105 (1992).
[18] *Id.*

obtained illegally or through improper means cannot be used at trial to convict a defendant. Such illegally or improperly obtained evidence is referred to as the "fruits of the poisonous tree."[19] Such evidence is disallowed because it is deemed to be just as tainted as the methods used to secure it. It is important for those who seek to mitigate the effect of DNA evidence to determine whether the DNA evidence that is central to the case at hand is possibly subject to suppression. In addition to suppression motions made because of the circumstances surrounding the collection of the evidence, counsel should also consider other novel approaches to suppression based on traditional evidence suppression arguments.[20]

6.3.1.1.1.1.7 DNA Profile Should Have Been Purged from Database Another important scenario that may present itself is that in which the defendant is linked to a crime scene because of a hit that occurred when a DNA database such as CODIS was used In such a case, the attorney must ask a number of questions. Why was the defendant's profile in the database? Was the profile supposed to be in the database? Is there law that may preclude the defendant's profile from being included in a database? The answers to the above questions are essential to determining whether an attack on the use of the defendant's DNA profile can be made. Defense counsel should be sure that the defendant was not entitled to have his or her profile expunged from the database. Defense counsel should also be aware of the law relevant to DNA collection and DNA databases.

A recent case (V. S. V. Kincade, No. 02-50380, 9th cir. October 02, 2003) in the Ninth Circuit invalidated a DNA collection statute. In the future, it is likely that other courts will weigh in on the debate surrounding the collection of DNA from persons who have some contact with the legal system for the purposes of creating a database. Counsel should be aware of the relevant considerations and arguments. Cases in which the placement of one's DNA profile in a database was disallowed might help to ensure that the DNA evidence in a later case is suppressed.

6.3.1.1.2 Admission of Statistics that Govern DNA Results

6.3.1.1.2.1 ROUTES OF ATTACK

6.3.1.1.2.1.1 Expert Not Qualified to Testify as to Statistics The admission of statistics governing the frequency with which a particular DNA profile appears in a particular population is important to the evaluation of whether the DNA evidence being presented can link the defendant to the scene of a crime The person who testifies at trial should be qualified to testify regarding such statistics. It is important that defense counsel challenge the witness's qualification to testify regarding population genetics and other statistical issues.

[19]*See Wong Sun et al. v. United States*, 371 U.S. 471; 83 S. Ct. 407; 9 L. Ed. 2d 441 (1963).
[20]*See Mapp v. Ohio*, 367 U.S. 643; 81 S. Ct. 1684; 6 L. Ed. 2d 1081 (1961).

6.3.1.1.2.1.2 Statistics Do Not Conform to Standards Accepted by the Scientific Community Defense counsel should always explore and possibly challenge the methods used by the witness to explain population genetics concepts that govern the DNA evidence in question Is the defendant familiar with the National Research Council Report?[21] Chapters 4 and 5 of the NRC report deal with population genetics and statistical issues relevant to the evaluation of DNA evidence.

Another important time at which attacks should be made upon the statistics being employed to explain the frequency with which a particular DNA profilc appcars in a particular population is when a novel DNA test is being employed. When a DNA testing method is new or without precedent, it is important to not only attack the acceptance of the testing method itself in the scientific community but also to attack the statistics that are purported to show the frequency with which a particular DNA profile appears in a particular population. These attacks are best made at a *Frye* or *Daubert* hearing.

6.3.1.1.2.1.3 Irrelevant/Improper Database Used The database used to link the defendant to the crime scene is also relevant and should be scrutinized What race is the defendant? Does the database include people of the defendant's particular group, tribe, or ethnicity?[22] The National Research Council makes an important recommendation in its report:

> If the person who contributed the evidence sample is from a group or tribe for which no adequate database exists, data from several other groups or tribes thought to be closely related to it should be used.

If the database being used is not a relevant database, an objection to its use should be made.

6.3.1.1.3 Admission of an Explanation of the DNA Results

6.3.1.1.3.1 ROUTES OF ATTACK

6.3.1.1.3.1.1 Expert Not Qualified to Testify as to Statistics in Context Opinion Is Being Offered One method of attack, as described above, is to object to the expert witness's qualification to testify regarding the statistics being offered at trial Where the statistics employed are particularly novel or there is some other factor that creates a unique circumstance in which the case and the evaluation of DNA evidence is different than that usually made in a particular type of case, an attack on the witness's qualification to testify in a particular context might be important. For instance, it is important to remember that DNA is a science of exclusion not inclusion. If the expert is trying to include the defendant in a particular group of possible contributors of the DNA found at a crime scene, object! The expert is

[21]*See The Evaluation of Forensic DNA Evidence* (NRC 1996).
[22]*See The Evaluation of Forensic DNA Evidence*, Recommendation 4.3, p. 123 (NRC 1996).

not qualified to testify as to statistics in that context. No one is! The science does not support such statements or conclusions!

6.3.1.1.3.1.2 Attacking Laboratory Techniques and Conditions

1. *Use of Accepted Techniques* Discovery requests should be made for laboratory protocols and handbooks that show the types of techniques employed by laboratory technicians testing DNA evidence. The techniques employed by the laboratory should be compared to the standards set forth by various accrediting agencies, including ASCLD-LAB.

2. *Quality Control and Quality Assurance* Discovery requests should also be made to obtain materials pertaining to quality control and quality assurance programs employed in the laboratory. "*Quality control* refers to measures that are taken to ensure that the product, in this case a DNA-typing result and its interpretation, meets a specified standard of quality .... *Quality assurance* refers to measures that are taken by a laboratory to monitor, verify, and document its performance."[23] The educational and training backgrounds of the technicians who handled the samples should be requested. Maintenance logs pertaining to the equipment used should be requested. The protocols and logs that pertain to reagents utilized should also be requested.

3. *Use of Proficiency Testing and Audits* Proficiency testing is required for laboratory accreditation by ASCLD-LAB.[24] A proficiency test is the testing of specimens submitted to a laboratory just as evidence samples are submitted.[25] Audits usually entail a review of a laboratory by an independent agency that evaluates the operations of the laboratory.[26] An audit usually ends in a determination of whether the laboratory is performing according to a predefined standard.[27]

It is important that records of proficiency testing and audits be requested during discovery. These discovery requests may uncover documents that show problems within the laboratory that analyzed the DNA evidence in a particular case. Audits usually take place approximately every 2 years.[28]

4. *Laboratory Error* Laboratory error, if it has possibly taken place, is one of the single most important factors that should be presented to a jury when attempting to mitigate the effect of DNA evidence.[29] Laboratory error can be present in the form of contamination due to sample mishandling, sample mixup, incorrectly recorded data, improper temperature controls, failure to perform duplicate tests, analyst bias, and failure to supervise technicians.[30] Make sure that discovery requests are calculated to obtain documents that may show that the above factors may be present

[23]*See The Evaluation of Forensic DNA Evidence*, p. 76 (NRC 1996).

[24]*See The Evaluation of Forensic DNA Evidence*, p. 78 (NRC 1996).

[25]*See Id.*

[26]*Id.*

[27]*Id.*

[28]*Id.* at 80.

[29]W. C. Thompson, F. Taroni, and C. G. D. Aitken, *How the Probability of a False Positive Affects the Value of DNA Evidence*, Journal of Forensic Sciences, 48(1), 47–54, (2003).

[30]*See generally The Evaluation of Forensic DNA Evidence*, Chapter Three (NRC 1996).

in your case. Request documents that pertain to programs put in place to prevent analyst bias. Does the laboratory have procedures in place to prevent such bias?

6.3.1.1.4 Attacking the DNA Test Used Every DNA testing method has its strengths and limitations. Make sure that you know the particular strengths and limitations of the test utilized in your case. Review the charts herein that point out the strengths and limitations of several DNA tests. Is your test listed? Are there factors in your particular case that make some limitation more prominent or pertinent? If so, make such factors known to the jury.[31]

6.3.1.1.5 Attacking Chain of Custody When seeking to mitigate the effect of DNA evidence at trial, attacking the chain of custody may be a good way of showing that factors exist that should be considered before conclusions are drawn with regard to a particular issue. The section herein on contamination explores this route of transmission further. Generally, however, through discovery requests, the defense attorney should request that a full chain of custody regarding pertinent evidence be presented. The temperature of the environments in which the evidence was stored should be a part of these requests. High temperatures can affect DNA evidence adversely. Was there moisture in the environment? Could that have affected the DNA? Was there extended and/or repeated exposure to these factors? Could there have possibly been contact with DNA from other sources? Question the witness regarding these items. Make sure that the chain of custody is complete. If it is possible that third parties had access to the evidence, make that known. If the evidence was obtained from or stored in a hostile environment, such a fact is also relevant. Breaks in the chain of custody should be explored, explained, and exploited!

6.3.1.1.6 Attacking Expert Witness The expert witness who will present the results of the testing that was done on the DNA evidence in a particular case is the voice of the evidence. It is important if you are seeking to mitigate the effect of that evidence to discredit the witness if possible. At the very least, an attorney seeking to mitigate the effect of DNA evidence should be seeking to create reasonable doubt.

Creating reasonable doubt through the cross-examination of the expert witness can be accomplished in a number of ways. If the witness was paid, this fact should be mentioned. The expert witness should be questioned regarding the fact that he or she is being paid a fee to testify. The fact that he or she may be dependent on such payment to earn a living should be established. If the witness testifies frequently for a fee, this should be established as well. The fact that the expert may be a hired gun may work to the defense attorney's advantage.

If the expert witness who is describing the DNA testing did not actually perform the tests, this should be explored at trial. Did the expert supervise the testing? If the witness did supervise the testing, how closely did he or she work with the technicians? Can the witness personally testify as to the chain of custody? Can

[31]R. Willing, Mismatch Calls DNA Tests into Question in Britain was 'To Be Expected' as Databases Include More Samples, *USA Today*, Feb. 8, 2000, p. 03A.

the witness actually testify to the fact that there was no possibility of cross-contamination or some other problem such as failure of equipment or mishandling?

The attorney who seeks to mitigate the effects of DNA evidence at trial should always search the Internet as well as local and national newspapers for articles that may mention the laboratory that performed the DNA testing, the laboratory's directors or technicians, and the particular expert witness who is to testify. If there has been some scandal or impropriety at the laboratory, question the witness about the article. Use anything you find to your advantage.[32]

Your background search should also include calls to other attorneys and law firms that may be acquainted with this particular expert's work. Did that attorney find anything useful? Have you found any previous testimony of the expert in your case? Is there anything that you find in that testimony to be useful in your case? Is there previous contradictory testimony? It is important to know the expert, the expert's credentials, and anything other attorneys may know about the expert before trial.

Have you questioned the expert's credentials? Should you question the expert's credentials? Is the expert just someone with a degree? If this is the case, you may want to stress the importance of having hands-on experience in order to discredit the witness. The way to do this might be to ask questions that only someone with hands-on experience would be able to answer and then to call your own expert, who might have hands-on experience to state that. This, however, is a loaded gun as such smear tactics may backfire and create a bad impression. However, this route of attack may be available in some cases.

6.3.1.1.7 Contamination Contamination is another key concern that must be addressed during the evaluation of DNA evidence in a legal proceeding. "Contamination has been used as an umbrella term to cover any situation in which a foreign material is mixed with an evidence sample."[33] If there is a possibility that the evidence could have been contaminated, the jury should be made aware of such a fact. It is for this reason that chain of custody must be explored at trial. Contamination could have taken place at many points during the chain of custody. Contamination could have taken place at the scene of the crime or the place at which the DNA evidence being tested was collected. Contamination could have taken place during transport and storage of the DNA evidence. During transport or storage, the evidence could also have been subjected to temperatures that could have led to degradation of the evidence. At the crime scene, explore who had access to the evidence. Explore how it was handled, how it was collected. Could a law enforcement official have contaminated the evidence? Could the sample taken have gotten to the crime scene in another manner? PCR is very sensitive. Explore the factors that would lead to a misleading result.

Could there have been contamination in the laboratory? Make discovery requests regarding contamination in the laboratory in recent months. Request all information

[32]*See* J. F. Kelly and P. K. Wearne, *Tainting Evidence: Inside the Scandals at the FBI Crime Lab*, Free Press, New York (1998).
[33]*The Evaluation of Forensic DNA Evidence*, p. 82 (NRC 1996).

that pertains to any instance of contamination involving the laboratory technicians or unit that handled the DNA evidence at issue. Check local and national newspaper reports to see if there are articles regarding contamination in the laboratory that conducted the testing in your case.[34] If there is, make the jury aware of such problems. Ask the witness about those cases. How are they different then the current case?

Was there inadvertent contamination due to sample mishandling? Was the sample problematic from the beginning as in the case of mixed samples? "Mixed samples are contaminated by their very nature."[35] For example, a DNA sample from a vaginal swab might contain both semen and vaginal fluids. Could blood from both parties and even a third party also have been present? Explore the possibility of such and the effects of such on the type of testing employed. Carryover contamination is also something to consider. Carryover contamination can occur during PCR testing "when a PCR amplification product finds its way into a reaction mix before the target template DNA is added."[36] "The carryover product can then be amplified along with the DNA from an evidence sample, and the result can be that an incorrect genetic type is assigned to the evidence sample."[37] In cases where the genetic type of the contaminant matches the genetic type of the person in question, such a coincidence can lead to a false match. In your discovery requests, ask for procedures and protocols that describe techniques used to prevent false matches and carryover contamination. Explore these possibilities when questioning the witness from the testing laboratory.

6.3.1.1.8 Attacking the Choice Not to Employ Several Different DNA Tests, Including Sequencing In a criminal prosecution, the prosecutor must prove his or her case beyond a reasonable doubt. One way to potentially prevent the prosecutor from doing so in a case where DNA evidence is being introduced is by attacking the decision not to employ several different kinds of DNA tests, including sequencing.

Each DNA testing method has its strengths and limitations. The limitations of a test, when pointed out, may increase the possibility of creating reasonable doubt in the minds of jurors. However, when several tests are employed, especially by different laboratories, and the results are reported to the jury, the conclusions may seem more credible and trustworthy to those jurors. Such being the case, it may be possible to create reasonable doubt by not only showing the limitations of the DNA test being employed but also by showing the jurors that those limitations might have been overcome by the use of multiple testing methods or by the use of two different laboratories to perform the tests. The fact that there may have been a way to overcome the limitations of a particular test and a choice was made not to conduct additional testing may help to create reasonable doubt in the minds of the jurors or, at the very least, may decrease the trustworthiness of the DNA test results

[34]*See* A. Liptak. The Nation; You Think DNA Evidence Is Foolproof? Try Again. *The New York Times*, Section 4, p. 5, column 1, March 16, 2003. *See also* N. Madigan, Houston's Troubled DNA Crime Lab Faces Growing Scrutiny, *The New York Times*, Section 1, p. 20, column 3, February 9, 2003.

[35]*The Evaluation of Forensic DNA Evidence*, p. 84 (NRC 1996).

[36]*Id.*

[37]*Id.*

being introduced. Chapter 4 contains tables in which the advantages and limitations of various DNA testing methods are set forth. It may be wise to consult these tables in a case in which only one DNA testing method was employed, despite clear factors that may lead a prudent scientist or attorney to employ other testing methods. In the future, new methods and technologies may also be relevant.[38]

6.3.1.1.9 Use of PCR and Sensitivity to Contamination Polymerase chain reaction is a powerful method that can amplify small amounts of DNA. This very strength can potentially be a weakness. Such is the case when the DNA sample that is subjected to PCR becomes contaminated. Contamination can occur at any point after the evidentiary specimen had been deposited at the crime scene. It can occur at the scene, during transport to the laboratory, or within the laboratory. It is possible for both the evidentiary DNA as well as the contaminating DNA to become amplified during the PCR process. Depending on the source of the contamination, this could adversely affect the defendant. If the contaminating DNA is that of the defendant, there is a chance that the amplified evidentiary sample would appear to be a mixture that contains the defendant's DNA. It is precisely for this reason that laboratories test exemplars and evidence at different times so as to minimize the possibility of lab-borne contamination. However, in general, "if contamination does occur, it will most likely result in an 'exclusion' or 'inconclusive' result and be in favor of the defendant."[39]

Both TWGDAM (now SWGDAM) and ASCLD have emphasized the need to minimize and avoid the possibility of DNA contamination during PCR amplification of casework specimens. Using the HLA DQA1 and PM test kits, it was demonstrated that contamination was never observed when nanogram quantities of DNA were mishandled or aerosolized. In fact, contamination was only observed when amplified product was carelessly manipulated or purposefully sprayed near or directly into open tubes containing water or genomic DNA. The attorney that can demonstrate contamination of a specimen can certainly use that to develop reasonable doubt in the minds of the jurors. It should be pointed out that the types of DNA testing currently employed are even more sensitive than those in usage in the early 1990s, so contamination remains a serious issue for forensic DNA testing. If contamination of an evidentiary specimen does occur, the interpretation of the resulting electropherogram becomes critically important.[40]

A case for contamination has to be made using clear evidence of the possibility of contamination. Simple carelessness in the laboratory will probably not rise to the level necessary to create reasonable doubt. To demonstrate that a test had been

[38] *See* National Commission on the Future of DNA Evidence, The Future of Forensic DNA Testing Predictions of the Research and Development Working Group (2000).

[39] *See* J. M. Butler, *Forensic DNA Typing: Biology & Technology behind STR Markers* Academic Press, N. Y., 2001.

[40] *See* C. A. Scherczinger, C. Ladd, M. T. Bourke, M. S. Adamowicz, P. M. Johannes, R. Scherczinger, T. Beesley, and H.C. Lee. *A Systematic Analysis of PCR Contamination.* Journal of Forensic Sciences, 44(5) 1042–1045 (1999). *See also* J. M. Butler, *Forensic DNA Typing: Biology & Technology behind STR Markers*, Academic Press, New York, 2001, pp. 102–103.

compromised, it is necessary to demonstrate serious carelessness, flagrant and/or gross deviation from protocols designed to prevent contamination, or a realistic potential for contamination or cross-contamination based on the behavior of laboratory personnel, investigators, and others at the crime scene or by any other person who may have come in contact with the DNA sample.

While the Scherczinger and Lee study makes it harder to create reasonable doubt by showing simple carelessness in the DNA laboratory, it does provide a valuable weapon for attorneys who are attempting to mitigate the effect of DNA evidence at trial using a contamination argument. The group found in its study that PCR contamination was not usually noted. If the study is introduced in an attempt to show that simple carelessness will not usually affect the results of DNA testing, the defense counsel may use the study to show that laboratory contamination during PCR analysis is often not noted and such is a possible reason that a more powerful argument could not be made in favor of possible contamination.

6.3.1.1.10 Preventing Testimony Regarding the Ultimate Issue A circumstance may arise in which there may be testimony at trial regarding the guilt of the defendant. This scenario may arise in a couple of ways. One way this may happen is when after the DNA test results are introduced, the expert makes a conclusion regarding the origin of the evidence and indirectly implies that the defendant is guilty. Conclusions regarding the ultimate issue to be decided by the trier of fact or jury may appear in laboratory reports and records of tests results. Such conclusions are improper because they are not supported by the tests results or the science of DNA analysis.

6.3.1.1.10.1 DNA EVIDENCE IS USEFUL FOR EXCLUSION, IT CANNOT IDENTIFY WITH CERTAINTY When used for identification, DNA is a science of exclusion, not a science of inclusion. This simply means that DNA can only be used to draw conclusions about whether one can be ruled out as a potential contributor of a particular DNA sample. DNA cannot be used to show that a defendant is absolutely the source or contributor of a particular DNA sample. Instead, conclusions must be drawn that state that the defendant cannot be ruled out as a contributor. Statistics are then introduced to show the probability that the defendant could be the source of the sample. However, there is no absolute certainty in the science of DNA analysis.

The public tends to believe that DNA can be used to show a defendant's guilt. However, nothing could be further from the truth. It is important to educate the public about what DNA testing can and cannot accomplish. DNA analysis is a great tool and is more reliable than some other forensic identification methodologies. This being the case, there is no need to misconstrue or misuse the conclusions that can be drawn from the results of DNA analysis.

6.3.1.1.10.2 OBJECTING TO TESTIMONY REGARDING DEFENDANT'S GUILT If a party attempts to give testimony regarding the defendant's guilt or innocence, or toward the ultimate issue of using DNA analysis as a foundation, OBJECT! If one attempts to introduce a report in which a conclusion is made regarding guilt based upon the results of DNA analysis, OBJECT! If insinuations are made that are not consistent with or in the spirit of the concept that DNA analysis is a science of exclusion, rather than inclusion, OBJECT! OBJECT! OBJECT! Making such objections is important. By not raising an objection, you may lose or waive the ability to appeal based on such testimony or the use of such evidence. It is important to get such objections on the record. The fact is, DNA is a science of exclusion, and claims regarding the ultimate issue are not supported by the science and are improper!

6.3.1.1.11 Addressing Relatives (Mitochondrial DNA) There are a number of important factors that must be considered when evaluating the possible contributors of DNA evidence. In general, where there is reason to believe that a close relative of the defendant could have been at the crime scene, such a possibility should be explored. The National Research Council recommends that nuclear DNA profiles of relatives be obtained when the possible contributors of the evidence sample include relatives of the suspect.[41]

Another scenario in which relatives must be considered is when mitochondrial DNA is being evaluated. The mitochondrial DNA profiles of siblings who share the same biological mother are identical. It is for this reason that such relatives must be included in the analysis of those who might have contributed a particular DNA sample.

[41]*See The Evaluation of Forensic DNA Evidence*, Recommendation 4.4, p. 123 (NRC 1996).

7

Exonerating the Innocent through DNA

7.1 POSTCONVICTION APPEALS BASED UPON DNA EVIDENCE

If you will accept bold ideas, new theories, courageous innovation, and disputed principles with an open and inquisitive mind and a renewed commitment to make the law an instrument of advantage for disadvantaged people, we will be a significant generation.

—E. Clinton Bamberger, first director of the OEO Legal Service Program, in a speech to the National Legal Aid and Defender Association (November 18, 1965)

THE INNOCENCE PROJECT

The Innocence Project was created in 1992 by Barry Scheck, a professor working at the Benjamin N. Cardozo Law School of Yeshiva University in New York City, and Peter Neufeld, his law partner. They realized the importance of DNA technology in identifying the source of biological evidence found at a crime scene. The analysis of DNA helps to determine guilt or innocence of suspects and defendants at trials where violent acts such as rape and/or homicide have been committed. Scheck and Neufeld reasoned that if evidence from old cases/trials could be found, new DNA technology could help to prove their clients' innocence. They worried about those innocent individuals who had been wrongfully convicted. Over the past 11 years, law students and attorneys from around the country have volunteered their time and knowledge to work with

DNA: Forensic and Legal Applications, by Lawrence Kobilinsky, Thomas F. Liotti, and Jamel Oeser-Sweat
ISBN 0-471-41478-6

the Innocence Project to help free the innocent. Hundreds of letters from inmates arrive regularly at the law school, each complaining that that they have been wrongfully convicted and that DNA could set them free. At the time of this writing, 138 inmates have been exonerated based on DNA testing in the United States. Despite this fact, many prosecutors still refuse to consent to DNA testing. On average, it takes up to 4 years for the Innocence Project to resolve a postconviction case. At the Innocence Project approximately 75% of the cases it accepts go unresolved as a result of loss or destruction of the biological evidence. Further, beyond the cases accepted by the Innocence Project, the majority of wrongful conviction cases have no biological evidence. However, in Innocence Project cases where such evidence is found, approximately 50% of convicts are found not to be the contributor of the evidence. In 2001 the number of exonerations reached a peak of 23 (see Fig. 7.1).

Now that it has been established that many individuals have been wrongfully convicted, we must ask ourselves why they were convicted in the first place. Were eyewitness identifications inaccurate? Were lineups or photo arrays conducted in less than a scientifically appropriate manner? Were false confessions obtained from suspects who were below normal intelligence or who were too young to understand what they were doing?[1] Were there analyses that were performed incorrectly? Was there testimony that was incorrect or misunderstood? Were there scientific reports containing false conclusions? Were the reports misinterpreted? It is important to examine the factors and causes of wrongful convictions so that such errors are minimized or eliminated in the future. Reforming the criminal justice system is on the top of the agenda for the Innocence Project leadership. They are trying to establish acceptable suspect identification guidelines, to establish methods to detect false confessions, to eliminate bad science, and to produce statutes that would make it easier for a convict to obtain postconviction testing. To date, only 38 states have statutes allowing for postconviction DNA testing. Furthermore, exonerees have no guarantees of compensation for time spent in prisons. The Innocence Project has been able to demonstrate the serious weaknesses of the criminal justice system. The weaknesses in our criminal justice system are a clear basis for establishing a death penalty moratorium. Such a moratorium has been established in Chicago where a number of individuals who were on death row have been exonerated.

Aliza B. Kaplan, Esq., Dep. Director, Innocence Project

Since the early 1990s our nation has been confronted by numerous cases of actual innocence where belated DNA testing revealed that an alleged perpetrator was in fact innocent. Numerous cases across the country including over 138 cases already documented by the Innocence Project of the Cardozo Law School in New York have

[1] *See* J. Hoffman, "Police Tactics Chipping Away at Suspects' Rights," *The New York Times*, March 29, 1998 at 1 and 40; and J. Hoffman, "Police Refine Methods So Potent, Even the Innocent Have Confessed," *The New York Times*, March 30, 1998 at 1 and B4.

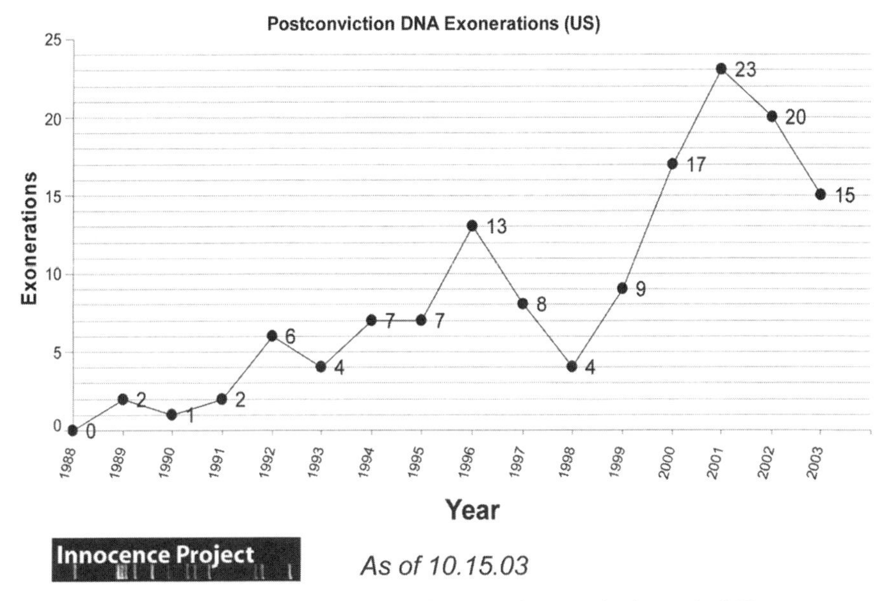

Fig. 7.1 In 2001 the number of exonerations reached a peak of 23.

shown that DNA testing is vital both for conviction and exoneration purposes. Many states have remedies available for postverdict testing of DNA evidence.[2] Obviously, evidence is not always available and many convictions are obtained without forensic evidence. But, where forensic evidence is available and DNA has not been conducted or there are possibilities that the DNA testing was flawed for other reasons, counsel may wish to consider DNA testing. This is also true of prosecutors who seek to conduct DNA testing to resolve or to solve cases that remain open. Databanks have now allowed prosecutors and police investigators to take DNA samples from defendants and check them against old and forgotten cases. There are ongoing issues concerning double jeopardy, the statute of limitations, probable cause, and *ex post facto* rule lawmaking; For example, when defendants have been convicted solely on the basis of cases that the prosecution was aware of during the prosecution of the case and DNA testing subsequently reveals other cases for which the defendant was not indicated when plea negotiations were taking place and/or the defendants trial was taking place. Obviously, double jeopardy and other claims may arise as a result of these new arrests. Given the fact that DNA testing may exonerate defendants who are falsely accused, many states have responded by enacting legislation that provides for postconviction DNA testing. New York is one of those states and since 1994 has developed a databank. A defendant in New York may move to set aside his or her verdict and conviction under CPL §440, a postconviction

[2]*See* Kathryn M. Kase, *Play It Again Sam, Sam Play It Again, Sam: Post-Conviction Motions in New York State*, New York State Bar Association Criminal Justice Section Journal, Vol. 9, No. 1 at p. 105 (Summer 2001) for a very well researched account of postconviction remedies in New York State.

statute that codifies applications for postconviction relief based upon new evidence and other factors not germane to this chapter. This statute will be discussed in more detail herein. In general, many postconviction DNA statutes are premised upon claims of actual innocence or complete exoneration. To obtain postconviction relief based on DNA evidence in New York, the state statute requires that the conviction has to have occurred before January 1, 1996, that testing is sought on specific biological evidence, and, lastly, that there is a reasonable probability the verdict would have been more favorable to the defendant had the DNA testing been performed earlier.[3]

In reviewing the prospects for DNA testing and whether it is a worthwhile endeavor, lawyers must speak at length with their clients and review all other evidence to see what corroborating facts or evidence may exist, or what was testified to, or upon which the jury based its verdict. A claim of actual innocence by virtue of DNA testing may also suggest to

DNA IN ACTION: DNA EVIDENCE PROVES MAN INNOCENT DESPITE MURDER CONFESSION

Mr. Eddie Joe Lloyd was a patient in a mental hospital when he was interrogated during an investigation into the murder of a 16-year-old girl in 1984. Mr. Lloyd signed a detailed confession to the killing of Michelle Jackson. Upon sentencing Mr. Lloyd to life in prison, Judge Townsend reportedly stated: "The sentence that the statute requires is inadequate. The only justifiable sentence, I would say, would be termination by extreme constriction." Fortunately, the state of Michigan, where the trial took place, did not have the death penalty.

After 17 years, 3 months, and 5 days, Mr. Eddie Joe Lloyd walked out of prison. DNA evidence proved he did not commit the crime he confessed to and helped to win his freedom. "Mr. Lloyd was the 110th person, and the first in Michigan, to be exonerated because of DNA evidence according to the Innocence Project at the Benjamin N. Cardozo School of Law in New York, which led the drive to overturn his conviction." According to Mr. Lloyd's lawyers, "the case highlights a national problem of false confessions, particularly with suspects who are minors or who suffer from mental illness". However, in the case of Mr. Lloyd, prosecutors and defense lawyers worked together to have the conviction overturned. DNA had once again helped an innocent man gain his freedom.

Source: Jodi Wilgoren, Man Freed after DNA Clears Him of Murder, *New York Times* (August 27, 2002). 10, Section A, page 10, column 4.

the court that the defendant is unconstitutionally detained in violation of his rights. Thus, the new DNA evidence may also suggest a due process claim whereby, if this evidence had been available or discovered or if there had been a proper police

[3]*See* NYCPL §440.30(1-a).

investigation and testing at the time of the allegations, the defendant would have been exonerated. Hence, his due process rights to a fair trial under the circumstances have been adversely effected. One of the first steps after interviewing the defendant is to do an investigation by reviewing everything available in the way of a trial record to determine what forensic evidence there was against the defendant and in other respects. Obviously, an admission or a confession must be closely scrutinized to determine whether it was voluntarily obtained, the age of the defendant at the time, as well as his or her mental status and capabilitics. But, ultimately, after those first two steps are undertaken, counsel may then have to determine whether biological evidence still exists.

As of 2001, Jane Segal Green, Executive Director of the Innocence Project of the Cardozo Law School, indicated that in fully 70% of postconviction cases investigated by her organization, biological evidence had been destroyed, making postconviction testing impossible. Prosecutors are required to keep evidence only until state appeals have been exhausted. The reality is that CPL §440 type of motions or motions to set aside convictions may occur well after all appeals have been exhausted through intermediate courts and the New York Court of Appeals. In fact, they often await a resolution of petitions for habeas corpus in federal courts, which now by law must be made within 1 year of the exhaustion of all state appeals. It goes without saying that in locating new evidence or biological evidence, the defense will have to speak with prosecutors, law enforcement laboratories, and, in the most drastic sense, may, under the public health law, have to move to exhume the body of the deceased. Public health laws require that such an application be made to the district attorney on motion, and a denial of that request may then cause the defense to initiate a New York Civil Practice Law and Rules, Article 78 proceeding in the nature of mandamus that would compel the exhumation of the body of the deceased for DNA testing. Such a case is currently pending in New York in the matter of *People v. Pacheco* (Nassau County Indictment No. 1027N-02).

In the event that the defense believes that there is evidence available that may exonerate the defendant and that DNA testing is planned, generally the first order of business would be to file a motion for an order to preserve and protect that evidence. On an intermediate level, defense counsel may consider serving the police department and others with letters asking that the evidence be preserved, but there is no guarantee of preservation in the absence of a court order to that effect. Accordingly, defense counsel should consider initiating an order to show cause, requesting emergency relief and a temporary restraining order preventing the destruction or mutilation or deterioration of any evidence in the possession of the district attorney or the police department or any laboratory that may have been involved in DNA testing. The counsel will then consider what type of DNA testing may be needed under the circumstances. For that purpose counsel may enlist the services of his or her own laboratory or experts to advise him or her as to what type of DNA testing may be indicated under the circumstances of a particular case. The cost of such testing has decreased over the years and will vary from lab to lab. Once again the Innocence Project of the Cardozo Law School may provide much needed information in that regard.

DNA RESULTS PROVE THREE MEN DID NOT COMMIT MURDER

Michael Jordan was one of the most amazing basketball players of this century. Calvin Ollins is not fortunate enough to truly appreciate why. "I've been locked up almost his whole career," Mr. Ollins reportedly stated while being interviewed about his imprisonment.

Omar Saunders, Calvin Ollins, and Larry Ollins were freed after a hearing in which charges against them and a fourth man were dropped, despite their confessions and convictions in the 1986 slaying of Lori Roscetti. Ms. Roscetti was a 23-year-old medical student who was brutally murdered and dumped on a desolate road near a Chicago housing project. Ms. Roscetti's face was severely traumatized as a result of being hit with a concrete block, her ribs were shattered, and there was a bloody shoe print found on her chest. During the investigation of the murder and at their trial, a forensic scientist may have given false testimony regarding blood grouping of the biological evidence in the case. It also appears that detectives may have coerced confessions based upon a profiler's hypothesis. The inmates' lawyer Kathleen Zellner pressed officials to reopen the case. Upon reinvestigation, officials discovered that semen and hairs found on Ms. Roscetti's body and in her car did not match any of the four men who had been convicted of the rape-homicide.

"This evidence appeared so overwhelming, and yet we know now, obviously, that it was underwhelming," Rob Warden, director of the Center on Wrongful Convictions at Northwestern School of Law reportedly stated. Ms. Celeste Stack, an assistant state's attorney who worked on the reinvestigation into the murder, petitioned Judge Dennis Porter of the Cook County Criminal Court in Chicago to vacate the charges based upon the new DNA evidence. The judge agreed and the charges were vacated. Upon being released from prison, the three men were given plaques etched with the date and the words "The first day of the rest of your life."

Source: Jodi Wilgoren, Three Cleared by DNA Tests Enjoy Liberty After 15 Years, *New York Times* (December 6, 2001). Section A, page 20, column 1.

7.2 POSTCONVICTION DNA TESTING: RECOMMENDATIONS FOR HANDLING REQUESTS

> It is better to risk saving a guilty person than to condemn an innocent one.
>
> —Voltaire (1694–1778)

In 1996, the National Institute of Justice published a report that revealed the stories of 28 men exonerated by DNA evidence. The report, entitled *Convicted by Juries, Exonerated by Science: Case Studies in the Use of DNA Evidence to Establish Innocence after Trial*, inspired then Attorney General Janet Reno to request that the National Institute of Justice (NIJ) identify ways to maximize the value of DNA in our criminal justice system.[4] The NIJ immediately went to work, and the result was the subsequent publication of another valuable resource, this one entitled *Postconviction DNA Testing: Recommendations for Handling Requests.*[5] This work is a great resource for people who might be affected by a postconviction DNA testing request.

In *Postconviction DNA Testing: Recommendations for Handling Requests*, the NIJ identifies five categories that should be used to evaluate cases that may potentially rcquirc DNA tcsting.[6] These categories are as follows:

1. Cases in which both the prosecutor and defense counsel concur on the need for DNA testing.
2. Cases in which the prosecutor and defense counsel cannot agree on whether an exclusion would amount to a demonstration of innocence, would establish reasonable doubt of guilt, or would merely constitute helpful evidence.
3. Cases in which testing will be inconclusive due to the present state of evidence or technology.
4. Cases in which it is impossible to do any testing because the crime scene evidence was either destroyed, never collected, or cannot be located despite best efforts.
5. Cases in which false claims of innocence are made. In such cases, prosecutors and defense counsel generally agree that no testing is warranted.

These categories cover virtually all of the possible scenarios that might inspire or accompany a request for postconviction DNA testing. It is especially important that when a category 1 case presents itself, both sides work to obtain the necessary postconviction DNA testing. Category 2 cases also present a unique opportunity to ensure that justice takes place. Despite disagreements with respect to the impact of the evidence, the fact that some exoneration, whether total or partial, is possible should be enough to convince the parties to work together.

7.2.1 Role and Response of the Prosecutor

Wrong must not win by technicalities.
—Aeschylus (?–456 BC), *The Eumenides* (N.D.), translated by Richmond Lattimore.

In category 1 cases, the results of DNA testing may show actual innocence and completely exonerate the defendant.[7] If the prosecutor feels comfortable with

[4]*See Conners, E., T. Lundregan, N. Miller, and T. McEwen Convicted by Juries, Exonerated by Science: Case Studies in the Use of DNA Evidence to Establish Innocence after Trial*, National Institute of Justice Research Report (1996). Available online: http://www.ncjrs.org/txtfiles/dnaevid.txt

[5]National Commission on The Future of DNA Evidence. *Postconviction DNA Testing: Recommendations for Handling Requests*, National Institute of Justice Research Report (September 1999). NCJ 177626. available online: http://www.nejrs.org/pdffiles1/nij/177626.pdf

[6]*Id.* at 3.

[7]*See Postconviction DNA Testing: Recommendations for Handling Requests*, National Institute of Justice Research Report, p. 6.

DNA testing and can do so, he or she should facilitate the testing of DNA in category 1 cases. Prosecutors should be cooperative and at the very least civil in category 2 cases.[8] It is important that prosecutors respond appropriately to DNA testing requests. The NIJ recommends that prosecutors not delay in responding to a request for DNA testing to ensure that testing is performed in a timely manner and is not barred by any statute of limitations.[9] In addition, prosecutors should take affirmative steps to ensure that relevant and important evidence, such as evidence obtained from the crime scene or the victim, is not destroyed or unnecessarily exhausted.[10] An awareness of institutional and statutory policies governing the destruction of such evidence is necessary. Prosecutors should be willing to use their best efforts to locate crime scene samples and other relevant evidence.[11] As officers of the court, it is important that prosecutors ensure that justice prevail.

Conferences with defense counsel at which the prosecution makes counsel aware of facts relevant to postconviction DNA testing are also helpful. The aforementioned Department of Justice recommendations point out that:

Defense counsel may be raising an issue about prior DNA testing that could be resolved if the prosecutor showed defense counsel underlying laboratory notebooks or other materials that the jurisdiction does not ordinarily disclose. In such cases, prompt disclosure will often save time and money.[12] Defense counsel may be unaware of facts that are known to the prosecutor.

7.2.2 Role and Response of the Defense Attorney

> As lawyers, our first responsibility is, of course, to see that the legal profession provides adequate representation for all people in our society.
>
> —Richard Nixon, in his speech to the National Legal Aid and Defender Association (October 1962)

The Department of Justice's postconviction testing recommendations describe the role of defense counsel in postconviction proceedings. The report stresses that defense counsel should be cordial to prosecutors during postconviction proceedings. Attorneys should always, as a matter of professionalism, show due deference to his or her adversary. However, the role of defense counsel in postconviction proceedings must be that of the zealous advocate. As an attorney, you step into the shoes of your client. You are his agent, and in many cases, his avatar. While you should always behave as a professional, you must also be ready to shake the judicial system to its very core if necessary. After all, if there is actual reason to believe your client is innocent—meanwhile, your client is sitting in a jail cell—the system has failed. It is up to you to ensure that our system of justice is working to its full potential. Think outside the box. When rules do not afford adequate relief and remedies, seek to have

[8] *Id.* at 7.
[9] *Id.*
[10] *Id.*
[11] *Id.*
[12] *Id.*

them changed. Your job as defense counsel is to demonstrate to the world that your client is innocent. There are many courts in which your fight must take place. State and federal courts provide statutes for postconviction relief. In the absence of such a statute, seek to have the court mandate DNA testing. If this fails, you must not give up. Bring the battle to the public's attention. The more egregious your situation, the more power it has to move others. Draft a press release and send it to the major media outlets. Politicians may be able to grant a pardon to your client. The judicial system is not the only game in town. Whatever you decide to do, do it like a professional. Do it for yourself. Do it for the love of the judicial system. Do it for your client. Irrespective of your inspiration, Just Do It!!!!

7.3 LEGAL STANDARDS GOVERNING POSTCONVICTION TESTING

Obtaining a favorable disposition subsequent to conviction is not an easy task. There are many ways in which one can go about seeking postconviction relief. However, many avenues of relief are barred by time. Other avenues of relief are limited to a certain number of hearings or proceedings.

7.3.1 Argument for a Constitutional Right to Postconviction DNA Testing

While there is no uniformity on the subject, an argument exists that there is a constitutional right to postconviction DNA testing.[13] There have been rulings by courts that seem to indicate that such a right to testing exists.[14] However, other cases do not seem to indicate that such a right exists.[15] Irrespective of where the individual courts stand on whether a constitutional right to postconviction testing exists, the argument for a constitutional right to testing, which stems from the Supreme Court's decision in *Brady v. Maryland*,[16] is a strong one. In *Brady*, the Court held that a defendant has a constitutional right at or before trial to be informed of exculpatory evidence of which the state is aware. The extension of *Brady* to posttrial proceedings in which exculpatory evidence exists seems to be a logical extension that is in the interest and spirit of justice.

In *Matter of Dabbs v. Vergari*[17] an inmate requested access to evidence to have DNA testing take place. This testing was to be a part of a motion to vacate the

[13]*See Postconviction DNA Testing: Recommendations for Handling Requests*, p. 11.
[14]*See People v. Callace*, 573 N.Y.S.2d 137, 138 (Suffolk County Ct., 1991) (holding that discovery right in statute authorized vacation of convictions on the basis of new evidence); *Jenkins v. Scully*, No. CIV-91-298E, 1992 WL 32342, at *1 (W.D.N.Y. Feb. 11, 1992) (holding state must produce evidence for DNA testing pursuant to rules governing habeas corpus).
[15]*See Ohio v. Wogenstahl*, No. C-970238, 1998 WL 306561, at *1 (Ohio Ct. App. Dist. 1 June 12, 1988) (Noting that request for DNA retesting because trial results were inconclusive is in the nature of a discovery request that the court is not required to grant in a postconviction proceeding).
[16]*See Brady*, 373 U.S. 83 (1963).
[17]*See Dabbs*, 570 N.Y.S. 2d 765 (Sup. Ct., Westchester County 1990).

conviction based on newly discovered evidence. Despite the prosecution's opposition to the defendant's request and arguments that no statutory right to such postconviction discovery existed at the time, that the potential results of such testing was speculative, and that granting such a request would inspire other sexual offenders to request that such DNA testing take place, the court granted the defendant's request for postconviction DNA testing. The court relied on *Brady* in making its decision to allow postconviction DNA testing.[18] The testing ultimately paved the way for the vacateur of the defendant's conviction.[19] Several other courts also allowed for postconviction DNA testing using *Brady* as the basis for providing those tests, despite often inflexible postconviction remedies that were in place during the time in which the courts made their decisions.[20] However, despite the numerous cases that allow testing pursuant to *Brady*, it is of particular note that some recent courts used the challenge and method in which such a *Brady* argument was being made in their decision-making process.[21] Despite the fact that many states have postconviction testing statutes in place, situations sometimes present in which no remedy exists under such a statute. When no remedy exists under a postconviction testing statute, arguing that a constitutional right to postconviction DNA testing may be a defendant's only hope.

7.3.2 Other Non-Postconviction Testing Statute Arguments

It is the spirit and not the form of law that keeps justice alive.

—U.S. Chief Justice Earl Warren (1891–1974)

7.3.2.1 Habeas Corpus Relief Another method that may be available to persons who seek postconviction DNA testing is federal habeas corpus relief. Habeas corpus is the designation of a number of writs that are used to bring a party before a court or judge.[22] The writ of habeas corpus dates back to old English law, where the writ of habeas corpora juratorum was used to command the sheriff to bring people before the court.[23] Prisoners who feel they are unlawfully being detained or imprisoned may which to consider the use of a writ of habeas corpus.[24] Prisoners who seek to challenge the constitutional grounds of a conviction in the absence of a

[18]*See Dabbs*, 570 N.Y.S.2d at 787–68.

[19]*See People v. Dabbs*, 587 N.Y.S.2d 90, 93 (N.Y. Sup. Ct. 1991).

[20]*See State v. Thomas*, 586 A.2d 250, 253-54 (1991) (noting that postconviction DNA testing can be appropriate even when procedural bars exist); *Sewell v. State*, 592 N.E.2d 705, 707–708 Ind. Ct. App. Dist. 3 1992) (allowing postconviction DNA testing 10 years after conviction); *Commonwealth v. Brison*, 618 A.2d 420, 423 (Pa. Super. Ct. 1992) (noting "where evidence has been preserved, which has high exculpatory potential, that evidence should be discoverable after conviction"); *Mebane v. State*, 902 P.2d 494, 497 (Kan. Cr. App. 1995) (noting that when a proper showing under *Brady* is made, requests for DNA testing can be granted).

[21]*See Harvey v. Horan*, No. 01-6703, 2002 WL 86874, at *7 (4th Cir. January 23, 2002) (holding that a prisoner seeking access to DNA evidence did not state a valid claim under *Brady* because he was "not challenging a prosecutor's failure to turn over material, exculpatory evidence that, if suppressed, would deprive the defendant of a fair trial.").

[22]*See Blacks Law Dictionary*, 6th ed. West (1990). West publishing co. Eagan. MN.

[23]*See U.S. v. Tod*, 263 U.S. 149, 44 S.Ct. 54, 57, 68 L.Ed.221.

[24]*See People ex rel. Luciano V. Murphy*, 160 Misc. 573, 290 N.Y.S. 1011.

method of statutory relief may have an ideal method in the form of a writ of habeas corpus.[25] Though the writ of habeas corpus has been an effective method for obtaining a judicial forum at which a prisoner may have the legality of his or her detention considered, it has become increasingly harder in recent years to obtain habeas corpus relief. The Supreme Court in *Herrera v. Collins* ruled that a defendant petitioning for habeas corpus relief could not obtain such relief, despite his claim of newly discovered evidence that might prove his innocence.[26] The Court went on to state that "claims of actual innocence based on newly discovered evidence have never been held to state a ground for federal habeas relief absent an independent constitutional violation occurring in the underlying state criminal proceeding."[27] The difficulty of obtaining postconviction DNA testing through normal channels or habeas corpus relief has troubled many legal scholars. Some states have begun to address the need for such testing by taking steps to afford avenues of relief.

7.4 POSTCONVICTION DNA TESTING STATUTES

In many jurisdictions, no statute providing for postconviction DNA testing exists. In such jurisdictions, persons who seek such testing must rely on the intervention of courts and/or prosecutors or attempt to obtain testing through the methods described above. This absence of statutory relief has been of concern to many legal commentators.[28]

The need to see that justice be served prompted many states to remedy the lack of avenues through which persons potentially wrongfully incarcerated could prove their innocence.[29] New York and Illinois were the first states to enact statutes that

[25]*See Fay v. Noia*, 372 U.S. 391, 83 S.Ct. 822, 9 L.Ed.2d 837. *See also* Liotti, Thomas F., *Lawyers and the First Amendment—Mutually Exclusive Terms?* The Champion, August 1992 at 23. The Champion is a publication of the National Association of Criminal Defense Lawyers (NACDL).
[26]506 U.S. 390, 396 (1993).
[27]*Id.* at 400.
[28]*See* Kathy Swedlow, Don't Believe Everything you Read: A Review of Modern "Post-Conviction" DNA Testing Statutes, 38 Cal. W. L. Rev. 355 (2002); Heidi C. Schmitt, Post-Conviction Remedies Involving the Use of DNA Evidence to Exonerate Wrongfully Convicted Prisoners: Various Approaches under Federal and State Law, 70 UMKC L. Rev. 1001 (2002); Holly Schaffter, Postconviction DNA Evidence: A 500 Pound Gorilla in State Courts, 50 Drake L. Rev. 695 (2002); Karen Christian, "And the DNA Shall Set You Free": Issues Surrounding Postconviction DNA Evidence and the Pursuit of Innocence, 62 Ohio St. L.J. 1195 (2001); Paul C. Giannelli, Serenity Now or Insanity Later?: The Impact of Post-Conviction DNA Testing on the Criminal Justice System: Panel Two: The Legal, Ethical, and Practical Issues of Post-Conviction DNA Testing: Impact of Post-Conviction DNA Testing on Forensic Science, 35 New Eng.L. Rev. 627 (2001); Rochelle L. Haller, The Innocence Protection Act: Why Federal Measures Requiring Post-Conviction DNA Testing and Preservation of Evidence are Needed in Order to Reduce the Risk of Wrongful Executions, 18 N.Y.L. Sch. J. Hum. Rts. 101 (2001).
[29]Indeed, the accomplishments of Barry Scheck, Peter Neufeld, and the Innocence Project at Cardozo Law School have served as a wake up call to lawmakers and jurists throughout the country. Numerous cases of innocent people wrongfully incarcerated have come to light thanks to the efforts of Mr. Scheck, Mr. Neufeld, the Innocence Project, and many other similar movements and programs. *See* Jim Dwyer, Peter Neufeld and Barry Scheck. *Actual Innocence: Five Days to Execution and Other Dispatches from the Wrongly Convicted*, Doubleday, New York (2000).

contained provisions regarding postconviction DNA testing.[30] Subsequent to the enactment of postconviction DNA testing statutes in New York and Illinois, many other jurisdictions enacted similar statutory schemes.[31] A list of postconviction DNA statutes and legislation appears herein as an appendix.

The New York postconviction testing statute is contained in N.Y. Crim. Proc. Law § 440.30. The statute specifically applies to cases in which the defendant requests performance of a forensic DNA test on specific evidence. DNA test results, if available, would have been admitted in the trial and would have likely resulted in a more favorable verdict to the defendant.[32] The statute states in relevant part:

> In cases of convictions occurring before January first, nineteen hundred ninety-six, where the defendant's motion requests the performance of a forensic DNA test on specified evidence, and upon the court's determination that any evidence containing deoxyribonucleic acid ("DNA") was secured in connection with the trial resulting in the judgment, the court shall grant the application for forensic DNA testing of such evidence upon its determination that if a DNA test had been conducted on such evidence, and if the results had been admitted in the trial resulting in the judgment, there exists a reasonable probability that the verdict would have been more favorable to the defendant.[33]

A number of things mentioned in the statute are of note. The first is that the New York statute only applies to older cases. Second, the statute applies in circumstances where the verdict would possibly have been more favorable to the defendant. This wording is particularly significant. It seems to indicate that irrespective of actual innocence, DNA testing may be possible. In other words, in New York, it seems that DNA testing is available for cases in which DNA evidence would either have lowered the defendant's possible sentence because of a more favorable verdict or in cases of actual innocence in which a verdict of not guilty would have probably been reached at trial.

The Illinois postconviction DNA testing statute is a remarkably different creature altogether. Section 116-3 of the Illinois Code of Criminal Procedure is specifically entitled: "Motion for Fingerprint or Forensic Testing Not Available at Trial

[30] *See* N.Y. Crim. Proc. Law 440.30 (Consol. 2001); 725 Ill. Comp. Stat. 5/116-3 (2002).

[31] Ariz. Rev. Stat. 13-4240 (2002); Ark. Code Ann. 16-112-124 to -129 (Michie 2001); Cal. Penal Code 1405 (West 2002); Del. Code Ann. tit. 11, 4504 (2001); D.C. Code Ann. 4031-4035 (2001 & Supp. 2002); Fla. Stat. ch. 925.11(1)(a) (2001); Idaho Code 19-2719 (Michie 2002); Idaho Code 19-4902 (Michie 2002); Ind. Code 35-38-7-1 to -19 (2001); La. Code Crim. Proc. Ann. art. 926.1 (2002); Me. Rev. Stat. Ann. tit. 15, 2137 (West 2001); Md. Code Ann., Crim. Proc. 8-201 (2001); Mich. Comp. Laws 770.16 (2002); Minn. Stat. 590.01 (2002); Mo. Rev. Stat. 547.035 (2002); Neb. Rev. Stat. 29-4117 to -4125 (2001); N.J. Stat. Ann. 2A:84A-32a (West 2002); N.M. Stat. Ann. 31-1A-1 (Michie 2001); N.C. Gen. Stat. 15A-269 (2002); Okla. Stat. tit. 22, 1371, 1371.1 (2002); Okla. Stat. tit. 22, 1372 (2001); S.B. 667, 2001 Leg., 71st Sess. (Or. 2001); Tenn. Code Ann. 40-30-401 to -413 (2002); Tex. Code Ann. Crim. Proc. art. 64.03 (2002); Utah Code Ann. 78-35a-301 to -304 (2001); Va. Code Ann. 19.2-327.1 (2001); Wash. Rev. Code 10.73.170 (2002); Wisc. Stat. 974.07 (2002).

[32] *See* N.Y. Crim.Proc.Law § 440.30(1)(a).

[33] N.Y. Crim.Proc.Law § 440.30(1)(a).

Regarding *Actual Innocence*" (emphasis added).[34] The Illinois statute specifically states that the DNA testing which is being requested must not have been available at the time of trial.[35] The Illinois statute also specifies that the DNA testing to be performed should be "on evidence that was secured in relation to the trial which resulted in [the defendant's] conviction."[36]

In a motion requesting postconviction DNA testing, a defendant in Illinois must present a prima facie case that:

1. Identity was the issue in the trial, which resulted in his or her conviction
2. Evidence to be tested has been subject to a chain of custody sufficient to establish that it has not been substituted, tampered with, replaced, or altered in any material aspect.[37]

In addition to the above, the statute also directs the trial court to allow testing as set forth above "under reasonable conditions designed to protect the State's interests in the integrity of the evidence and the testing process" upon making certain determinations.[38] The trial court must allow DNA testing to take place if it determines that:

1. Result of the testing has the scientific potential to produce new, noncumulative evidence materially relevant to the defendant's assertion of actual innocence.
2. Testing requested employs a scientific method generally accepted within the relevant scientific community.[39]

In essence, the Illinois statute does two things: (1) It authorizes postconviction DNA testing on evidence collected in relation to trial in cases where testing such evidence would possibly show the actual innocence of the defendant, and (2) the statute also delegates a gatekeeping function to the trial court judge. It is of note that the standard of admissibility set forth seems to be that set forth in *Frye*. New York also follows the *Frye* standard of admissibility.

7.5 PREVENTING POSTCONVICTION DNA TESTING THROUGH WAIVER

> What debt did she owe to a social order which had condemned and banished her without a trial? She had never been heard in her own defense; she was innocent of the charge on which she had been found guilty . . .
>
> —Lily Bart, Character in Edith Wharton's *The House of Mirth* (1905)

[34]*See* 725 Ill. Comp. Stat. 5/116-3 (2002).
[35]*See* 725 Ill. Comp. Stat. 5/116-3(a).
[36]*Id.*
[37]725 Ill. Comp. Stat. 5/116-3(b).
[38]725 Ill. Comp. Stat. 5/116-3(c).
[39]*Id.*

Recent visitors to the California Attorneys for Criminal Justice website may have noticed an article entitled "AG Draft Aims at Preventing Postconviction DNA Testing."[40] The article/page warns of potential attempts by prosecutors to obtain waivers from defendants of their right to postconviction DNA testing under California law during plea agreements. Attorneys on both sides of a case should be aware of the problems that can be associated with waivers of the right to postconviction DNA testing. Such a waiver does not seem to be in the interests of justice. As pointed out herein, there are substantial requirements that must be met before a motion requesting postconviction DNA testing will be considered. These requirements help to weed out potentially frivolous cases. If a defendant is able to demonstrate to a court that postconviction DNA testing is in the interests of justice and fairness, and that the results of such testing might have led a jury had the case been tried to conclude that the defendant was actually innocent, it would be horrible for such a fact to be suppressed because of the pressures that accompany pretrial plea negotiations and dispositions.

POSTCONVICTION APPEALS BASED UPON DNA EVIDENCE

Dennis Maher is "number 127" of 144 individuals (as of this writing) who were convicted of various felonies and who served lengthy prison sentences. However, thanks to postconviction DNA testing, these people were eventually exonerated. Mr. Maher served more than 19 years in Massachusetts prisons, first at Walpole and then in the Massachusetts Treatment Center for sexual offenders, located in Bridgewater, MA. He was convicted of 2 rapes and an attempted rape. He was arrested in 1983 because a police officer had spotted him walking down the street wearing a red sweatshirt with an attached hood. He was then 23 years old. One of the rape victims had indicated that her assailant wore a sweatshirt matching this description. Throughout his almost two-decade incarceration, Maher continued to proclaim his innocence. In 1993, the Innocence Project became involved in his case. Years of searching for the evidence and filing motions in Maher's case produced no results or evidence in his case. In 2000, Aliza Kaplan, an attorney for the Project, took responsibility for his case by serving as his counsel. It was not until 2001 that an intern working for the Project discovered two boxes of evidence located in the basement of the Middlesex courthouse in Cambridge. The underwear and blue corduroy slacks worn by the first victim were found in these boxes. The clothing was tested for the presence of semen and then DNA typed. It was determined that the genetic profile was not that of Mr. Maher. In 2003, a slide obtained from a rape kit prepared from the second rape victim was found in a police station in Ayer, MA. DNA obtained from the sperm cells on the slide was tested. Once again the genetic profile excluded Maher as the source. At the age of 42, more than 19 years after his arrest, he was cleared of all charges

[40]*See* AG Draft Aims at Preventing Post-Conviction DNA Testing: Working Group 2 Draft Recommendations, available at *http://www.cacj.org/whats_new_ag_draft.htm.*

and is now trying to return to a normal life. Were it not for DNA testing and the fact that biological evidence was discovered, he would still be languishing in prison today.

7.6 THE FUTURE OF DNA TECHNOLOGY

The development of forensic DNA techniques in the mid-1980s was followed by constant advancements in both science and technology. We have seen RFLP technology come and go. The use of reverse dot blot analysis and even AmpFLP analysis has fallen out of favor. The techniques that are used in today's forensic labs are quite different than those used earlier. They are more sensitive, more rapid, more specific, more reliable, more economical, and less labor intensive. Science is not static. It is constantly changing, bringing new improvements and advancements. Science is closely linked to technology and therefore scientific advancement leads to technological benefits. Improvements in forensic methodology over time are expected and welcome. They will be met with a new wave of validation studies, peer-reviewed published articles, education, and training and also with debate and deliberation by legal practitioners. Scientific breakthroughs are often reflected in advancements in technology. Who would have guessed that the development of PCR would have had such a profound impact on so many scientific and medical fields? Forensic DNA scientists have benefited greatly from computerization of instrumentation, the establishment of state and national databases, and the production of more sophisticated equipment and more powerful multiplex kits. In particular, the increased funding of DNA related programs by federal and state legislatures has had a profound impact on criminalistics laboratories in the United States. There are huge costs associated with education and training of lab personnel, with building forensic laboratories, with purchasing major equipment and supplies, and with maintaining and improving quality control and quality assurance in the laboratory. Improvements in funding have certainly been a blessing for DNA-related forensic science. With all of these advancements in science and technology and with better funding for crime laboratories, we must still ask about the future of DNA technology.

Without a crystal ball we can only guess at what the near future will bring to forensic science. In the laboratory, the trend will be toward automation and robotic systems that are capable of reducing the labor involved in testing. Laboratory automation workstations are already in use in some large forensic laboratories. They have been used primarily for DNA extraction from database samples, but it is likely that the same instrumentation will be used to extract DNA from authentic casework specimens. Given the trend of miniaturization of analytical devices, it is likely that in the future, crime scene investigators will be able to perform DNA analysis of evidence found at the scene using portable equipment. The analysis will be done rapidly, and the determined genetic profile(s) will be submitted to the state's main DNA laboratory via encrypted telecommunications lines to query the offender database. This could result in the development of a suspect early in the investigation conducted by law enforcement. It is also likely that the exciting new technology known as SNP analysis will become more favored over time and may not just complement the existing tests but may even supplant the PCR-STR analysis that is commonly in use today. The ability to determine the gender of the source of DNA evidence was a remarkable advancement resulting from a knowledge of the amelogenin locus. In the future, it is very likely that analysts will be able to determine the ethnicity of the individual as well as the gender. A company named DNA Print Genomics, Inc., provides its customers with information regarding their "ancestry proportions" or admixture ratios. Most individuals have mixed racial backgrounds. The results of SNP testing using their Ancestry kit, indicate an individual's ancestry proportions in terms of Indo-European, sub-Saharan African, Native American, and Asian. They test a number of "population-specific alleles" that can be identified using SNP detection. Perhaps, it may one day be possible to test DNA to determine one's physical appearance.

Another likelihood is that more individuals will find their DNA profiles added to the national DNA database. It is likely that legislation will be passed that will require juveniles who commit serious crimes to be included in the database. It is possible that eventually every citizen will be required to have his or her profile in a national database despite concerns about privacy issues and constitutional protections. As the database grows, it is more likely that repeat offenders will be caught.

While funding of DNA-related programs is welcome and extraordinarily beneficial, it is hoped that there will be renewed interest in improving the many other disciplines and technologies of forensic science, for while DNA has advanced rapidly over the past two decades, many of the other technologies have not. It is also hoped that more attention will be paid to the education of individuals seeking employment in crime laboratories, continuing education and training for those who work at crime scenes and for those who already work in crime labs, and that more money for forensic-related research will be provided. In this way, science and technology can better be used to solve the problem of crime in our society.[41]

IMPORTANT CASES INVOLVING THE USE OF DNA IN THE COURTROOM

United States v. Jakobetz, 955 F. 2d at 786, 791, 799–800 (2d Cir. 1992), Cert. denied 506 U.S. 834 (1992)

United States v. Shea, 957 F. Supp. 331, 333, 345 (D.N.H. 1997)

Aff'd, 159 F. 3d 37 (1998), Cert. denied, 526 U.S. 1077 (1999)

Daubert v. Merrill Dow Pharmaceuticals, 509 U.S. 579 (1993)

Federal Rules of Evidence, Rule 702

[41]*See* V. W. Weedn, and J. Hicks, *The Unrealized Potential of DNA Testing*. U.S. Department of Justice, Office of Justice Programs, National Institute of Justice, Washington, D.C., (1998).

United States v. Hicks, 103 F. 3d 837, 844-5 (9th Cir. 1996), Cert denied, 520 U.S. 1193 (1997)

United States v. Beasley, 102 F. 3d 1440, 1445, 1447 (8th Cir. 1996), Cert denied, 520 U.S. 1246 (1997)

United States v. Trala, 162 F. Supp. 2d 336, 341-342 (U.S.D.C. 2001)

Harvey v. Horan, 285 F. 3d 398, 305 n.1 (4th Cir. 2002)

United States v. Cuff, 37 F. Supp 2nd 279, 282 (S.D.N.Y. 1999)

United States v. Wright, 215 F. 3d 1020, 1027, (9th Cir. 2000)

United States v. Gains, 979 F. Supp 1429, 1433 at n.4 and 1435 (S.D. Fla. 1997)

People v. Owens, 187 Misc. 2d 838 (N.Y. Sup. Ct. 2001)

People v. Kelly, 288 AD2d 695, 732 NYS2d 484 (3d Dept. 2001)

Missouri v. Salmon, 89 S.W. 3d 540 (Missouri Court of Appeals, 2002)

People v. Allen, 72 Cal. App. 4th 1093 (1999)

People v. Shreck, 22 P. 3d 68, 83 (Colo. 2001)

Lemour v. State, 802 So. 2d 402, 408 (Fla. Dist. Ct. App. 2001)

Commonwealth v. Rosier, 425 Mass. 807 (1997)

State v. Butterfield, 27 P. 3d 1133, 1143 (Utah 2001)

ARTICLES ABOUT LEGAL ASPECTS OF DNA TESTING

DNA Database Searches and the Legal Consumption of Scientific Evidence. *Michigan Law Review*, February 1999 Vol. 97, No. 4 pp. 931–984.

Commonwealth v. Joseph O'Dell: Truth and Justice or Confuse the Courts? The DNA Controversy. *New England Journal on Criminal and Civil Confinement*, Winter 1999 Vol. 25, No. 1 pp. 311–331.

Genetic Justice: A Lawyer's Guide to the Science of DNA Testing. *Illinois Bar Journal*, January 1999 Vol. 87, No. 1 pp. 18–26.

The DNA Database: Civil Liberty and Evidentiary Issues. *The Criminal Law Review*, July 1998 pp. 437–454.

Juries and Crime Labs: Correcting the Weak Links in the DNA Chain. *Journal of Law and Medicine*, 1998 Vol. 24, Nos. 2 & 3 pp. 345–363.

A Comeback for Hair Evidence. *ABA Journal*, May 1998 pp. 66–69.

Defense Access to State-Funded DNA Experts: Considerations of Due Process. *California Law Review*, December 1997 Vol. 85, No. 6 pp. 1803–1839.

A Second Chance for Justice: Illinois' Post-Trial Forensic Testing Law. *Judicature*, November/December 1997 Vol. 81, No. 3 pp. 114–117.

DNA Identification Act of 1994. P.L. 103-322, 108 STAT. 2071 sections 210301 to 210306.

Genes and Justice. *Judicature*, November–December 1999, Vol. 83, No. 3.

Tough Luck for the Innocent Man. *ABA Journal*, March 1999 pp. 46–52.

Appendix A

Bibliography: Selected by Topic Area

DNA

Elaine Johnson Mange and Arthur P. Mange, *Basic Human Genetics*, 2d ed., 1998. Sinaver Publishing, Sunderland, MA

Forensic DNA Analysis

John M. Butler, *Forensic DNA Typing: Biology and Technology Behind STR Markers*, 2001. Academic Press, New York

Ian W. Evett and Bruce Weir, *Interpreting DNA Evidence: Statistical Genetics for Forensic Scientists*, 1998. Sinaver Associates, Sunderland, MA

National Research Council Committee on DNA Forensic Science: *An Update, The Evaluation of Forensic DNA Evidence*, 1996. National Academy Press, Washington D.C.

National Research Council Committee on DNA Technology in Forensic Science, *DNA Technology in Forensic Science*, 1992. National Academy Press, Washington D.C.

Grand Jury Law and Procedures

Sara Sun Beale, William C. Bryson, James E. Felman, and Michael J. Elston, *Grand Jury Law and Practice*, West, 2002. Eagan New York

DNA: Forensic and Legal Applications, by Lawrence Kobilinsky, Thomas F. Liotti, and Jamel Oeser-Sweat
ISBN 0-471-41478-6

Jury Selection

Cathy E. Bennett and Robert B. Hirschhorn, *Bennett's Guide to Jury Selection and Trial Dynamics*, West, 1995. Eagan MN

Ted A. Donner and Richard K. Gabriel, *Jury Selection Strategy & Science*, 3rd. ed., West, 2002. Eagan MN

Trial Techniques

Alfred S. Julien and Dominic Gianna, *Opening Statements*: Winning in the Beginning by Winning The Beginning, Clark Boardman Callaghan, 2004. Eagan MN

Roberto Aron and Jonathan L. Rosner, *How to Prepare Witnesses for Trial*, 2nd ed., West, 1998. Eagan MN

Mark A. Dombroff, *Dombroff on Unfair Tactics*, Aspen, 2001. Aspen Publishing Co. N.Y.

Evidence and Discovery

Thomas A. Mauet, *Pretrial*, 5th ed., Aspen Publishing, New York, 2002.

Jane Campbell Moriarty, *Psychological and Scientific Evidence in Criminal Trials*, West, Eagan MN 1996

Joe S. Cecil Carol E. Drew Marie Cordisco Dean P. MileTich *Reference Manual on Scientific Evidence*, 2nd ed., Federal Judicial Center, West Group. Eagan MN 1994

Paul C. Giannelli and Edward J. Imwinkelried, Scientific Evidence Vols I & II, 3rd ed., Michie (now Lexis Law Publishing) Miamisburg, Ohio 2000

David L. Faigman, David H. Kaye, Michael J. Saks, and Joseph Sanders. *Modern Scientific Evidence—The Law and Science of Expert Testimony*, Vol. 3, West, 2002. Eagan MN

Postconviction Remedies and Considerations

Larry W. Yackle, *Postconviction Remedies*, West, 2002. Eagan MN

Josephine R. Potuto, *Prisoner Collateral Attacks: Federal Habeas Corpus and Federal Motion Practice*, 1991 Publisher: Deerfield, IL

Lissa Griffin, *Federal Criminal Appeals*, West, Eagan MN 1991

Ira P. Robbins, *Habeas Corpus Checklists*, West, Eagan MN 1999

John M. Burkoff and Hope Hudson, *Ineffective Assistance of Counsel*, West, Eagan MN 1993

Randy Hertz and James S. Liebman, *Federal Habeas Corpus Practice and Procedure*, Lexis-Nexis, 2002. 4th ed Dayton, Ohio

Chester J. Antieau, *The Practice of Extraordinary Remedies: Habeas Corpus and The Other Common Law Writs*, 1987. Oceana, New York

Appendix B

Cases Involving the Admissibility of DNA Evidence

DECISIONS THAT GOVERN OR EXPLAIN THE ADMISSIBILITY OF SHORT TANDEM REPEAT (STR) TESTING

State Courts

Arizona

State v. Lynch, No. CR 98-11390 (Ariz. Super. Ct. 1999)

California

People v. Baylor, No. INFO29736 (Cal. Super Ct. 2000)
People v. Elizarraras, No. 50651 (Cal. Super Ct. 2000)
People v. Hackney, No. 97F02466 (Cal. Super Ct. 1999)
People v. Hill, No. 232982 (Cal. Super. Ct. 2000)
People v. Hunt, No. SA034500 (Cal. Super Ct. 2000)
People v. Moveo, No. 168277 (Cal. Super. Ct. 2000)
People v. Allen, 85 Cal. Reprt.2d 655 (Cal. App. 2 Dist., 1999)
People v. Bertsch and Hronis, No. 94F07255 (Cal. Super. Ct. 1999)
People v. Bokin, No. 168461 (Cal. Super Ct. 1999)

Colorado

Shreck v. People, 00SA105 (Co. Sup. Ct. 2001)
Flores v. People, 99 CR2022 (Co. Dist. Ct. 2000)
Shreck v. People, 98 CR 2475 (Co. Dist. Ct. 2000)

DNA: Forensic and Legal Applications, by Lawrence Kobilinsky, Thomas F. Liotti, and Jamel Oeser-Sweat
ISBN 0-471-41478-6

Delaware

State v. Roth, No. 9901000330 (Del. Super. Cr. 2000)

Florida

Yisrael v. State, No. 99-20176CF10A (Fl. Dist. Ct. 2000)

Maryland

Commonwealth v. Gaynor, No. 98-0965-0966 (Mass. Super. Ct. 2000)
Commonwealth v. Rosier, 685 N.E.2d 739, 743 (Mass. Super. Ct., 1997)

Michigan

People v. Calvin, No. 00-4395-FY (Mich. Dist. Ct. 2000)
People v. Phillips, No. 00-02025-FC (Mich. Dist. Ct. 2000)
People v. Kopp et al., No. 00-04014-FH (Mich. Dist. Ct. 2000)

Minnesota

State v. Kirkendahl, No. 00044987 (Minn. Dist. Ct. 2001)
State v. Dishmon, No. 99047345 (Minn. Dist. Ct. 2000)

Mississippi

State v. Staples, No. CR1999-03841 (Mo. Dist. Ct. 2000)
State v. Boyd, No. 991-3613 (Mo. Dist. Ct. 2000)

Nebraska

State v. Champ, No. A-00-617 (Neb. App., 2001)
State v. Jackson, No. S-97-522. 582 N.W.2d 317, 325 (Neb. App., 1998)

New York

RFLP

People v. Wesley, 83 N.Y.2d 417, 611 N.Y.S.2d 97, 633 N.E.2d 451 (1994) (RFLP sufficiently reliable to be admissible)

PCR

People v. Owens, No. IND 547/99 (N.Y. Sup. Ct. 2001)
People v. Hamilton, 255 A.D.2d 693, 681 N.Y.S.2d 117 (3d Dept. 1998) (PCR evidence admissible under *Frye* subject to appropriate chain-of-custody evidence.)

Rhode Island

State v. Motyka, No. N1-1999-0341A (R.I. Super. 2001)

Utah

State v. Butterfield, No. 990654 (2001 UT 59) Supreme Court

mitochindrial DNA analysis and description makes the inference that the results of such DNA analysis were considered.)

State v. Williams, Cir Ct, Anne Arundel County, Md. (May 6, 1998, North, J., *affd* Ct Spec App, Md, Apr. 12, 2000.)

Michigan

People v. Holtzer, 2003 Mich. App. LEXIS 523 (Mich. Ct. App. Feb. 25, 2003). (Affirming the judgment of the lower court and holding that mtDNA evidence was properly admitted.)

Mississippi

Adams v. State, 794 So. 2d 1049, 2001 Miss. App. LEXIS 170 (Miss. Ct. App. 2001). (mtDNA was admissible at trial; see concurring opinion on the admissibility of mtDNA.)

New York

People v. Wise, 752 N.Y.S.2d 837, 2002 N.Y. Misc. LEXIS 1708 (N.Y. Sup. Ct. 2002). (Court granted defendants' motion to vacate judgment based upon newly discovered mtDNA evidence that established that only the DNA of a nonparty to the case and not that of the defendants was present in the samples taken from the victim or at the crime scene.)

People v. Klinger, 185 Misc. 2d 574, 713 N.Y.S.2d 823, 2000 N.Y. Misc. LEXIS 390 (N.Y. County Ct. 2000). (Holding that evidence adducted at a *Frye* hearing established that mtDNA analysis and interpretations were generally accepted as reliable and that mtDNA evidence was admissible.)

North Carolina

State v. Underwood, 134 N.C. App. 533, 518 S.E.2d 231, 1999 N.C. App. LEXIS 856 (1999). (Holding that mtDNA was sufficiently reliable to warrant its admissibility into evidence.)

Pennsylvania

Commonwealth v. Dillon, Ct Common Pleas, Lackawanna County, Pa, Jan. 28, 1998.

Commonwealth v. Rorrer, Ct Common Pleas, Lehigh County, Pa, Jan. 20, 1998, *affd 748 A2d 776, 1999 Pa Super LEXIS 3913.*

South Carolina

State v. Council, 335 S.C. 1, 515 S.E.2d 508, 1999 S.C. LEXIS 76 (S.C. 1999). (Holding that mtDNA evidence was admissible.)

Tennessee

State v. Ware, 1999 Tenn. Crim. App. LEXIS 370 (Tenn. Crim. App. Apr. 20, 1999). (Admissibility of mtDNA was within the lower court's discretion and was properly admitted.)

State v. Scott, 1999 Tenn. Crim. App. LEXIS 758, 1999 WL 547460 (Tenn. Crim. App. July 28, 1999), reversed in part on other grounds, 33 S.W.3d 746 (Tenn. 2000). (Reversing for new trial based on trial court's failure to approve DNA expert assistance for the defense.)

State v. Dean, 76 S.W.3d 352, 2001 Tenn. Crim. App. LEXIS 791 (Tenn. Crim. App. 2001). (Holding that mtDNA was properly admitted and that chain of custody was established.)

Texas

Sheckells v. State, 2001 Tex. App. LEXIS 6730 (Tex. App. Dallas Oct. 8, 2001). (Affirming the admission of testimony regarding mitochondrial DNA evidence at trial on the grounds that "evidence was to be viewed in the light most favorable to the trial court's decision and, absent an abuse of discretion" the trial court's decision would not be disturbed.)

Washington

State v. Smith, 2000 Wash. App. LEXIS 795 (Wash. Ct. App. May 26, 2000). [Noting that the trial court found mtDNA admissible, generally upholding the evidentiary rulings by the trial court and stating that "the admission of evidence of mitochondrial DNA (mtDNA), if error, was harmless because evidence of nuclear DNA showed that the victim could not have been excluded as the donor of blood stains found on Smith's watch and in his truck".]

DECISIONS DESCRIBING THE TESTING OF ANIMAL DNA

Alabama

United States v. Guthrie, No. 93 6508. 50 F.3d 936, 944 (11th Cir. 1995).

Iowa

U.S. v. Boswell, No. 00-4005.—F.3d—, 2001 WL 1223128 (8th Cir. Iowa 2001). (PCR testing conducted on swine DNA held to be admissible.)

Vermont

State v. Demers, No. 96-452. 167 Vt. 349, 707 A.2d 276 (Vt., 1997). (Results from DNA analysis of a deer were used to obtain a search warrant.)

DECISIONS DESCRIBING THE TESTING OF PLANT DNA AND VIRAL DNA

Arizona

State v. Bogan, 905 P.2d 515, 519-20 (Ariz. Ct. App. 1995), *app. dismissed*, 920 P.2d 320 (Ariz. 1996).

California

State v. Schmidt, N. 99-1412. 771 So.2d 131 (La.App. 3 Cir., 2000). (DNA testing of HIV virus ruled admissible in an attempted murder case.)

DECISIONS IN WHICH COURTS TAKE DNA STATISTICS INTO ACCOUNT

Alaska

Dayton v. State, 2002 WL 1352496 (Alaska App., 2002).

Arizona

Arizona v. Garcia, 197 Ariz. 79, 3 P.3d 999 (Ariz. App. Div. 1999). (Holding that DNA mixture calculations were admissible under the *Frye* standard.)

California

People v. Funston, No. C032472. (Cal.App. 3 Dist. 2002). (Holding that error-rate statistics were not admissible.)

Connecticut

Connecticut v. Sivri, 646 A.2d 169 (Conn. 1994).

Florida

Butler v. State, No. SC95158. 2002 WL 926283 (Fla., 2002).

Darling v. State, No. SC94691. 808 So.2d 145, 27 Fla. L. Weekly S41 (Fla., 2002).

Brim v. Florida, 2000 WL 1568741 (Fla.App. 2 Dist. 2000).

Iowa

State v. Belken, No. 99-2001. 633 N.W.2d 786 (Iowa, 2001).

State v. Williams, 574 N.W.2d 293 (Iowa, 1998). (Holding trial court erred in admitting matching evidence without statistics.)

Massachusetts

Massachusetts v. Curnin, 565 N.E.2d 440 (Mass. 1991).

Minnesota

Minnesota v. Thomas, 2002 WL 142074 (Minnesota Court of Appeals 2002).

Mississippi

Hull v. State, 687 So. 2d 708 (Miss. 1996). (Holding that trial court erred in admitting matching DNA evidence without statistics.)

Nebraska

Nebraska v. Carter, 524 N.W.2d 763 (Neb. 1994). (The racial group of the perpetrator was unknown and court held limiting evidence on statistical frequency to two racial groups was prejudicial.)

New Hampshire

State v. Vandebogart, 616 A.2d 483 (N.H. 1992).

Pennsylvania

Com. v. Jones, 2002 WL 31630346 (Pa.Super., 2002).

Virginia

Hills v. Com., 33 Va.App. 442, 534 S.E.2d 337 (Va.App., 2000) (No. 0367-99-4).

Washington

Washington v. Cauthron, 846 P.2d 502 (Wash. 1993).

DECISIONS GOVERNING OR EXPLAINING MITOCHONDRIAL DNA TESTING (MTDNA)

Federal Courts

Missouri

United States v. Coleman, 202 F.Supp. 2d 962, 2002 U.S. Dist. LEXIS 12563 (E.D. Mo. 2002). [Citing United States v. Beasley, 102 F.3d 1440, 1445 (8th Cir. 1996) (PCR) and *United States v. Martinez*, 3 F.2d 1191 (8th Cir. 1993) (RFLP) for the proposition that the Eighth Circuit has "taken judicial notice of reliability of the general theory and techniques of DNA profiling . . .". The Court allowed the admission of mitochondrial DNA Analysis.]

Texas

Blair v. Johnson, 2001 U.S. Dist. LEXIS 6149 (W.D. Tex. Apr. 19, 2001). (Court ordered mitochondrial DNA testing in a postconviction/habeas corpus proceeding and evidence appeared to exclude petitioner as a donor of hair samples that were characterized at trial as belonging to him.)

Virginia

Cherrix v. True, 205 F. Supp. 2d 525, 2002 U.S. Dist. LEXIS 10217 (E.D. Va. 2002). (Postconviction/habeas corpus proceeding in which the court adopted the petitioner's DNA testing plan which set forth the terms of DNA testing including mitochondrial DNA testing. No indication of admissibility.)

State Courts

Arkansas

Ware v. State, 348 Ark. 181, 75 S.W.3d 165, 2002 Ark. LEXIS 188 (Ark. 2002). (Describing the admission of mitochondrial DNA evidence in trial at issue.)

Connecticut

State v. Pappas, 256 Conn. 854, 776A.2d 1091, 2001 Conn. LEXIS 284 (2001). (Trial court did not abuse its discretion in denying defendant's motion to exclude mtDNA evidence on the basis that such evidence was sufficiently reliable under the *Daubert* standard.)

State v. Grant, 2002 Conn. Super. LEXIS 1127 (Conn. Super. Cr. Apr. 9, 2002). [Noting that "Pappas holds that mitochondrial DNA ("mtDNA") evidence satisfies the validity requirement of Porter."]

Delaware

Vanlier v. State, 813 A.2d 1142, 2002 Del. LEXIS 793 (Del. 2002). [Noting that "the State sought a Daubert hearing to validate mitochondrial DNA as a relevant and credible form of identification (and) (i)ronically, the test the State wished validated . . . resulted in an expert opinion helpful to the defense."]

State v. Hammons, 2001 Del. Super. LEXIS 545 (Del. Super. Ct. Sept. 19, 2001). (Holding that mtDNA evidence was scientifically reliable and relevant.)

Florida

Magaletti v. State, 2003 Fla. App. LEXIS 4511, 28 Fla. L. Weekly D 884 (Fla, Dist. Ct. App. 2d Dist. Apr. 4, 2003). (Affirming the trial court's determination that mtDNA was admissible.)

Illinois

People v. Kliner, 2002 Ill. LEXIS 1641 (Ill. Dec. 19, 2002). (Opinion notes that trial court ordered that mitochondrial DNA testing be performed on evidence in a postconviction proceeding.)

Indiana

Anderson v. State, 718 N.E.2d 1101, 1999 Ind. LEXIS 1029 (Ind. 1999). (Court describes mitochondrial DNA analysis in facts. It is possible that mtDNA was indeed admitted at the underlying trial.)

Maryland

Clark v. State, 140 Md. App. 540, 781 A.2d 913, 2001 Md. App. LEXIS 156 (2001). (Opinion describes the fact that defense witness was an expert in mitochondrial DNA analysis and description makes the inference that the results of such DNA analysis were considered.)

Michigan

People v. Holtzer, 2003 Mich. App. LEXIS 523 (Mich. Ct. App. Feb. 25, 2003). (Affirming the judgment of the lower court and holding that mtDNA evidence was properly admitted.)

Mississippi

Adams v. State, 794 So. 2d 1049, 2001 Miss. App. LEXIS 170 (Miss. Ct. App. 2001). (mtDNA was admissible at trial; see concurring opinion on the admissibility of mtDNA.)

New York

People v. Wise, 752 N.Y.S.2d 837, 2002 N.Y. Misc. LEXIS 1708 (N.Y. Sup. Ct. 2002). (Court granted defendants' motion to vacate judgment based upon newly discovered mtDNA evidence that established that only the DNA of a nonparty to the case and not that of the defendants was present in the samples taken from the victim or at the crime scene.)

People v. Klinger, 185 Misc. 2d 574, 713 N.Y.S.2d 823, 2000 N.Y. Misc. LEXIS 390 (N.Y. County Ct. 2000). (Holding that evidence adducted at a *Frye* hearing established that mtDNA analysis and interpretations were generally accepted as reliable and that mtDNA evidence was admissible.)

North Carolina

State v. Underwood, 134 N.C. App. 533, 518 S.E.2d 231, 1999 N.C. App. LEXIS 856 (1999). (Holding that mtDNA was sufficiently reliable to warrant its admissibility into evidence.)

South Carolina

State v. Council, 335 S.C. 1, 515 S.E.2d 508, 1999 S.C. LEXIS 76 (S.C. 1999). (Holding that mtDNA evidence was admissible.)

Tennessee

State v. Ware, 1999 Tenn. Crim. App. LEXIS 370 (Tenn. Crim. App. Apr. 20, 1999). (Admissibility of mtDNA was within the lower court's discretion and was properly admitted.)

State v. Scott, 1999 Tenn. Crim. App. LEXIS 758, 1999 WL 547460 (Tenn. Crim. App. July 28, 1999), reversed in part on other grounds, 33 S.W.3d 746 (Tenn. 2000). (Reversing for new trial based on trial court's failure to approve DNA expert assistance for the defense.)

State v. Dean, 76 S.W.3d 352, 2001 Tenn. Crim. App. LEXIS 791 (Tenn. Crim. App. 2001). (Holding that mtDNA was properly admitted and that chain of custody was established.)

Texas

Sheckells v. State, 2001 Tex. App. LEXIS 6730 (Tex. App. Dallas Oct. 8, 2001). (Affirming the admission of testimony regarding mitochondrial DNA evidence at trial on the grounds that "evidence was to be viewed in the light most favorable to the trial court's decision and, absent an abuse of discretion" the trial court's decision would not be disturbed.)

Washington

State v. Smith, 2000 Wash. App. LEXIS 795 (Wash. Ct. App. May 26, 2000). (Noting that the trial court found mtDNA admissible, generally upholding the evidentiary rulings by the trial court and stating that "the admission of evidence of mitochondrial DNA (mtDNA), if error, was harmless because evidence of nuclear DNA showed that the victim could not have been excluded as the donor of blood stains found on Smith's watch and in his truck.")

Appendix C

Information Pertinent to Attempts to Overturn Convictions Based Upon DNA Evidence

INNOCENCE PROJECTS

The following list contains entities that may provide assistance to those who are seeking to have their convictions overturned because they are innocent, were in no way involved with the crimes for which they were convicted and seek to prove their innocence using DNA evidence.

Foreign Innocence Projects

Australia

Australian Innocence Project
Lynne Weathered, Director
Griffith University Innocence Project
Griffith Law School
GU PMB 50
GCMC 9726
Australia

DNA: Forensic and Legal Applications, by Lawrence Kobilinsky, Thomas F. Liotti, and Jamel Oeser-Sweat
ISBN 0-471-41478-6

UTS Innocence Project
Kirsten Edwards
Faculty of Law
PO Box 123
Broadway NSW 2007
Sydney, Australia

National Innocence Projects

Centurion Ministries
221 Witherspoon Street
Princeton, New Jersey 08542
*Only handles cases in which a prisoner has been sentenced to either death or life in prison without parole, cases in which a prisoner has exhausted most or all appeals, and cases in which a prisoner is claiming absolute innocence—no self-defense or accidental death cases.

Innocence Project
Benjamin N. Cardozo School of Law
55 Fifth Avenue, 11th Floor
New York, New York 10003
*Only handles cases in which physical or biological evidence could prove innocence.

State Innocence Projects

Alabama

University of Houston Law Center Innocence Network
100 Law Center
Houston, Texas 77204-6371

Alaska

Innocence Project Northwest
University of Washington Law School
1100 NE Campus Parkway
Seattle, Washington 98105

Arizona

Justice Project of Arizona
Larry A. Hammond
Osbord Maledon, P.A.
2929 North Central Avenue, Suite 2100
Phoenix, Arizona 85012-2794

Professor Andy Silverman
University of Arizona School of Law
Tucson, AZ 85721

Professor Robert Bartels
Arizona State University College of Law
Tempe, AZ 85287

Arizona Attorneys for Criminal Justice
3737 North Seventh Street, Suite 105
Phoenix, AZ 85014

Arkansas

Midwest Innocence Project
UMKC School of Law
5100 Rockhill Road
Kansas City, MO 64110
Contact: Professor Ellen Suni

James E. Hensley
Attorney at Law
P.O. Box 639
Cabot, AR 72023

California

California Innocence Project
California Western School of Law
225 Cedar Street
San Diego, California 92102
Contact: Justin Brooks

Northern California Innocence Project
874 Lafayette Street
Santa Clara, CA 92102
Contact: Professor Linda Starr
Professor Cookie Ridolfi
Professor Mary Likins

Professor Robert Weisberg
Stanford University Law School
Nathan Abbot Way at Alvarado Row
Stanford, CA 94305

Professor Susan Rutberg
Golden Gate University School of Law
536 Mission Street
San Francisco CA 94105-2968

Colorado

Jim Scarboro
Arnold & Porter
1700 Lincoln Street, Suite 4000
Denver, Colorado 80203

Connecticut

New England Innocence Project
Testa Hurwitz & Thiebeault
125 High Street
Boston, Massachusetts 02110
*Only handles cases in which physical or biological evidence could prove innocence.

Delaware

Innocence Project of the National Capital Region
American University, Washington College of Law
4801 Massachusetts Avenue, N.W.
Washington, D.C. 20016

Lisa Schwind
Office of the Public Defender
Carvel State Building
820 French Street, 3rd Floor
Wilimington, DE 19801

District of Columbia

Innocence Project of the National Capital Region
American University, Washington College of Law
4801 Massachusetts Avenue, N.W.
Washington, D.C. 20016

Florida

Professor Catherine Arcabascio
Nova Southeastern University
Shepard Broad Law Center
Fort Lauderdale, Florida 33314

Georgia

Aimee Maxwell, Executive Director
Georgia Innocence Project
730 Peachtree Street, N.E.
Suite 705
Atlanta, GA 30308

Hawaii
No Innocence Project

Idaho
Innocence Project Northwest
University of Washington Law School
1100 NE Campus Parkway
Seattle, Washington 98105

Diana Cavigliano
Idaho Innocence Project
P.O. Box 442321
Moscow, ID 83844-2321

Illinois
Center on Wrongful Convictions
Northwestern University School of Law
357 East Chicago Avenue
Chicago, Illinois 60611

Indiana
Frances W. Hardy
Law School Clinic
Indiana University School of Law at Indianapolis
735 West New York Street
Indianapolis, Indiana 46202

Iowa
Professor Ellen Suni
Wisconsin Innocence Project
975 Bascom Mall
Madison, Wisconsin 53706
Midwest Innocence Project
UMKC School of Law
5100 Rockhill Road
Kansas City, MO 64110

Kansas
Professor Ellen Suni
Midwest Innocence Project
UMKC School of Law
5100 Rockhill Road
Kansas City, MO 64110

Kentucky

University of Houston Law Center Innocence Network
100 Law Center
Houston, Texas 77204-6371

Roberta M. Harding, Professor of Law
University of Kentucky Innocence Project Externship
University of Kentucky College of Law
209 Law Building
Lexington, KY 40506-0048

Gordon Rahn
Project Coordinator
Kentucky Innocence Project
Department of Public Advocacy
P.O. Box 555
Eddyville, KY 42038

Louisiana

Emily Bouton
Innocence Project New Orleans
636 Baronne Street
New Orleans, LA 70113
*Handles only cases arising in Orleans or Jefferson Parishes.

Maine

New England Innocence Project
Testa Hurwitz & Thiebeault
125 High Street
Boston, Massachusetts 02110
*Only handles cases in which physical or biological evidence could prove innocence.

Maryland

Innocence Project of the National Capital Region
American University, Washington College of Law
4801 Massachusetts Avenue, N.W.
Washington, D.C. 20016

Massachusetts

New England Innocence Project
Testa Hurwitz & Thiebeault
125 High Street
Boston, Massachusetts 02110
*Only handles cases in which physical or biological evidence could prove innocence.

Michigan

Thomas M. Cooley Law School Innocence Project
200 South Capital Avenue
Lansing, Michigan 48901

Minnesota

Erika Applebaum
Innocence Project of Minnesota
Hamline University School of Law
1536 Hewitt Avenue
St. Paul, MN 55104

Wisconsin Innocence Project
975 Bascom Mall
Madison, Wisconsin 53706

Mississippi

University of Houston Law Center Innocence Network
100 Law Center
Houston, Texas 77204-6371

Missouri

Professor Ellen Suni
Midwest Innocence Project
UMKC School of Law
5100 Rockhill Road
Kansas City, MO 64110

Montana

Innocence Project Northwest
University of Washington Law School
1100 NE Campus Parkway
Seattle, Washington 98105

Rocky Mountain Innocence Center
358 South 700 East, B235
Salt Lake City, Utah 84102

Nebraska

Professor Ellen Suni
Midwest Innocence Project
UMKC School of Law
5100 Rockhill Road
Kansas City, MO 64110

Nevada

Rocky Mountain Innocence Center
358 South 700 East, B235
Salt Lake City, Utah 84102

New Hampshire

New England Innocence Project
Testa Hurwitz & Thiebeault
125 High Street
Boston, Massachusetts 02110
*Only handles cases in which physical or biological evidence could prove innocence.

New Jersey

Centurion Ministries
221 Witherspoon Street
Princeton, New Jersey 08542
*Only handles cases in which a prisoner has been sentenced to either death or life in prison without parole, cases in which a prisoner has exhausted most or all appeals, and cases in which a prisoner is claiming absolute innocence—no self-defense or accidental death cases.

Rutgers University School of Law Innocence Project
Rutgers University Law School
Innocence Project for Justice Constitutional Litigation Clinic
123 Washington Street
Newark, NJ 07102

New Mexico

Professor April Land
Professor Barbara Bergman
New Mexico Innocence and Justice Project
University of New Mexico School of Law
1117 Stanford NE
Albuquerque, NM 87131

Rocky Mountain Innocence Center
358 South 700 East, B235
Salt Lake City, Utah 84102

New York

Innocence Project
Benjamin N. Cardozo School of Law
55 Fifth Avenue, 11th Floor
New York, New York 10003
*Only handles cases in which physical or biological evidence could prove innocence.

New York State Defenders Association
New York State Defenders Association
194 Washington Street, Suite 500
Albany, NY 12210

Professor Will Hellerstein
Daniel Hedwed
Second Look Program
Brooklyn Law School
250 Joralemon Street
Brooklyn, NY 11201

North Carolina

North Carolina Center on Actual Innocence
P.O. Box 52446
Shannon Plaza Station
Durham, North Carolina 27727-2446

Professor Therea Newman
Professor Rich Rosen
Pete Weitzel, Executive Director
Chris Mumma, Legal Counsel
North Carolina Center on Actual Innocence
Duke/University of North Carolina Law Schools
Duke University School of Law
Durham, NC 27708-0360

North Dakota

No Innocence Project

Ohio

Professor Bryan Ward
Ohio Northern University
College of Law
525 S. Main Street
Ada, Ohio 45810

Martin Yant
P.O. Box 14306
Columbus, Ohio 43214

William Gallagher
Ohio Innocence Project
Ohio Association of Criminal Defense Lawyers
2320 Chagrin Boulevard, Suite 525
Cleveland, OH 44122

Oklahoma

University of Houston Law Center Innocence Network
100 Law Center
Houston, Texas 77204-6371

Oklahoma Indigent Defense Systems
1660 Cross Center
Norman, Oklahoma 73019
*Only handles cases in which physical or biological evidence could prove innocence.

Oregon

Innocence Project Northwest
University of Washington Law School
1100 NE Campus Parkway
Seattle, Washington 98105

Pennsylvania

Bill Moushey
The Innocence Institute of Western Pennsylvania
c/o Point Park College Department of Journalism and Mass Communications
201 Wood Street
Pittsburgh, PA 15222-1984

Rhode Island

New England Innocence Project
Testa Hurwitz & Thiebeault
125 High Street
Boston, Massachusetts 02110
*Only handles cases in which physical or biological evidence could prove innocence.

South Carolina

Innocence Project
Benjamin N. Cardozo School of Law
55 Fifth Avenue, 11th Floor
New York, New York 10003
*Only handles cases in which physical or biological evidence could prove innocence.

South Dakota

No Innocence Project

Tennessee

Tennessee Innocent Project
C/O UP Pro Bono
1505 W. Cumberland
Knoxville, TN 37996

Texas

Professor David Dow
University of Houston Law Center Innocence Network
100 Law Center
Houston, Texas 77204-6371

Catherine Greene Burnett
South Texas Innocence Project
South Texas College of Law
1303 San Jacinto
Houston, TX 77002

Joyce Ann Brown
MASS, Inc. (Mothers for the Advancement of Social Services)
6301 Gaston Avenue, Suite 300
Dallas, TX 75214

Utah

Rocky Mountain Innocence Center
358 South 700 East, B235
Salt Lake City, Utah 84102

Vermont

New England Innocence Project
Testa Hurwitz & Thiebeault
125 High Street
Boston, Massachusetts 02110
*Only handles cases in which physical or biological evidence could prove innocence.

Virginia

Innocence Project of the National Capital Region
American University, Washington College of Law
4801 Massachusetts Avenue, N.W.
Washington, D.C. 20016

Washington

Innocence Project Northwest
University of Washington Law School
1100 NE Campus Parkway
Seattle, Washington 98105

West Virginia

Innocence Project of the National Capital Region
American University, Washington College of Law
4801 Massachusetts Avenue, N.W.
Washington, D.C. 20016

Wisconsin

Wisconsin Innocence Project
975 Bascom Mall
Madison, Wisconsin 53706

Keith Findley, Associate Clinical Professor
John Pray, Associate Clinical Professor
Wisconsin Innocence Project
University of Wisconsin Law School
Remington Center
Madison, WI 53706

Wyoming

Rocky Mountain Innocence Center
358 South 700 East, B235
Salt Lake City, Utah 84102

INTERNET RESOURCES*: INNOCENCE PROJECTS

California Innocence Project: http://www.cwsl.edu/icda/i_Innocence.html

Center on Wrongful Convictions: http://www.law.northwestern.edu/wrongfulconvictions/

Equal Justice Initiative of Alabama: http://www.eji.org/

Georgia Innocence Project: http://www.ga-innocenceproject.org/

University of Houston Law: www.law.uh.edu/faculty/ddow2/dpage2/innocence.html

The Innocence Project: http://www.innocenceproject.org/

Innocence Project New Orleans: http://www.ip-no.org/

North Carolina Center on Actual Innocence: http://ncinnocencecenter.law.duke.edu/

Northern California Innocence Project: http://www.ncip.scu.edu/

*Information current as of February 5, 2003.

Innocence Project Northwest: http://www.law.washington.edu/ipnw/index.shtml

Thomas M. Cooley Innocence Project: http://www.cooleylaw.edu/innocence/home.htm

Truth in Justice: http://www.truthinjustice.org/

Wisconsin Innocence Project: http://www.law.wisc.edu/FJR/innocence/

Sources

Center on Wrongful Convictions
Northwestern University School of Law
357 East Chicago Avenue
Chicago, Illinois 60611
http://www.law.northwestern.edu/wrongfulconvictions/

Innocence Project
Benjamin N. Cardozo School of Law
55 5th Avenue, 11th Floor
New York, NY 10003
http://www.innocenceproject.org/

Appendix D

Offenses in New York State Resulting in Mandatory DNA Testing for Database Inclusion

A sample appropriate for DNA testing to be included in the New York State DNA Identification Index shall be provided by a person (1) convicted of and sentenced for any one of the following offenses committed on or after December 1, 1999, OR (2) convicted of and sentenced for any one of the following offenses committed prior to December 1, 1999 where service of the sentence imposed thereon has not been completed prior to December 1, 1999:

120.05—Assault in the 2d degree
120.06—Gang Assault in the 2d degree
110.00/120.06—Attempted Gang Assault in the 2d degree
120.07—Gang Assault in the 1st degree
110.00/120.07—Attempted Gang Assault in the 1st degree
120.08—Assault on a Peace, Police Officer, Fireman or EMS Professional
110.00/120.08—Attempted Assault on a Peace, Police Officer, Fireman or EMS Professional
120.10—Assault in the 1st degree
110.00/120.10—Attempted Assault in the 1st degree

DNA: Forensic and Legal Applications, by Lawrence Kobilinsky, Thomas F. Liotti, and Jamel Oeser-Sweat
ISBN 0-471-41478-6

120.11—Aggravated Assault upon a Peace or Police Officer
110.00/120.11—Attempted Aggravated Assault upon a Peace or Police Officer
120.60(1)—Stalking in the 1st degree
125.15—Manslaughter in the 2d degree
125.20—Manslaughter in the 1st degree
110.00/125.20—Attempted Manslaughter in the 1st degree
125.25—Murder in the 2d degree
110.00/125.25—Attempted Murder in the 2d degree
125.27—Murder in the 1st degree
110.00/125.27—Attempted Murder in the 1st degree
130.25—Rape in the 3rd degree
130.30—Rape in the 2d degree
130.35—Rape in the 1st degree
110.00/130.35—Attempted Rape in the 1st degree
130.40—Sodomy in the 3rd degree
130.45—Sodomy in the 2d degree
130.50—Sodomy in the 1st degree
110.00/130.50—Attempted Sodomy in the 1st degree
130.65—Sexual Abuse in the 1st degree
130.66—Aggravated Sexual Abuse in the 3rd degree
130.67—Aggravated Sexual Abuse in the 2d degree
110.00/130.67—Attempted Aggravated Sexual Abuse in the 2d degree
130.70—Aggravated Sexual Abuse in the 1st degree
110.00/130.70—Attempted Aggravated Sexual Abuse in the 1st degree
130.75—Course of Sexual Conduct Against a Child in the 1st degree
110.00/130.75—Attempted Course of Sexual Conduct Against a Child in the 1st degree
130.80—Course of Sexual Conduct Against a Child in the 2d degree
135.20—Kidnapping in the 2d degree
110.00/135.20—Attempted Kidnapping in the 2d degree
135.25—Kidnapping in the 1st degree
110.00/135.25—Attempted Kidnapping in the 1st degree
140.20—Burglary in the 3rd degree
110.00/140.20—Attempted Burglary in the 3rd degree
140.25—Burglary in the 2d degree
110.00/140.25—Attempted Burglary in the 2d degree
140.30—Burglary in the 1st degree
110.00/140.30—Attempted Burglary in the 1st degree

150.15—Arson in the 2d degree
110.00/150.15—Attempted Arson in the 2d degree
150.20—Arson in the 1st degree
110.00/150.20—Attempted Arson in the 1st degree
160.10—Robbery in the 2d degree
110.00/160.10—Attempted Robbery in the 2d degree
160.15—Robbery in the 1st degree
110.00/160.15—Attempted Robbery in the 1st degree
215.16—Intimidating a Victim or Witness in the 2d degree
215.17—Intimidating a Victim or Witness in the 1st degree
110.00/215.17—Attempted Intimidating a Victim or Witness in the 1st degree
240.55—Falsely reporting an incident in the 2d degree
240.60—Falsely reporting an incident in the 1st degree
240.61—Placing a False Bomb in the 2d degree
240.62—Placing a False Bomb in the 1st degree
240.63—Placing a False Bomb in a Sports Stadium or Arena, Mass Transportation Facility, or Enclosed Shopping Mall
255.25—Incest
265.02(4), (5), (6), (7), and (8)—Criminal Possession of a Weapon in the 3rd degree
110.00/265.02 (4), (5), (6), (7), and (8)—Attempted Criminal Possession of a Weapon in the 3rd degree as a lesser included offense of that section as defined in 220.20 of the Criminal Procedure Law
265.03—Criminal Possession of a Weapon in the 2d degree
110.00/265.03—Attempted Criminal Possession of a Weapon in the 2d degree
265.04—Criminal Possession of a Dangerous Weapon in the 1st degree
110.00/265.04—Attempted Criminal Possession of a Dangerous Weapon in the 1st degree
265.08—Criminal Use of a Firearm in the 2d degree
110.00/265.08—Attempted Criminal Use of a Firearm in the 2d degree
265.09—Criminal Use of a Firearm in the 1st degree
110.00/265.09—Attempted Criminal Use of a Firearm in the 1st degree
265.12—Criminal Sale of a Firearm in the 2d degree
110.00/265.12—Attempted Criminal Sale of a Firearm in the 2d degree
265.13—Criminal Sale of a Firearm in the 1st degree
110.00/265.13—Attempted Criminal Sale of a Firearm in the 1st degree
265.14—Criminal Sale of a Firearm with the Aid of a Minor
110.00/265.14—Attempted Criminal Sale of a Firearm with the Aid of a Minor
490.10—Soliciting or Providing Support for an Act of Terrorism in the 2d degree

490.15—Soliciting or Providing Support for an Act of Terrorism in the 1st degree
110/490.15—Attempted Soliciting or Providing Support for an Act of Terrorism in the 1st degree
490.20—Making a Terroristic Threat
490.30—Hindering Prosecution of Terrorism in the 2d degree
110/490.30—Attempted Hindering Prosecution of Terrorism in the 2d degree
490.35—Hindering Prosecution of Terrorism in the 1st degree
110/490.35—Attempted Hindering Prosecution of Terrorism in the 1st degree

A sample appropriate for DNA testing shall also be provided by a person convicted of and sentenced for any one of the following offenses on or after December 1, 1999:

220.18—Criminal Possession of a Controlled Substance in the 2d degree
220.21—Criminal Possession of a Controlled Substance in the 1st degree
220.31—Criminal Sale of a Controlled Substance in the 5th degree
220.34—Criminal Sale of a Controlled Substance in the 4th degree
220.39—Criminal Sale of a Controlled Substance in the 3rd degree
220.41—Criminal Sale of a Controlled Substance in the 2d degree
220.43—Criminal Sale of a Controlled Substance in the 1st degree
220.44—Criminal Sale of a Controlled Substance in or near School Grounds
155.30(5)—Grand Larceny in the 4th degree

A sample appropriate for DNA testing shall also be provided by a person (1) convicted of and sentenced for any one of the following offenses committed on or after December 1, 1999, OR (2) convicted of and sentenced for any one of the following offenses committed prior to December 1, 1999 where service of the sentence imposed thereon has not been completed prior to December 1, 1999, but in either case (1) or (2) only where the offender has been convicted within the previous five (5) years of one of the offenses listed previously:

205.10—Escape in the 2d degree
205.15—Escape in the 1st degree
205.17—Absconding from Temporary Release in the 1st degree
205.19—Absconding from a Community Treatment Facility

Source: NYS Division of Criminal Justice Services, Office of Forensic Services Revised: 02/26/2003.

Appendix E

Postconviction DNA Testing, Preservation of Evidence and Compensation for Wrongful Convictions: Relevant Legislative Information

Jurisdiction	*Area of Relevance*	Status of Legislation	Number of Bill/ Statute
Arizona	Postconviction DNA testing	Enacted	Ariz. Rev. Stat §13-4240 (2000)
Arkansas	Postconviction DNA testing; preservation of evidence	Enacted	Ark. Code Ann. §16-112-201, 205, and 207 (2001)

DNA: Forensic and Legal Applications, by Lawrence Kobilinsky, Thomas F. Liotti, and Jamel Oeser-Sweat
ISBN 0-471-41478-6

Jurisdiction	*Area of Relevance*	Status of Legislation	Number of Bill/ Statute
California	Postconviction DNA testing; preservation of evidence; compensation for wrongful conviction	Enacted	Cal. Penal Code §1405 (2000)
Delaware	Postconviction DNA testing; preservation of evidence	Enacted	Del. Code Ann. tit. 11 §4504
Federal	Postconviction DNA testing; preservation of evidence; compensation for wrongful conviction	Pending	Senate Bill 486
Federal	Postconviction DNA testing; preservation of evidence; compensation for wrongful conviction	Pending	H.R. 912 (The Innocence Protection Act of 2001)
Florida	Postconviction DNA testing	Enacted	Fla. Stat. Ch. 925.11 (2001)
Hawaii	Postconviction DNA testing; preservation of evidence	Pending	HB 42
Idaho	Postconviction DNA testing	Enacted	Idaho Code §19-4902 (b) (2001)
Illinois	Postconviction DNA testing; preservation of evidence	Enacted	725 Ill. Comp. Stat. §5/116-3 (1997)
Indiana	Postconviction DNA testing	Enacted	Ind. Code 35-38-7-(§1-19) (2001)

Jurisdiction	*Area of Relevance*	Status of Legislation	Number of Bill/ Statute
Iowa	Postconviction DNA testing; preservation of evidence	Pending	SF 229
Kentucky	Postconviction DNA testing; preservation of evidence	Enacted	Chapter 422 - Kentucky Revised
Louisiana	Postconviction DNA testing; preservation of evidence	Enacted	2001 La. Acts 1020 (2001)
Maine	Postconviction DNA testing; preservation of evidence	Enacted	2001 Me. Laws 469 (2001)
Maryland	Postconviction DNA testing; preservation of evidence; compensation for wrongful conviction	Enacted	2001 Md. Laws 418 (2001)
Michigan	Postconviction DNA testing; preservation of evidence	Enacted	Mich. Comp. Laws §770.16 (2001)
Minnesota	Postconviction DNA testing; preservation of evidence	Enacted	Minn. Stat. §590.01–590.06 (1999)
Missouri	Postconviction DNA testing; preservation of evidence	Enacted	Mo. Rev. Stat. §547.035 (2001)
Nebraska	Postconviction DNA testing; preservation of evidence	Enacted	Neb. Rev. Stat. §29-4120–§29-4125 (2001)

Jurisdiction	*Area of Relevance*	Status of Legislation	Number of Bill/ Statute
New Hampshire	Postconviction DNA testing; preservation of evidence	Pending	HB 1258
New Jersey	Postconviction DNA testing; preservation of evidence	Enacted	New Jersey Permanent Law 2A:84A-32a
New Mexico	Postconviction DNA testing; preservation of evidence	Enacted	N.M. Stat. Ann. §31-1A-1 (2001)
New York	Postconviction DNA testing; preservation of evidence	Pending	A09250
New York	Postconviction DNA testing; preservation of evidence; compensation for wrongful conviction	Enacted	N.Y. Crim Proc. Law §440.30 (1-a)
North Carolina	Postconviction DNA testing; preservation of evidence; compensation for wrongful conviction	Pending	SB 164
Ohio	Postconviction DNA testing	Pending	SB 7
Oklahoma	Postconviction DNA testing; preservation of evidence	Enacted	Okla. Stat. tit. 22 §1371, 1371.1 1371.2 (2000)
Oregon	Postconviction DNA testing	Enacted	2001 Or. Laws 697 (2001)

Jurisdiction	*Area of Relevance*	Status of Legislation	Number of Bill/ Statute
Pennsylvania	Postconviction DNA testing; preservation of evidence	Enacted	SB 589; §9543.1
Rhode Island	Postconviction DNA testing; preservation of evidence	Enacted	Chapter 10-9.1-10 (RI General Laws)
Tennessee	Postconviction DNA testing	Enacted	Ten. Code Ann. §40-30-401 through 413 (2001)
Texas	Postconviction DNA testing; preservation of evidence; compensation for wrongful conviction	Enacted	2001 Tex. Gen Laws 64 (2001)
Utah	Postconviction DNA testing	Enacted	Utah Code Ann. §78-35a-301–304 (2001)
Virginia	Postconviction DNA testing; preservation of evidence	Enacted	Va. Code Ann. §19.2-327.1 (2001)
Washington	Postconviction DNA testing; preservation of evidence	Enacted	2001 Wash. Laws 301 (2001)
Wisconsin	Postconviction DNA testing; preservation of evidence	Enacted	AB 291

Source: Benjamin N. Cardozo School of Law, 55 5th Avenue, 11th Floor, New York, NY 10003; http://www.innocenceproject.org/.

Appendix F

Items Obtained through Discovery

I. INTRODUCTION

Discovery is one of the most important parts of a DNA case. During the discovery phase, information is requested that will help the defense (and/or the prosecution in some instances) to develop the theory of the case.

Discovery requests are made in many different ways. Methods used by the defense to obtain information include but are not limited to letters, motions, and oral requests for information. This appendix includes information that may be useful to a defense team seeking to obtain exculpatory information or other information that may be useful in defending a suspect at trial or in pretrial proceedings.

II. ITEMS A PARTY MAY WISH TO REQUEST DURING DISCOVERY

The following items should be requested by the defense in discovery requests:

A. General

- All material known to the district attorney or any other law enforcement officer, or agency, that, by due diligence, could be learned from other governmental agents or prospective witnesses in this case, that is arguably exculpatory in nature, favorable to the accused, or that may tend to support a theory of defense

DNA: Forensic and Legal Applications, by Lawrence Kobilinsky, Thomas F. Liotti, and Jamel Oeser-Sweat
ISBN 0-471-41478-6

  - (This is a catch-all provision.)
- List of witnesses
- The exact location of the alleged criminal conduct
- The exact time and date of the alleged criminal conduct
- The names, addresses, and birth dates of any witnesses to or informants about the alleged criminal conduct
- Information pertaining to whether any person whom the prosecution intends to call as a witness at trial has participated in a corporeal viewing of the defendant. This sort of viewing includes show-ups and lineups
- Information pertaining to any statements or actions that were made or performed by the person viewing the defendant that may be deemed to link the defendant with the crime or that can be construed as being sufficient enough to identify the defendant as the perpetrator of the offense charged
- Information pertaining to all instances of uncharged criminal, vicious, or immoral conduct that the prosecution intends to introduce for impeachment

Court Proceedings

- Grand jury synopsis sheet
- Indictment
- Omni form complaint
- Complaint room screening sheet
- Complaint report worksheet
- Complaint report and follow-up
- Order(s) to submit to DNA testing (if any) and the basis for such order

Police Reports and Procedures

- The exact time, date, and location within said premises of the arrest
- The names, addresses, and birth dates of any witnesses to defendant's arrest
- The predicate for the initial seizure of the defendant
- Miranda warnings
- Any written, recorded, or oral statement made by defendant to any law enforcement official including, but not limited to, police officers, district attorneys, corrections officers
- Any written, recorded, or oral statement made by defendant to any person acting under the direction of, or in cooperation with a law enforcement official.
- Arrest photos
- Omni form arrest sheet
- Booking arrest worksheet

- Arrest report
- Videotapes/audiotapes of confessions if relevant
- Police memo book entries
- Police DD5s
- Crime scene photographs
- Crime scene sketches and handwritten notes
- Crime scene reports
- Request for laboratory examination
- Telephone recordings and radio runs (when relevant including 911 tapes or recordings)

If Dog Used

- Canine report
- Canine certification

If Hair Was Tested

- Hair analysis report
- Bench notes

If Other Trace Evidence Tested

- Trace evidence report
- Bench notes
- Any written report or document, concerning a physical or mental examination, or scientific test or experiment, relating to the criminal action or proceeding that was made by, or at the request or direction of, a public servant engaged in law enforcement activity, or which was made by a person whom the prosecutor intends to call as a witness at trial, or which the people intend to introduce at trial

Cases in which Alleged Victim Was Injured in General

- Information regarding the nature of the injury received and the nature of any medical attention required or received by the alleged victim
- Information pertaining to whether the alleged victim has had any lasting after-effects as a result of the injury
- Information pertaining to how the alleged victim came to be injured
- Whether the injury was an aggravation of a preexisting medical condition, and, if so, describe the nature of such prior condition and the nature and extent of prior treatment received

Homicide Cases

- Hospital records (if any)
- Police identification of body
- Autopsy report(s)

Rape Cases

- Hospital records
- Latent print report (if relevant)

DNA Related Requests

- Laboratory "bench" notes
- Summary reports (evidence and suspect files)

For Both Serology and DNA Analysis

- Evidence vouchers
- Property clerk records
- Evidence tracking sheets and other relevant chain-of-custody information
- Curriculum vitae of expert testifying
- List of expert witnesses
- Proficiency test records for that expert
- Laboratory accreditation status
- Laboratory protocols for serological or DNA procedures used
- Procedures for withdrawal, collection, and transmission of DNA samples
- Authority to withdraw DNA from defendant (if evidence obtained without consent)
- Copy of warrant or order to obtain bodily substances for forensic DNA analysis
- Protocols and procedures for the storage of DNA samples
- Laboratory's database for all loci tested (usually 13) in major ethnic/racial groups
- Statistical calculations of genotypes and overall genetic profile
- Calculations for siblings and close relatives
- Information pertaining to the locations and whereabouts of any siblings or close relatives who reside or may have been in the vicinity of the victim
- Evidence and/or DNA samples for independent testing by the defense

- Certification of laboratory equipment calibrations including:
  - Temperature systems (freezers, refrigerators, incubators, heating blocks, thermal cycler, thermometers, thermocouple temperature monitors)
  - Weight (scales, balances)
  - Centrifuges (speedvac, refrigerated centrifuges, microfuges, etc.)

Where the SDIS or CODIS have been employed

- Certification from state DNA databank coordinator
- DNA databank specimen submission form
- DNA databank report of analysis

Appendix G

Glossary

A Single-letter representation of the purine base adenine. The letter *A* is also used in diagrams herein to represent a nucleotide containing adenine.

ABI 310 Genetic Analyzer A capillary electrophoresis instrument used to analyze amplified products and to produce gene scans of STR loci as well as to sequence DNA fragments.

Adenine A purine base. It is one of the molecules present in nucleic acids DNA and RNA. It is herein represented by the letter *A*.

Adventitious DNA DNA, either amplified or not, which finds its way into an amplification tube and becomes amplified along with the template DNA obtained from the evidentiary or exemplar samples.

Allele One of two or more alternative forms of a gene at a specific locus (chromosomal location).

Allele Frequency The proportion of a particular allele occurring among the chromosomes carried by individuals in a population.

Allelic Ladders A set of the most common alleles for each locus being tested. Used to determine sizes of unknown alleles based on comparison with those alleles found in the ladder.

Sources: National Research Council, Committee on DNA Technology and Forensic Science, DNA Technology and Forensic Science (1992); Paul Berg and Maxine Singer, *Dealing with Genes: The Language of Heredity*, University Science Books, 1992.

DNA: Forensic and Legal Applications, by Lawrence Kobilinsky, Thomas F. Liotti, and Jamel Oeser-Sweat
ISBN 0-471-41478-6

Alphoid DNA A class of satellite DNA found at the centromeres of chromosomes. It is complementary to primate DNA at the D17Z1 locus. Used to quantify human DNA in the QuantiBlot assay.

AmpFLP Amplified fragment-length polymorphism.

Amplicons The products of PCR amplification.

Amplification The process of increasing the number of copies of a specific region of DNA by polymerase chain reaction.

Annealing Binding of complementary single strands of DNA based on sequence recognition.

ASCLD American Society of Crime Laboratory Directors: ASCLD-LAB is involved with the accreditation of crime labs.

Autoradiogram (Autoradiograph, Autorad) A photographic recording of the positions on a film where radioactive decay of isotopes has occurred.

Autosome Any of the chromosomes other than the sex chromosomes X and Y. Humans have 22 pairs of autosomes.

Band The visual image representing a particular DNA fragment on an autoradiogram.

Band Shift The phenomenon in which DNA fragments in one lane of a gel migrate at a rate different from that of identical fragments in other lanes of the same gel.

Base Pair Two complementary nucleotides held together by hydrogen bonds; basepairing occurs between A and T and between G and C. It is also a unit used to describe nucleic acid fragment length.

Biallelic *See* Diallelic.

Bleed Through (Pull-Up) An artifact in the electropherogram for loci analyzed by multiplex kits using multiple fluorochromes for primer labeling. A high-intensity peak developed in the green channel appears in the blue channel.

Blind Proficiency Test A proficiency test in which the analyst is unaware that the evidence he is analyzing is part of a test.

Blot *See* Southern Blot.

C Single-letter representation of the pyrimidine base cytosine. The letter *C* is also used in diagrams herein to represent a nucleotide containing cytosine.

Capillary Electrophoresis A procedure in which fragments of DNA within a mixture are separated based on size as they migrate through a narrow polymer-filled tube as a result of a voltage difference across the ends of the tube.

CCD Camera A charge-coupled device that detects and records emission spectra of amplified DNA fragments. It collects fluorescence at four different emission wavelengths that correspond to blue, green, yellow, and red.

Ceiling Principle A conservative method of estimating the frequency of a genetic profile within the major racial groups knowing that each is composed of a number of subpopulations.

Chelex A resin composed of styrene divinylbenzene copolymers containing paired iminodiacetate ions each of which acts as a chelating group. Used to extract DNA in single-stranded form from biological specimens. Chelex binds to polyvalent metal ions that would normally catalyze the breakdown of DNA at elevated temperatures in solutions having low ionic strength.

Chromosome The structure by which hereditary information in the form of genes and noncoding DNA is physically transmitted from one generation to the next; the organelles within the cell nucleus that carries the gene.

CODIS Combined DNA Index System, the national DNA database established in 1998 and containing genetic profiles obtained from felons as well as profiles obtained from crime scene evidence.

COFiler A PCR multiplex kit that can be used to provide genetic information at the following loci: CSF1PO, D16S539, THO1, TPOX, D3S1358, D7S820, and Amelogenin.

Controls Tests performed in parallel with experimental samples and designed to demonstrate that a procedure worked correctly.

Cytosine A pyrimidine base. It is one of the molecules present in nucleic acids DNA and RNA. It is herein represented by the letter *C*.

dNTPs The building blocks of DNA that are incorporated into the newly synthesized DNA strands as directed by the Taq polymerase enzyme and template DNA.

Degradation The breaking down or fragmentation of DNA by chemical or physical means.

Denaturation The process of converting a molecule from its active form to an inactive form. In the case of DNA the molecule is converted from a double-stranded to a single-stranded form. This can happen by heating DNA to 95°C or by treatment with NaOH that will break hydrogen bonds.

Deoxyribonucleic Acid (DNA) The genetic material of organisms. DNA is usually double stranded. It is composed of two complementary chains of nucleotides in the form of a double helix. The four chemical bases that make up DNA are adenine, cytosine, guanine, and thymine.

Diallelic Referring to DNA variation showing only two forms with a frequency of more than 1%.

Diploid Having two sets of chromosomes, in pairs (compare haploid).

DNA Deoxyribonucleic acid.

DNA Band The visual image representing a particular DNA fragment on an autoradiogram.

DNA Databank (Database) A collection of DNA typing profiles of selected or randomly chosen individuals.

DNA Polymerase An enzyme that catalyzes the synthesis of double-stranded DNA.

DNA Probe A short segment of single-stranded DNA labeled with a radioactive or chemical tag that is used to detect the presence of a particular DNA sequence through hybridization to its complementary sequence.

Dye Blobs Occasionally, fluorescent dyes will come off of their associated primers and move throughout the capillary tube during fragment separation and detection.

Electrophoresis A technique in which different molecules are separated by their rate of movement in an electric field.

Electropherogram A graphical display of the fragments that are detected by the CCD camera of the ABI prism. Peaks illustrate relative fluorescence units (height) and fragment length (time).

Exons Sequences of DNA within a gene that actually code for the protein to be synthesized.

Enzyme A biological catalyst. An enzyme is a protein capable of speeding up a specific chemical reaction but that itself is not changed or consumed in the process.

Ethidium Bromide An organic molecule that binds to DNA and fluoresces under ultraviolet light and is used to identify DNA.

Extraction Control A negative control containing all reagents used to extract DNA from a biological specimen. No DNA is present in this sample. This tests for the presence of adventitious DNA within the extraction reagents. This negative control is treated like all other samples and amplified and analyzed to see if any DNA becomes amplified.

Fluorescence The emission of light after a molecule is excited by light energy. The emitted light is always at a wavelength longer than the wavelength of the exciting light. Fluorescence testing is usually more sensitive than tests based on color changes.

G Single-letter representation of the purine base guanine. The letter *G* is also used in diagrams herein to represent a nucleotide containing guanine.

Gamete A haploid reproductive cell.

Gametic (Phase) Equilibrium The state at loci on different chromosomes when the allele at one locus in the gamete varies independently of that at the other loci; in gametic (phase) disequilibrium, a specific allele at one locus is associated with an allele at another locus on a different chromosome with a frequency greater than expected by chance (see linkage disequilibrium).

Gel Semisolid matrix (usually agarose or acrylamide) used in electrophoresis to separate molecules.

Gene The basic unit of heredity. A gene is a sequence of DNA nucleotides on a chromosome.

Gene Frequency The relative percent occurrence of a particular allele in a population.

Genetic Drift Random fluctuation in allele frequencies usually due to small population size or sampling error.

Genome The total genetic makeup of an organism.

Genotype The genetic makeup of an organism, as distinguished from its appearance or phenotype. It is expressed as the two alleles at a single locus.

Guanine A purine base. It is one of the molecules present in nucleic acids DNA and RNA. It is herein represented by the letter *G*.

Haploid Having one set of chromosomes (compare diploid). The number of chromosomes found in gametes (sperm and egg).

Hardy–Weinberg Equilibrium The condition, for a particular genetic locus and a particular population, with the following properties: Allele frequencies at the locus are constant in the population over time, and there is no statistical correlation between the two alleles possessed by individuals in the populations in the absence of selection, migration, and mutation.

Heredity The transmission of characteristics from parent to offspring.

Heterozygote A diploid organism that carries different alleles at one or more genetic loci on its homologous chromosomes.

Heterozygous Having different alleles at a particular locus. For most forensic DNA probes the autoradiogram displays two bands if the person is heterozygous at the locus (compare to homozygous).

HLA *See* Human Leukocyte Antigen.

Homologous Chromosomes The two members of a chromosome pair. Chromosomes that pair during meiosis. The chromosomes of the pair have the same loci, have the same centromere positions, have responsibility for the same traits. One member of the pair comes from mother and the other from father.

Homozygote A diploid organism that carries identical alleles at one or more genetic loci on its homologous chromosomes.

Homozygous Having the same allele at a particular locus. For most forensic DNA probes, the autoradiogram displays a single band if the person is homozygous at the locus (compare with heterozygous).

Human Leukocyte Antigen (HLA) Protein–sugar structures on the surface of most cells, except blood cells, that differ among individuals and are important for acceptance or rejection of tissue grafts or organ transplants; the locus of one particular class, HLA DQα, is used for forensic analysis with PCR.

Hybridization The reassociation of complementary strands of nucleic acids, nucleotides, or probes.

Identifiler A multiplex kit that can be used to produce genetic information for 15 STR loci plus amelogenin.

Introns The intervening sequences between exons. Introns consist of noncoding DNA. During transcription both introns and exons are transcribed into RNA, however, the introns are excised, and the messenger RNA that is produced contains only the sequences coded by introns.

Isotope An alternative form of a chemical element. Isotopes are used in reference to the radioactive alternative forms, or radioisotopes.

Kilobase One thousand bases (or nucleotide base pairs).

Linkage Genes that are located on the same chromosome are inherited as a unit and are said to be linked. Such genes do not sort independently during mitosis and therefore cannot be used together when applying the product rule to determine gene profile frequencies.

Linkage Disequilibrium The phenomenon in which a specific allele at one locus is nonrandomly associated with an allele at another locus. Genes located on the same chromosome but at a distance can behave as if they are not linked.

Locus (*pl.* Loci) The specific physical location of a gene on a chromosome.

Marker A gene with a known location on a chromosome and a clear-cut phenotype that is used as a point of reference in the mapping of other loci.

Matrix File This is a file used to adjust for the spectral overlap between the fluorescent dyes used. Thus all spectral data is corrected during analysis.

Matrix Failure or Pull-up The failure of the detection instrumentation to properly resolve the dye colors used to label the STR amplicons.

Membrane The matrix (usually nylon) to which DNA is transferred during the Southern blotting procedure.

Microsatellite DNA A form of repetitious DNA that includes STRs. STRs consist of repeat units 2–7 bp long.

Minisatellite DNA A form of repetitious DNA that includes repeat units that are longer than STRs. The fragments containing these core repeats may be in the range of 0.5–20 kb.

Minus A, n−1, −A Incomplete addition of adenine. This artifact manifests as a split peak and results from excess DNA in the amplification reaction mix.

Molecular-Weight Size Marker A DNA fragment of known size from which the size of an unknown DNA sample can be determined.

Monomorphic Probe A probe that detects the same allele and hence the same pattern in everyone.

Multilocus Probe A DNA probe that detects genetic variation at multiple sites in the genome. An autoradiogram of a multilocus probe yields a complex, stripe-like pattern of 30 or more bands per individual.

Mutagen A physical agent (e.g., X-rays) or chemical agent that induces changes in DNA.

Nucleic Acid A nucleotide polymer of which major types are DNA and RNA.

Nucleoside A purine or pyrimidine base covalently linked to a five-carbon sugar (ribose or deoxyribose).

Nucleotide A unit of nucleic acid composed of a phosphate, a five-carbon sugar (ribose or deoxyribose), and a purine or a pyrimidine base.

Null Allele An allele that is not detected in an individual although it is present at the appropriate locus. This occurs due to mutation in the primer binding site. PCR results in calling a heterozygote a homozygote. In RFLP, a very small allele could run off the electrophoretic gel very rapidly and go undetected. Again a heterozygote might be observed as a homozygote.

Off-Scale Data When inputting too much template DNA into the PCR reaction mixture, the fluorescence emitted by the PCR products will be excessive and as a result there will be elevated baselines and pull-up of one or more colors corresponding to the peak size in other channels.

PCR Polymerase chain reaction.

Peak Height Relative fluorescence intensity of an allele that reflects the quantity of sample in the original PCR reagent mix.

Peak Height Imbalance A large difference (greater than 30%) in peak height for alleles detected at a single locus that suggests that a mixture of samples exists.

Phenotype The physical appearance or functional expression of a trait.

Plus A, n + 1, +A The Taq polymerase will at times add another base, usually an adenine nucleotide to the 3′ ends of double-stranded PCR amplified product. This is a nontemplate addition that results in an amplified target one base larger than the target DNA sequence. Multiplex kits are designed to promote this fully for all loci. This is done by primer design and by extending the final thermal cycle step at 60°C.

Point Mutation An alteration of one nucleotide in chromosomal DNA that consists of addition, deletion, or substitution of nucleotides.

Polymerase Chain Reaction (PCR) An *in vitro* process that yields millions of copies of desired DNA through repeated cycling of a reaction that involves the enzyme DNA polymerase. Cycles include denaturation, annealing or primers, and DNA polymerase-directed extension and ligation.

Polymorphism The existence of two or more discontinuous, segregating phenotypes in a population. The presence of more than one allele of a gene in a population at a frequency greater than that of a newly arising mutation. Operationally, a population in which the most common allele at a locus has a frequency of less than 99%.

Preferential Amplification The phenomenon of producing more amplified product from the shorter allele of a heterozygote than from the longer allele.

Primer A short sequence of DNA, generally a synthetic, single-stranded, oligonucleotide that binds specifically to a target DNA sequence and initiates synthesis together with a Taq polymerase and other reagent molecules, of a new strand of DNA complementary to the template DNA.

Probability The ratio of the frequency of a given event to the frequency of all possible events.

Probe A short segment of single-stranded DNA tagged with a group or radioactive atom that is used to detect a particular complementary DNA sequence.

Product Rule The probability of two independent events occurring simultaneously is calculated as the multiplicand (product) of their independent probabilities.

Proficiency Tests Tests to evaluate the competence of technicians and the quality performance of a laboratory. There are different types of proficiency tests. In open tests, the technicians are aware that they are being tested. In blind tests, the technicians are not aware that they are being tested. Internal proficiency tests are conducted by the laboratory itself. External proficiency tests are conducted by an agency independent of the laboratory being tested.

Profiler Plus A PCR multiplex kit that can be used to provide genetic information at the following nine loci: D3S1358, VWA, FGA, D8S1179, D21S11, D18S51, D5S818, D13S317, and D7S820 and amelogenin.

Protein A chain of amino acids, the building subunits of proteins, joined by peptide bonds.

Pull-up Matrix failure—the result of color bleeding from one spectral channel to another resulting in artifactual peaks.

Purine The larger of two kinds of bases found in DNA and RNA. A purine is a nitrogenous base with a double-ring structure, such as adenine and guanine. (compare pyrimidine).

Pyrimidine The smaller of two kinds of bases found in DNA and RNA. A pyrimidine is a nitrogenous base with a double-ring structure, such as cytosine, thymine, and uracil (compare purine).

Quality Assurance A program conducted by a laboratory to ensure accuracy and reliability of tests performed.

Quality Control Internal activities performed according to externally established standards used to monitor the quality of DNA typing to meet and satisfy specified criteria.

Random Match Probability The chance of a random match, that is, a match between the genetic profile obtained from the evidence and that from a random sample from the population.

Recombinant DNA Fragments of DNA from two different species, such as a bacterium and a mammal, spliced into a single molecule.

Relative Fluorescence Unit (RFU) A measure of the quantity of a particular color and size fragment. Visually the *y* axis on an electropherogram.

Replication The synthesis of new DNA from existing DNA.

Restriction Endonuclease, Restriction Enzyme An enzyme that cleaves DNA molecules at particular base sequences.

Restriction Fragment-Length Polymorphism (RFLP) Variation in the length of DNA fragments produced by a restriction endonuclease (restriction enzyme) that cuts at a polymorphic locus.

RFLP *See* Restriction Fragment-Length Polymorphism.

RFLP Analysis Technique that uses single-locus or multilocus probes to detect variation in a DNA sequence according to differences in the length of fragments created by cutting DNA with a restriction enzyme.

Ribonucleic Acid (RNA) A class of nucleic acids characterized by the presence of the sugar ribose and the pyrimidine uracil, as opposed to the thymine of DNA.

RNA Ribonucleic acid.

Scan Number Describes the location of a data point along the *x* axis on an electropherogram.

Serology The discipline concerned with the immunological study of body fluids.

Serum The liquid that separates from blood after the blood forms a clot.

Sex Chromosomes (X and Y Chromosomes) Chromosomes that are different in the two sexes and that are involved in sex determination.

Sex-Linked Characteristic A genetic characteristic, such as color blindness, that is determined by a gene on a sex chromosome and shows a different pattern of inheritance in males and females. X-linked is a more specific term.

Short Tandem Repeat (STR) The short tandem repeat (STR) consists of short repeat units, 2–7 bases long, tandemly connected to each other producing highly polymorphic loci. STRs usually are located in noncoding regions of DNA, either between genes or within introns.

Single-Locus Probe A DNA probe that detects genetic variation at only one site in the genome. It is an autoradiogram that uses one single-locus probe usually displays one band in homozygotes and two bands in heterozygotes.

Somatic cells The differentiated cells that make up the body tissues of multicellular plants and animals.

Southern Blot The nylon membrane to which DNA adheres after the process of Southern blotting.

Southern Blotting The technique for transferring DNA fragments that have been separated by electrophoresis from the gel to a nylon membrane.

Standards Criteria established for quality control assurance. This word is also used to describe established or know test reagents, such as molecular weight standards.

Stutter Bands ($n - 4$) An artifact in which a minor peak usually 4 bases shorter than the corresponding main allele peak is produced during the PCR process. Stutter results from slippage of the polymerase during the extension phase of the amplification process. Stutter can be found one or two core repeat units shorter or longer than the correct allele size.

SWGDAM Formerly known as TWGDAM—Scientific Working Group on DNA Analysis Methods.

T Single-letter representation of the purine base thymine. The letter *T* is also used in diagrams herein to represent a nucleotide containing thymine.

Tandem Repeats Multiple copies of an identical DNA sequence arranged in direct succession in a particular region of a chromosome.

Target DNA Template DNA used in the PCR amplification process.

Template DNA The DNA that is put into the PCR reaction mix to be copied. This template DNA is at first denatured by elevating the temperature and then annealed to the primers. The Taq polymerase extends the primers by directing dNTPs to position themselves in a complementary manner to the sequence of bases on the template.

Thymine A purine base. It is one of the molecules present in nucleic acids DNA and RNA. It is herein represented by the letter *T*.

Uracil A pyrimidine base in RNA.

Variable Number Tandem Repeats (VTNR) A polymorphic locus where alleles differ primarily in the number of times that a stretch of nucleotides (core repeat) are tandemly repeated. The size of the core differentiates the VNTR from the STR. The former is usually in the range of 15–30 bases long while the latter is in the range of 2–7 bases long.

VNTR Variable number tandem repeats.

Zygote Diploid cell that results from the fusion of male and female gametes.

Index

DNA: Forensic and Legal Applications, by Lawrence Kobilinsky, Thomas F. Liotti, and Jamel Oeser-Sweat
ISBN 0-471-41478-6